G000044750

Task/Subtask	Title
301.2.6	Updates
302	Support System Alternatives
302.2.1	Alternate Support Concepts
302.2.2	Support Concepts Updates
302.2.3	Alternate Support Plans
302.2.4	Support Plan Update
302.2.5	Risks
303	Evaluation of Alternatives and Tradeoff Analysis
303.2.1	Tradeoff Criteria
303.2.2	Support System Tradeoff
303.2.3	System Tradeoffs
303.2.4	Readiness Sensitivities
303.2.5	Manpower and Personnel Tradeoffs
303.2.6	Training Tradeoffs
303.2.7	Repair Level Analysis
303.2.8	Diagnostic Tradeoffs
303.2.9	Comparative Evaluations
303.2.10	Energy Tradeoffs
303.2.11	Survivability Tradeoffs
303.2.12	Transportability Tradeoffs
401	Task Analysis
401.2.1	Task Analysis
401.2.2	Analysis Documentation
401.2.3	New/Critical Support Resources
401.2.4	Training Requirements and Recommendations
401.2.5	Design Improvements

Task/Subtask	Title
401.2.6	Management Plans
401.2.7	Transportability Analysis
401.2.8	Provisioning Requirements
401.2.9	Validation
401.2.10	ILS Output Products
401.2.11	LSAR Updates
402	Early Fielding Analysis
402.2.1	New System Impact
402.2.2	Sources of Manpower and Personnel Skills
402.2.3	Impact of Resource Shortfalls
402.2.4	Combat Resource Requirements
402.2.5	Plans for Problem Resources
403	Post Production Support Analysis
403.2	Post Production Support Plan
501	Supportability Test, Evaluation, and Verification
501.2.1	Test and Evaluation Strategy
501.2.2	Objectives and Criteria
501.2.3	Updates and Corrective Actions
501.2.4	Supportability Assessment Plan (Post Deployment)
501.2.5	Supportability Assessment

The ILS Manager's
LSA Toolkit

McGraw-Hill Logistics Series
Jim Jones, Series Editor

BARNES • *Logistics Support Training: Design and Development*

JONES • *Integrated Logistics Support Handbook*

JONES • *Engineering Design*

JONES • *Logistics Support Analysis Handbook*

ORSBURN • *Spares Management Handbook*

SAYLOR • *TQM Field Manual*

LACY • *Systems Engineering Management*

The ILS Manager's LSA Toolkit

Availability Engineering

Dick Biedenbender

Florence Vryn

John Eisaman

McGraw-Hill, Inc.

New York   San Francisco   Washington, D.C.   Auckland   Bogotá
Caracas   Lisbon   London   Madrid   Mexico City   Milan
Montreal   New Delhi   San Juan   Singapore
Sydney   Tokyo   Toronto

Library of Congress Cataloging-in-Publication Data

Biedenbender, Richard E.
    The ILS manager's LSA toolkit : availability engineering / Richard
E. Biedenbender, Florence Vryn, John Eisaman.
        p.   cm. — (McGraw-Hill Logistics series)
    ISBN 0-07-005220-4
    1. Integrated logistics support.   2. United States—Armed Forces—
Procurement.   I. Vryn, Florence.   II. Eisaman, John.   III. Title.
IV. Series: Logistics series.
U168.B54   1993
355.4'11—dc20   92-36755
                                                                        CIP

1 2 3 4 5 6 7 8 9 0   DOH/DOH 9 8 7 6 5 4 3 2

ISBN   0-07-005220-4

The sponsoring editor for this book was Larry S. Hager, the editing
supervisor was Joseph Bertuna, and the production supervisor was
Donald F. Schmidt. It was set in Century Schoolbook by
McGraw-Hill's Professional Book Group composition unit.

Printed and bound by R. R. Donnelley & Sons Company.

Contents

Appendixes

Preface

As this book goes to print, the 1A LSA standard is slowly maturing and the 2A standard has been replaced by the 2B standard. CALS is beginning to have a major impact on LSA/LSAR.

As we survey the evolution, practice, and use of the LSA standards, two conflicting messages are found. First, anyone who reviews the history of the two standards can honestly point with pride to significant progress over the past 15 years. However, despite this progress, one also encounters considerable dissatisfaction with the current standards, particularly in terms of practice. I as well as the authors have encountered this numerous times. In late 1990, a knowledgeable senior OSD official was publicly stating in presentations that for some reason LSA had problems in implementation. From the industry end of the spectrum, the industry CALS LSA subgroup received two completely independent formal suggestions that the tasks in the 1A standard need to be better defined.

Making LSA/LSAR work is what this book is all about. It provides detailed suggestions on how to write better LSA contract requirements and alternatives on how to execute the LSA tasks. Thus this book is unique in that it discusses both sides of the coin. Understanding the other side can be of significant benefit in developing cost-effective responses.

The principal and secondary authors of the book have a unique range of experience, and this book covers development of the LSA standards, development of government guidance for application, review of contractor responses to LSA requirements, and last but not least, actual performance of the LSA tasks. Again, this range of experience provides the reader with a unique view across both requirer and performer perspectives.

The book is also unique in that it presents many new developments in LSA/LSAR which are not yet incorporated generally into DOD guidance. The book purposely includes significant amounts of detail, with numerous examples and checklists. Most readers will find a lot of ideas in the material on how to actually do some of the LSA tasks, a subject not really addressed in depth in the existing literature.

Some readers will find parts of the book at variance with their convictions in the sense that they disagree with some parts or feel they have a superior approach. Experts in all disciplines hold a variety of opinions on appropriate techniques. LSA is no exception. Hopefully, the book will foster further dialogue exploring differences that will lead to advances in LSA. I am confident that for every point of disagreement, something useful will be found.

This book has been developed for government as well as contractor personnel. The tips on tailoring and focusing, or further refining, the requirements in the 1A standard should help anyone write more cost-effective requirements. The responses should reflect both better understanding and improved results.

Lastly, a word about commercial products. Most people in the ILS world understand that commercial product development is a suitable candidate for the selective application DOD ILS/LSA. Linda Green, in her book, *Logistics Engineering*, mentions several companies who have actually done this. The example of the development of a commercial lawn mower was chosen because the product is familiar to most readers and it illustrates a simple commercial application.

There is a need for the information in this book. We have expended great effort to make the format user-friendly. The result of more specific information and detailed guidance will be decreased costs of analysis, increased productivity, and products that better meet the needs of the consumer, both government and civilian. It will be a useful addition to libraries. Readers will find in it ready solutions/departure points for problems which suddenly confront them in practice.

Jim Jones

Acknowledgments

This book would obviously not be possible without the cooperation of many people. It also would have been impractical, if not impossible, without the technology of the word processor.

In terms of motivation to do this book, I would like to express my appreciation to Walter Finkelstein and Jim Fulton.

In terms of book preparation, obviously, Florence Vryn and John Eisaman, my collaborators, deserve thanks. John lent his LSAR expertise to the preparation of Chapter 6 and to four of the appendixes as well as some of the graphics material. Flo contributed in a major way to Chapters 2 and 3 and the appendix on the new DOD acquisition policies. Equally important, she was especially helpful to me in structuring the book and in ensuring that the industry perspective is adequately represented.

I would also like to mention George Desiderio, OSD ILS/LSA Focal Point, John Peere, chief of LSA/LSAR at MRSA, and Bert Upton, head of ILS at NAVSEA, for providing some of the material used in the book preparation.

Special thanks go to the people at Interlog. John Brown, Del Dalpra, and Rich Richardson lent management support. Jim Campbell assisted in numerous ways. Joe Murray helped prepare some of the material for some LSA guidance prepared for a DOD program office which was adapted for use in this book. Jim Heidmous did much of the lawn mower example used in Chapter 5. C. C. Johnson was unbelievably good in rounding up and preparing graphics.

Last, but not least, I would like to thank my wife Lucy, who has helped with the word processing and who, on more than five occasions, saved my "life" in terms of saving "lost" work or finding and unlocking hidden word processing codes which threatened to halt all progress.

While I have tried to acknowledge sources, recognizing that these were drawn from a library collected over 15 years, I apologize for any omissions.

Dick Biedenbender

Introduction and Background

1.0   Introduction

This chapter presents an overview. It specifically discusses

Objectives   (1.1)

Intended audience   (1.2)

Synopsis   (1.3)

Intended use   (1.4)

Anticipated benefits   (1.5)

Requirer versus executor   (1.6)

Consider supportability, not just ILS/LSA   (1.7)

Suggested library   (1.8)

Updates   (1.9)

This book is primarily about MIL-STD-1388-1A, "Logistics Support Analysis" (LSA),[2] and its commercial counterpart, Availability Engineering. This book is intended to supplement and augment current books, such as those by Jim Jones[3] and Ben Blanchard.[4] Although there is some unavoidable overlap, we have tried to keep this to a minimum. While various management aspects of MIL-STD-1388-2A/2B, "DOD Requirements for a LSA Record" (LSAR),[5] are discussed, this is not a book about minute details of entering LSAR data. It does extensively discuss approaches to LSAR requirements and execution and key management concerns in developing the LSAR.

The basic purposes of the book are to

- Relate LSA to major new management initiatives.

- Integrate in one book currently fragmented information.

- Provide an understanding of both government and contractor perspectives.

- Provide "how to" ideas to both parties for preparing more cost-effective

  - LSA requirements for Request For Proposals (RFPs).
  - LSA task analyses.
  - LSA proposals/plans/procedures manuals.

■ Provide new ideas and departure point tools.

In short, the overall purpose of this book is to provide government and contractors with management information to manage LSA (and its commercial counterpart, Availability Engineering) more cost-effectively. It has become increasingly clear that the principles and methods of defense integrated logistics support/logistics support analysis (ILS/LSA) are adaptable to commercial projects. Dick Webster, one of the pioneers in applying ILS commercially, mentioned applications to commercial products such as transportation, electronic systems, power generation, information processing, heavy equipment, communications, light equipment manufacturing, the equipment service industry, and medical equipment in a presentation to the 1991 Society of Logistics Engineers Symposium. Examples of specific techniques used are Operational Availability Analysis, Life Cycle Costing (LCC), operation and maintenance feedback systems, built-in remote diagnostics, and readability/understandability. Reasons why defense logistics practices are increasingly of use commercially are discussed in Chap. 2.

Another trend of interest to many readers is the growing use of software and its implications logistically. This issue is probably a key factor to the near- and long-term future of the logistics profession. We discuss evolving software/ILS/LSA relationships in various parts of this book.

1.1 Objectives

We have been involved in the current LSA and LSAR standards since their inception. As Jim Jones says in the Preface, although a lot of progress has been made, many practicers on both sides of the LSA equation feel that further improvements in the standards and applications are needed. This is particularly true of MIL-STD-1388-1A. Thus, one of the objectives of this book is to consolidate in one place many of the "lessons learned" at the working level. In many cases this involves more "how to" information about what to require as well as "how to" execute. This book presents both new and more in-depth material on tailoring requirements and new approaches to LSA task performance and execution. The book also contains a more in-depth discussion of system readiness and availability than found elsewhere. While the book contains significant amounts of detail, it is not a "cookbook." Experiences with MIL-STD-1388-1A are not uniform. Some readers may find some parts of this book at variance with their experience. The majority, hopefully, will find parts that are immediately and directly applicable or that offer enough information to enable them to remove a mental block and/or devise superior approaches.

We have deliberately addressed both the requirer and the executor aspects, since we have found that it helps both parties to perform more effectively if they better understand the other's perspective.

In short, this book

Summarizes the best of government guidance in one place.

Contains new material not addressed anywhere in government guidance or the literature.

Incorporates numerous lessons learned but not yet reflected generally in widely available LSA documentation.

Contains numerous examples, checklists, and possible management tools potentially useful to the reader.

Discusses the application of LSA to software-intensive programs, since the use of software is rapidly growing.

1.2   Intended Audience

While the primary intended users of this book are obviously ILS/LSA managers in both government and industry, numerous other specialists could use it as a reference. A partial list of such specialists includes

- System engineering managers/specialists
- LSA and availability engineering analysts
- R&M engineers/specialists
- Spares/repair managers/specialists
- Training managers
- Publication managers/engineers
- Maintenance engineers
- Test equipment engineers
- Software managers
- Configuration management specialists
- Contracts and marketing managers

The material presented incorporates both government and industry experiences and practices. It has been updated to reflect evolving management and acquisition practices and technology trends. Much of the material has been used successfully as the basis for LSA evening courses at the university level and in courses conducted at both defense and contractor facilities. This book could be used as a basis for one of the courses in a logistics major at the junior, senior, or graduate level.

1.3 Synopsis

This section summarizes what is in this book. Before doing this, a few words need to be said about why this book is organized as it is. Obviously, some structure is necessary. LSA is complex and has many complex interfaces. In 15 years of teaching, doing, and writing, we have heard some students suggest that an alternate sequence or structure would have been more useful. However, the majority have felt comfortable with the sequence used. A detailed table of contents and an index of the LSA tasks and subtasks are provided to help the user quickly locate material of interest.

Chapter 2 briefly describes various management processes, such as the Department of Defense (DOD) acquisition processes, and various major management initiatives, such as total quality management (TQM) and computer-aided acquisition and logistics systems (CALS), that influence LSA activity. LSA must interface with such changes and trends if it is to be effective. Chapter 2 also discusses the probable impact of the changes and trends on LSA.

Chapter 3 discusses various major management considerations in LSA/LSAR. The perceptions of both the requirer and executor are presented. Challenges peculiar to each are discussed. Chapter 3 also presents various key management principles that are commonly recognized as essential to both requirer and executor management for effective LSA/LSAR.

Chapter 4 discusses what the individual LSA tasks and subtasks require and major considerations in the preparation of LSA solicitation requirements for a variety of procurement scenarios [nondevelopment items (NDI), product improvement programs (PIP), and so on], as well as the full development case, since many programs do not utilize all the phases of the acquisition process. Chapter 4 heavily emphasizes both tailoring and focusing (defining the tasks and subtasks to be performed more explicitly) of LSA requirements. While most requests for proposals (RFPs) do a decent job of tailoring to the subtask level, focusing is seldom seen in practice. Other considerations, such as competition, use of RFP Sec. 4 ("Instructions to Offerers"), and possible supportability contract incentives, are also discussed.

Chapter 5 is an in-depth discussion of *how to* execute the LSA tasks by phase. This subject is currently not generally covered elsewhere in a systematic fashion in the available literature. Yet, as Jones says in the Preface, there is growing recognition of the need to better define "how to." Chapter 5 is a start in this direction. The chapter includes an example application to a simple commercial product, a lawn mower, to illustrate application to both defense and commercial products.

Chapter 6 specifically addresses management of the LSAR, since generation of this data base is generally viewed as a significant cost element. If not properly managed, LSAR costs can overwhelm the budget for all LSA analysis.

Chapter 7 is devoted to analysis methods, with emphasis on readiness modeling and system engineering tradeoff approaches, since these come up frequently in practice but are seldom covered in the literature. LSA integration with the systems engineering process will grow as the concurrent engineering initiative gains momentum in the coming years. Several "quick and dirty" analysis methods for both readiness and operating and support (O&S) cost estimates are included to give the reader a departure point for development of such tools for local circumstances.

1.4 Intended Use

This book is intended primarily for use as a source of guidance and ideas and as a handy, one-stop service reference book. Table 1.1 summarizes major uses and the location of the material in the book.

The reader can see from the table that this book contains a wealth

TABLE 1.1   Possible Uses of the Material in This Book

Use	Chapter/Appendix
1. Relating LSA to key management processes	Chap. 2 Appendixes 2 and 3
2. Management, general	Chap. 3 Appendixes 15 and 17
3. Prepare requirements	Chaps. 3 and 4 Appendixes 6, 8, 9, 10, 11, 12, 13, 14, 19, 22, and 26
4. Prepare proposals	Chaps. 3, 5, and 7 Appendixes 4, 8, 9, 13, 15, 17, 18, 19, 20, 22, 25, and 27
5. Prepare plans	Chaps. 3, 5, and 7 Appendixes: See 4 above
6. Prepare procedures manual	Chaps. 3, 5, and 7 Appendixes: See 4 above
7. Prepare task inputs	Chap. 3 Appendix 14
8. Execute tasks	Chap. 5 Appendixes 6, 18, 19, 20, 21, 22, 23, and 24
9. Review LSA products	Chap. 3 Appendixes 15, 16, and 25
10. Select analysis methods	Chap. 7 Appendixes 18, 20, and 21
11. Focusing/new ideas	Chaps. 4 and 5 Appendixes 6, 8, 9, 18, 20, 21, and 22
12. Use of LSA products as input to program documents	Chaps. 3, 4, and 5

of material useful for preparing LSA requirements, proposals, and plans. The material can be used directly or as a departure point for task execution. For example, if either the government or the contractor wished to review a particular LSA task report, he or she could use the checklists in Appendix 15. A government ILS manager preparing an RFP could use the table in Chap. 3 to identify the various actions he or she must initiate. He or she could consult Appendix 5 to identify which specialists to consult to obtain the needed requirer inputs. The possibilities for use of the material are obviously too extensive to provide a complete list.

1.5   Anticipated Benefits

Benefits from the use of this book will, of course, vary with the individual user. Those who have significant background in ILS/LSA will recognize from this overview sections of immediate use in augmenting personal experience. Those with less experience will find Chaps. 1 through 3 useful in understanding the context of LSA. This will significantly enhance the benefits of the "how to" in the succeeding chapters.

Both government and industry managers are urged to review Chaps. 2 and 3, since the LSA interrelationships are as complex as any in the RFP or contract. Mastering these aspects of LSA yields savings in both time and money, improves job performance, aids in decision making, and stimulates new and innovative approaches. Potential benefits include

Improved systems/equipments in terms of system readiness, availability, and cost.

Reduced program supportability risk.

Higher-quality, more timely, and/or cost-effective LSA/LSAR products.

Time savings in terms of preparing, reviewing, executing, or planning LSA/LSAR efforts.

1.6   Requirer versus Executor

The requiring organization (normally the government or prime contractor) and the performing organization (normally a hardware contractor) frequently change with the phase, type of program, nature of the industry, whether or not there is competition, and other program variables. In preconcept, no hardware contractor exists, so the requiring organization is the performing organization or its field organizations or service contractor. In concept exploration/validation, the requirer may do a system-level LSA, while the major hardware contractors analyze in greater depth their proposed approach.

1.7   Consider Supportability, Not Just ILS/LSA

The term *supportability* in this book is defined as the combination of support-related design requirements and ILS management, LSA, and ILS element requirements. The ILS/LSA proponent in both the requirer and bidder organizations should realize that supportability importance is the sum of these components. It is not unusual to see an RFP in which ILS is not ranked as important as performance but in which support-related design parameters are given a high rating in terms of performance criteria. Thus, when the two sets of factors are combined, supportability is in fact given a high priority in the source selection. The conclusion is obvious—Read the entire proposal before reaching conclusions about relative importance.

1.8   Suggested Library

A decent library is essential to truly cost-effective LSA. Basic documents in this regard are

MIL-STD-1388-1A[2]

MIL-STD-1388-2A/2B[5]

This book

Jim Jones' book, *Logistics Support Analysis*[3]

Ben Blanchard's book, *Logistics Engineering and Management*[4]

AMC Pamphlet 700-4, "LSA Techniques Guide"[6]

AMC Pamphlet, "Cost Estimating Methodology for LSA"[7]

Numerous additional documents also could be useful, including

Systems Engineering Guide[8]

System Engineering Management[9]

Reliability Centered Maintenance (RCM) and level of repair analysis (LORA) standards and other documents, including the army graphics technique

Defense Systems Management College's "Cost Analysis Strategy Assessment" life cycle cost model[10]

Other LSA-related documentation used by your customer and/or referenced in Requests for Proposals (RFP) in the past.

Another source of information is the recently established U.S. Army Material Readiness Support Agency LSA Electronic Bulletin Board. Details of how to access this service are presented in Appendix 1.

1.9 Updates

LSA and the environments in which it takes place are continuously evolving. As the depth of experience with this subject has grown, it has become increasingly clear that there are variances in opinion and that greater specificity is needed in parts of the LSA standard. You are therefore invited to submit suggestions for improvement to this book (please be as specific as possible):

McGraw-Hill, Inc.
Science and Engineering Books
11 West 19th Street
New York, NY 10011

The LSA and Related Processes

2.0 Introduction

LSA and Availability Engineering are not objectives in themselves. To be effective, they must be properly related to the management environment in which they are applied. This chapter provides an overview of the LSA process and other related processes/initiatives. It specifically discusses

An overview of the LSA process (2.1)

Management development/acquisition processes (2.2)

The system engineering process and supportability design influence (2.3)

The ILS, HARDMAN/MANPRINT, Human Systems Integration (HSI), and Maintenance Planning Analysis processes (2.4)

Related management initiatives/trends [such as Computer-aided Acquisition and Logistics support, [CALS Concurrent Engineering (CE)] and the growing software intensity of programs] (2.5)

Summary (2.6)

2.1 An Overview of the LSA Process

2.1.1 LSA objectives

Although stated slightly differently in various documents, the generally accepted objectives of LSA are

Exert readiness and economic influence on requirements and design.

Optimally integrate the ILS elements.

Identify detailed ILS element resource requirements.

One or all of these objectives may apply to a given situation. A fourth objective, risk assessment/reduction, although mentioned less frequently, is embedded as an integral part into many of the LSA tasks. Risk assessment is now a major part of the new DOD acquisition policy and procedures. Risk-reduction techniques can include technology demonstrations, aggressive prototyping (including manufacturing processes and hardware and software systems), early operational assessments, and schedule tradeoffs.

2.1.2 The LSA standards

The United States Department of Defense (DOD) LSA process is embodied in two military standards. MIL-STD-1388-1A, "Logistics Support Analysis," establishes a structured engineering approach to the achievement of LSA objectives. MIL-STD-1388-2A (now 2B), "DOD Requirements for an LSA Record" (LSAR), establishes a standard medium to systematically record, store, process, and report various data that supports the LSA objectives.

The LSA standard has a series of 15 tasks which are listed in Table 2.1. The 15 tasks are subdivided into about 79 subtasks to facilitate tailoring of requirements for specific programs. A useful index of tasks and subtasks is in Table 1 of the standard. This table shows the purpose of the tasks and the LSA objectives that each subtask supports. The inside cover of your book also lists the task and subtask titles to

TABLE 2.1   Index of LSA Tasks

Task sections	Tasks
100 Program Planning and Control: Provides for Formal Program Planning and Review	101 Development of an Early Logistic Support Analysis Strategy 102 Logistic Support Analysis Plan 103 Program and Design Reviews
200 Mission and Support Systems Definition: To Determine Logistic Requirements	201 Use Study 202 Mission Hardware, Software, and Support System Standardization 203 Comparative Analysis 204 Technological Opportunities 205 Supportability and Supportability-Related Design Factors
300 Preparation and Evaluation of Alternatives: To Identify Operations and Maintenance Tasks to be Performed, Associated Safety Risks, and Optimal Maintenance Level	301 Function Requirements Identification 302 Support System Alternatives 303 Tradeoff Analysis
400 Determination of Logistic Support Resource Requirements: To Identify All Support Resources for the New System and Its Impact on Existing Resources and to Develop Plans for Postproduction Support	401 Task Analysis 402 Early Fielding Analysis 403 Postproduction Support Analysis
500 Supportability Assessment: To Ensure that Specified Requirements Are Achieved and Deficiencies Corrected	501 Supportability Test, Evaluation, and Verification

SOURCE: Navy LSA Applications Guide, May 1986.

TABLE 2.2   LSA Task/LSA Objective Relations

Task	Design influence	Support system influence	Logistic resource determination
201 Use Study	Δ	□	□
202 Standardization	Δ	Δ	□
203 Comparative Analysis	Δ	□	
204 Technological Opportunities	Δ	□	
205 Supportability and Related Design Factors	Δ	□	
301 Functional Requirements Identification	□	Δ	□
302 Support Systems Alternatives		Δ	
303 Tradeoff Analysis	Δ	Δ	□
401 Task Analysis	□	□	Δ
402 Early Fielding Analysis			Δ
403 Postproduction Support Analysis		□	Δ
501 Supportability Test, Evaluation, and Verification	Δ	Δ	Δ

□ = primary objective.
Δ = secondary objective.

make it easy for you to find them when you read this book. Table 2.2 summarizes the primary and secondary objectives each task serves. The LSA tasks provide answers to basic questions about the support needed for a specific program, as shown in Table 2.3.

Some of the LSA subtasks result in the generation of data to be stored in the LSAR. MIL-STD-1388-2A employed the various data

TABLE 2.3   The LSA Tasks Answer These Questions

Task no.	Task title	Question answered
101	LSA Strategy	How can I tailor to best invest my scarce LSA resources for my program?
102	LSA Plan	How does the contractor propose to execute LSA requirements?
103	Reviews	What is the status of contractor execution?
201	Use Study	How will the system be deployed, and how will it be used?
202	Standardization	How can standardization benefit the program?
203	Comparative Analysis	What are the readiness, support, manpower, and operating and support cost drivers and problem areas on similar systems?
204	Technological Opportunities	What can design do to reduce support resources needed and preclude repetition of past problems on this program?
205	Design Factors	What are reasonable support and support-related design goals for the program?
301	Functional Requirements	What functions must be performed by the support system of the new program?
302	Support Alternatives	What are the alternatives available to the support system?
303	Tradeoffs	What is the best support concept and best performance, support, and operating concept for the program?
401	Task Analysis	What ILS element resources are needed?
402	Early Fielding	What is the impact of this program on the logistics infrastructure?
403	Postproduction	What needs to be done to ensure that my program will be supportable after production ceases?
501	Assessment and Test	What support problems are indicated by T&E and other assessments?

TABLE 2.4   MIL-STD-1388-2A Data Sheets

Data sheet	Title	Scope
A	Operation and Maintenance Requirements	Identification of Hardware Assembly, Source, Quantity Required Frequency and Duration of Operational Employment Allocation of Preventive and Corrective Maintenance Actions Among Organizational, Intermediate, and Depot Levels Availability Requirements
B	Item Reliability and Maintainability Characteristics	Identification, Source, Quantity Required Availability Requirements Maintainability Considerations Function of Item Maintenance Concepts and Qualitative Maintainability Requirements
B1	Failure Modes and Effects Analysis	Failure Modes and Effects Analysis Damage Mode and Effects Analysis Survivability and Vulnerability Requirements
B2	Criticality and Maintainability Analyses	Criticality Analyses Maintainability Analyses High-Risk Items
C	Operation and Maintenance Task Summary	Identification, Source, Quantity Required Maintenance Task, Level, Time Required, Manning Skills, Support Equipment Requirements Task Sequence Frequency of Occurrence
D	Operation Maintenance Task Analysis	Identification, Source, Quantity Required Task Identification and Description, Time Required, Skills Identification of Common and Special Tools, Parts, Material Required Per Task
D1	Personnel and Support Requirements	Training Requirements Per Task Personnel Required Per Task Support Equipment Required Per Task Supply Support Required Per Task
E	Support Equipment and Training Material Description and Justification	Identification, Source, Quantity Required Size, Weight, Storage Volume, Nonrecurring and Recurring Costs Functions to Be Performed Characteristics and Installation Factors Justification for the New Material and Skill Requirements
E1	Unit Under Test and Automatic Test Program and Training Material Description	Test Program Set Elements Hardware and Software Required for Testing
F	Facility Description and Justification	Identification of New Facility Functions and Tasks to Be Performed Requirements, Design Criteria, Lead Times, Construction, Required Utilities Facility Utilization, Rate, Cost Justification

TABLE 2.4   MIL-STD-1388-2A Data Sheets (Continued)

Data sheet	Title	Scope
G	Skill Evaluation and Justification	Identification, Source, Quantity Required Skill Specialty Costs Functions to Be Performed Additional Skill and Training Requirements Selection Criteria, e.g., Physical, Mental, Educational Justification
H	Support Items Identification	Spare Parts Data Provisioning Screening Data Packaging Data
H1	Support Items Identification (Application-Related)	Application Data of Each Item on H Sheet Data Required for Initial Support Requirements Determination, Repair Parts Manual, etc.
J	Transportability Engineering Characteristics	Identification of Transportability Requirements

records shown in Table 2.4. These records have been changed to relational data-base tables in MIL-STD-1388-2B. Table 2.5 shows the general relationships between the 1A subtasks, the 2A records, and the 2B data tables. Historically, the non-LSAR-related LSA tasks have been submitted in hard-copy (paper) form, but there is now a trend to require these in digital format, such as magnetic disks, using mutually agreeable software programs.

TABLE 2.5   LSAR Data Record/Tables Related to MIL-STD-1388-1A Tasks

MIL-STD-1388-1A task/subtask	Applicable LSAR data records MIL-STD-1388-2A	Applicable LSAR tables MIL-STD-1388-2B
205.2.2	A, B	All A tables and GA
205.2.3	A, B	All A tables and GA
205.2.5	A, B	All A tables and GA
301.2.4.1	B, B1, B2, C, D, D1	All B and C tables and XI
301.2.4.2	B, B1, B2, C, D, D1	All B and C tables and XI
301.2.4.3	B, C, D, D1	All B and C tables and XI, GA
301.2.5	B, C, D, D1	All B and C tables and XI, GA
401.2.1	C, D, D1	All C tables and XI
401.2.2	C, D, D1	All C tables and XI
401.2.3	E, E1, E2, F, G, J, as applicable	All E and U tables and GB, GC, and GD
401.2.4	D1, G	Tables EE, GA, GB, GC, GD
401.2.5	D1, G	All C and F tables and XI
401.2.7	J	All J tables
401.2.8	H, H1	All H and X tables
401.2.9	B, C, D, D1, E, E1, E2, F, G, H, H1, J, as applicable	All tables as applicable
401.2.10	B, C, D, D1, E, E1, E2, F, G, H, H1, J, as applicable	All tables as applicable
401.2.11	B, C, D, D1, E, E1, E2, F, G, H, H1, J, as applicable	All tables as applicable
501.2.3	B, C, D, D1, E, E1, E2, F, G, H, H1, J, as applicable	All tables as applicable

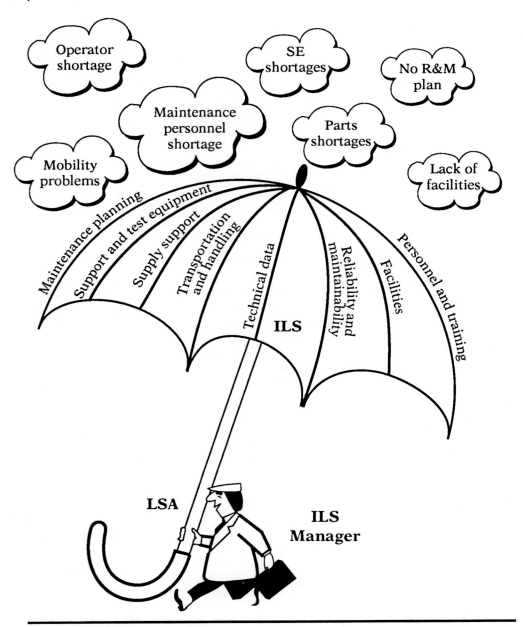

Figure 2.1   LSA helps integrate ILS.

The two LSA standards are integrators. They provide the ILS manager with information useful to integrate support considerations into design, to integrate the ILS elements into an optimized support system design, and to minimize duplication in the generation of detailed ILS element requirements. Figure 2.1 illustrates this integration role. The LSA integrates the ILS elements into the support system design in the 300 series tasks. The LSAR performs this function at the detailed ILS element support resource requirement level.

2.2 Management Development/Acquisition Processes

2.2.1 General

All military and commercial products undergo some management process that usually includes requirements (needs or desires) determination, development, production, and support. All such products normally have management reviews/decision points on whether to cancel or continue the program and some sort of strategy by which the project will be pursued. This section discusses these aspects in more detail.

2.2.2 Requirements Determination

A requirement may be a real need (daily food), a desire (I'd like a new car, although the old one is still acceptable), or a product marketing may feel is sellable to consumers (e.g., a desire will be created). It is important to realize that requirements determination interacts with technology capabilities. This fact is explicitly recognized by U.S. DOD acquisition policy. Numerous examples of this exist both commercially and militarily. Commercially, the electronics and computer revolution has resulted in many new products not conceived of as a desire or a need a few years ago. Militarily, "smart" bombs and ammunition were not a requirement until technology for these was developed.

Commercially, the decision to develop a new product normally is based on a combination of marketing judgments and technical input. Military requirements go through a slightly different process. Requirements determination in the U.S. DOD is a formal process.

For major defense systems, i.e., those above certain dollar thresholds, requirements determination is embedded in DOD Directive 5000.1, "Defense Acquisition,"[12] and DOD Instruction 5000.2, "Defense Acquisition Management Policies and Procedures,"[13] both dated February 23, 1991. Since many readers of this book will be interested in the changes in these documents, a summary discussion is contained in Appendix 2. Reference 14 presents another management perspective on this process.

The guidance in these documents states that needs will be expressed initially in broad operational terms and progressively translated into system-specific performance requirements. Assessments are to be made to determine if the need/deficiency can be satisfied by

A change in doctrine, tactics, training, or organization.

Changes (product improvement) to existing material.

Initiation of a new development.

It is noteworthy that such assessments also may identify opportunities created by possible technological breakthroughs (as exemplified by the "smart" bomb).

2.2.3 New trends in development production processes

The use of nondevelopment items (NDIs), including commercial items and product-improvement efforts, has long been an acquisition strategy employed in both the DOD and commercial industry in the United States as well as abroad. Obvious examples commercially include the automobile industry, where product improvement occurs annually while totally new designs are introduced less frequently. We noted in Sec. 2.2.2 that the U.S. defense establishment has recently made a major change in its acquisition process (see Appendix 2 for a discussion of this new process). While this has occurred in part because of dissatisfaction with the former process, this change also was fostered by the need to shrink the defense budget and to adapt new management approaches to the process. It is a premise of this book that these interests and trends are causing the commercial and defense development production processes to more closely resemble each other. Jacques Gansler recommended that "first, companies must shift from the separation of engineering, manufacturing, and support functions to an integrated operation in which process and product engineering are managed as a single unit."[15] He went on to note that "Japan is a nation whose economy is focused on the civilian sector...they stress in their defense industrial sector the same characteristics found in their civilian sector: high volume (with a manufacturing-process orientation in engineering), low cost, and high quality." Evidence of this trend toward greater commonality follows.

Adaptations are occurring in both defense and commercial industries of such management initiatives as total quality management (TQM), computer-aided design, manufacturing, and support [termed computer-aided acquisition and logistic support (CALS) in the DOD], and concurrent engineering (CE). This trend is reflected commercially in advertisements by automobile manufacturers that illustrate the integrated engineering approach by showing the single location from which the design operates. There are numerous TV commercials showing management commitment to quality.

At the same time, there is growing commercial industry recognition that more emphasis is needed on such systems engineering specialties as reliability, maintainability, and safety. These characteristics of products are receiving more and more emphasis in the commercial sector. Consumers expect an automobile to fail infrequently and repairs to be effected swiftly and as inexpensively as possible. With integrated circuits at the heart of control systems as diverse as personal computers and automobile ignition systems, these considerations rank high in the priorities of consumers of commercial products.

A third point substantiating our main theme is the increasing awareness in commercial industry of the defense acquisition emphasis on the advantages in cost and schedule to be gained in using NDIs when such items fit military needs. The new DOD acquisition guidance is very clear in this regard. Where commercial products have potential application to defense or government, it can be to the benefit of the manufacturer to examine how much extra it would take, in the way of supportability, to meet defense needs.

What does all this mean to logisticians and to ILS and LSA? We mentioned in Chap. 1 that ILS/LSA concepts were being increasingly applied to a widening range of commercial products. Examples of such techniques crossing over to the commercial sector include analyses of operational availability, life cycle cost, CALS, readability/understandability, maintenance feedback systems, and built-in diagnostics. This should not be surprising. Just as LSA has followed reliability and maintainability (R&M) as a recognized military need, so LSA techniques are following the need for R&M into the commercial sector. Some commercial product objectives are the same as those of the DOD. The incorporation of LSA-type analysis both serves the commercial market and enhances the potential for commercial products to be appropriate for DOD use.

A last point worth mentioning is that these trends are obviously not unique to the United States, but, in part, are international in scope.

2.2.4   LSA impact

In the U.S. DOD, LSA requirements and execution must be tailored to fit with the requirements-determination decision mentioned in Sec. 2.2.2, the type (NDI, PIPs, etc.) of acquisition, and the various strategies, such as TQM, CE, CALS, competition, etc., cited above. LSA tailoring by the type of acquisition is discussed in more detail in Chap. 4.

TABLE 2.6 Logistic Support Analysis Process Overview

	Preconcept	Concept exploration	Demonstration and validation	Engineering manufacturing development	Production deployment post production
Program documentation		Justification for major system new start	System concept paper	Decision coordination paper and integrated program summary	
Baseline Specifications			Functional	Allocated	Product
			System specification →		
				Development specification →	
					Product process material specification →
Reviews and audits	System replacement review			System design review · Preliminary design review · Clinical design review · Functional config. audit	Formal qualifications review · Physical config. audit · Production readiness review · Post production support review
ILS planning		← ILS plan integrated support system →			
Logistic, support, analysis, effort	Development of an Early Logistic Support Analysis Strategy (101)\n\nUse Study (201)\n\nComparative Analysis (203)	Development of an Early Logistic Support Analysis Strategy (101)\n\nLogistic Support Analysis Plan (102)\n\nProgram and Design Reviews (105)\n\nUse Study (201)\n\nMission Hardware, Software and Support System Standardization (202)\n\nComparative Analysis (203)\n\nTechnological Opportunities (204)\n\nSupportability and Supportability-Related Design Factors (205)\n\nFunctional Requirements Identification (301)\n\nSupport System Alternatives (302)\n\nEvaluation of Alternatives and Tradeoff Analysis (303)\n\nSupportability Test, Evaluation, and Verification (301)	Development of an Early Logistic Support Analysis Strategy (101)\n\nLogistic Support Analysis Plan (102)\n\nProgram and Design Reviews (105)\n\nUse Study (201)\n\nMission Hardware, Software and Support System Standardization (202)\n\nComparative Analysis (203)\n\nTechnological Opportunities (204)\n\nSupportability and Supportability-Related Design Factors (205)\n\nFunctional Requirements Identification (301)\n\nSupport System Alternatives (302)\n\nEvaluation of Alternatives and Tradeoff Analysis (303)\n\nTask Analysis (401)\n\nSupportability Test, Evaluation, and Verification (301)	Logistic Support Analysis Plan (102)\n\nProgram and Design Reviews (105)\n\nUse Study (201)\n\nMission Hardware, Software and Support System Standardization (202)\n\nComparative Analysis (203)\n\nSupportability and Supportability-Related Design Factors (205)\n\nFunctional Requirements Identification (301)\n\nSupport System Alternatives (302)\n\nEvaluation of Alternatives and Tradeoff Analysis (303)\n\nTask Analysis (401)\n\nEarly Fielding Analysis (402)\n\nSupportability Test, Evaluation, and Verification (301)	Logistic Support Analysis Plan (102)\n\nProgram and Design Reviews (105)\n\nPost Production Support Analysis (403)\n\nSupportability Test, Evaluation, and Verification (501)\n\nFollowing LSA Tasks Applicable to Design Changes Only\n\nMission Hardware, Software and Support System Standardization (202)\n\nSupportability and Supportability-Related Design Factors (205)\n\nFunctional Requirements Identification (301)\n\nEvaluation of Alternatives and Tradeoff Analysis (303)\n\nTask Analysis (401)\n\nEarly Fielding Analysis (402)
Test and evaluation		← Developmental T&E	initial operational T&E		Production acceptance T&E follow on T&E
Logistics outputs for subsequent phase	Supportability constraints LSA strategy	Cost and readiness improvement targets\n\nSupport concept alternatives\n\nSupportability-related objectives	Support concept\n\nSupportability-related goals and thresholds	Logistic support resource requirements\n\nCompetitive action plan	Assessment of supportability and readiness\n\nPlans for hardware and support improvement\n\nPost production support

SOURCE: MIL-STD-1388-1A, "Logistic Support Analysis."

The application of the LSA tasks shown in Table 2.6 for new developments obviously must be modified to fit these variables. Types of acquisitions and various acquisition strategies may be combined as illustrated in the following March 1991 example from a joint program office:

Harmonizing operational requirements among the Services and Unified and Specified Combatant Commands.

Procuring off-the-shelf technologies for initial systems, thereby reducing cost, risk, and duration of development.

Developing improved specifications for systems after the Services have acquired hands-on operational experience, using, for example, available NDI. Operational experience is essential for reducing costs by providing users the basis for establishing specific performance specification upgrades.

Conducting advanced research and development that enhances the system future capabilities. Advanced technologies are incorporated through block upgrades to the systems.

Maintaining all equipment interfaces, interface control documents, and specifications to ensure effective block upgrades and interchangeability of systems and subsystems.

Ensuring interoperability among all systems and subsystems with the command, control, communications, and intelligence (C^3I) systems of the services.

Employing a competitive and evolutionary acquisition process to incorporate block upgrades to air vehicles, payloads, data links, mission planning and control stations, launch and recovery, and logistics support systems.

These acquisition strategies have a number of LSA implications. Specifically, these are

Need for LSA NDI approaches to selection of off-the-shelf technologies.

Collection of support-related data during initial field use.

Need for LSA approaches for application to off-line subsystem/ component development if design is to be influenced.

Need for LSA approaches to optimize logistics resources when competition exists.

2.3   The System Engineering (SE) Process and Supportability Design Influence

2.3.1   The SE process

The complications of management of the design, development, production, and fielding of complex items have resulted in the development

TABLE 2.7   Table of Contents: DSMC SE Management Guide

of the Systems Engineering process. A good discussion of this process commercially or generically is presented in Blanchard, *System Engineering Management.*[9] Blanchard defines *system engineering* to constitute the application of scientific and engineering effort to

1. Transform an operational need into a description of system performance parameters and a preferred system configuration through the use of an iterative process of functional analysis, synthesis, optimization, definition, design, test, and evaluation.

2. Integrate related technical parameters and ensure compatibility of all physical, functional, and program interfaces in a manner that optimizes the total system definition and design.

3. Integrate reliability, maintainability, logistics support, human factors, safety, security, structural integrity, producibility, and other related specialties into the total engineering effort.

For U.S. DOD needs, this process is embodied in MIL-STD-499. A good discussion of the process is in the Defense System Management College (DSMC), "System Engineering Management Guide."[8] Table 2.7 shows the table of contents of this guide to illustrate the scope of the SE process. Figure 2.2 illustrates the steps in the process, and Fig. 2.3 provides suggested documentation and briefly describes the steps in the process.

SE documentation normally includes

- The System Engineering Management Plan* (SEMP)
- System, Segment, Prime Item, and Computer Software Configuration Item Specifications
- Interface Control Documents
- Trade Study Reports/Directions*
- Risk-Analysis Management Plan
- Survivability/Hardness Plan*

Note the ILS/LSA content of the special documentation in Fig. 2.3.

---

*Possible LSA content in each of these documents.

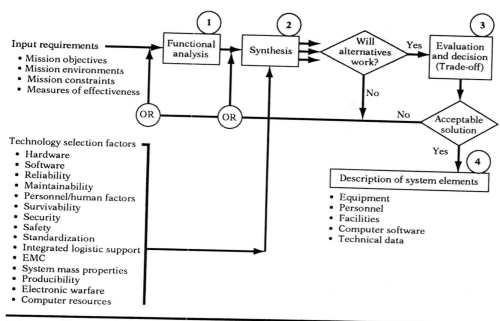

Figure 2.2 The SE process.

2.3.2 SE-LSA integration

Most experts consider LSA (some also say ILS) a part of the SE process. A common view is shown in Fig. 2.4. The new U.S. DOD acquisition policy states that LSA is a part of the SE process. The tasks in the LSA standard were designed by applying the SE concept to logistics analysis needs.

LSA is clearly related in part to the SE process. This is illustrated by Tables 2.8 and 2.9, which portray the relationships of the LSA tasks and the LSA records to the SE process and documentation previously described. These figures were derived from more detailed figures in Ref. 8.

2.3.3 LSA-SE Specialty Interfaces

There is agreement that LSA is related to various SE specialties or techniques. The most obvious of these are R&M, Human Engineering (HE), safety, and configuration management. These relationships are discussed in detail in Jones[3] and Blanchard.[4]

2.3.4 SEMP LSA content example

An example of inclusion of LSA in a multiprogram draft SEMP, dated October 1990, contains the following material: "The LSA tasks pro-

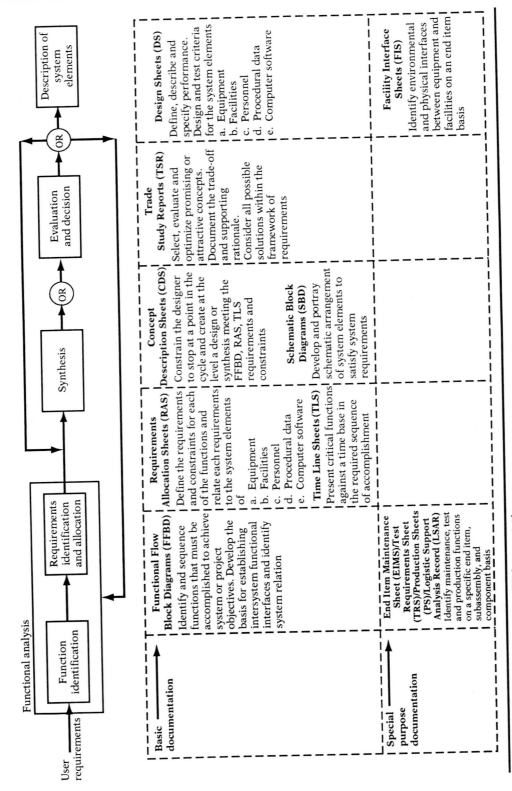

Figure 2.3 Basic and special-purpose SE documentation.

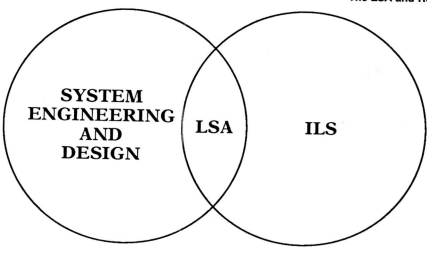

Figure 2.4 Relations.

TABLE 2.8 SE-LSA Task Relations

SE Process step	Functional analysis						Synthesis		Evaluation and decision	Description of system elements	
	Function identification				Requirements identification and allocation						
SE Document / LSA Task	F F B D	E I M S	T S R	P S	R A S	T L S	C D S	S B D	T S R	D S	F I S
200 Series		X	X	X	X	X	X		X		
300 Series		X			X				X		
400 Series 401, 402		X	X					X	X	X	X
403				X							X
500 Series			X	X					X		

vide a major source of supportability and readiness information for the system engineering studies. Table 2.10 lists specific LSA tasks useful in this regard. Since these tasks involve engineering as well as support considerations, engineering must collaborate with ILS/LSA on these tasks."

The SEMP requirement above and other program LSA policy/ guidance provide the multiprogram ILS/LSA communities the authority to obtain the system and design engineering assistance necessary to ensure high-quality LSA products. The "team" principle underlying

TABLE 2.9   SE-LSAR Relation

SE Process step	Functional analysis						Synthesis		Evaluation and decision	Description of system elements	
	Function identification				Requirements identification and allocation						
SE Document	F F B D	E I M S	T S R	P S	R A S	T L S	C D S	S B D	T S R	D S	F I S
LSAR Table A	X				X	X	X		X		
B		X			X				X	X	
C		X			X	X			X		X
E	X	X			X		X		X	X	
F		X			X			X	X	X	X
G					X	X			X		
H		X			X				X		
J					X				X		

TABLE 2.10   LSA Engineering-Related Tasks

Task 201	Use Study
Task 202	Standardization
Task 203	Comparative Analysis
Task 204	Technology Opportunities
Task 205	Supportability and Supportability-Related Design Factors
Task 301	Functional Requirements
Task 302	Support System Alternatives
Task 303	Evaluation of Alternatives and Tradeoff Analysis
Task 501	Test and Evaluation

this requirement is consistent with the increasing use of concurrent engineering concepts.

2.3.5   Requirements/design influence

The LSA tasks produce information to be used to influence operating and design requirements. Operating requirements, as well as design requirements, are explicitly included because there are cases where the initially specified desired operating requirements are impossible or too costly to support. As a function of decrease in unit support cost attributable to economies of scale, the ratio of support cost per mission to equipment cost decreases dramatically as the number of mission equipments supported at a given location increases.

A major use of LSA task results is to affect design requirements. This is particularly true of the 200 series tasks. To achieve this goal, the system performance and design specifications must include support-related requirements. In general, most designers, contrary to what a lot of logisticians believe, wish to make their designs "supportable." One problem is that *supportable* is a general term that has different meanings to different people. One approach to making its meaning more concrete is to define *supportability* by identifying those paragraphs of the specification which heavily affect supportability.

2.3.6 An example of supportability in system specifications

This draft system specification example, dated June 1991, contains both generic family and specific program performance specifications. Since this is a key place to influence requirements and design, inclusion of supportability requirements is essential. Table 2.11 includes current paragraphs about the "family" requirements, generally expanded in the individual program sections, which are of particular interest to ILS/LSA. While these requirements of interest are too lengthy to repeat here, several of these—readiness, the family "design to maintenance" concept, and computer resource support—are key to supportability and are therefore included in this example.

TABLE 2.11   Support-Related System Specification Requirements

Paragraph no.	Title
3.1.1.2	Commonality Requirements
3.1.1.2.1	Common Core Development Items
3.1.2.1	Readiness
3.1.2.2	Operational Modes
3.1.2.3	Manning Requirements
3.1.3.2.1	System Reliability
3.1.3.2.2	Mission Reliability
3.1.2.2.3	Recovery Reliability
3.1.3.2.4	Operational Availability
3.1.4.2	Storage Environments
3.1.4.3	Transportation Environments
3.1.5.1	Design for Modularization
3.1.5.2	Design for Standardization
3.1.5.3	Design for Maintainability
3.1.5.4	Accessibility
3.1.5.5	Human Factors
3.1.5.6	Configuration Management
3.1.5.8	Service Life
3.1.5.10	Identification Marking
3.1.5.11	Packaging Design
3.1.6	Integrated Logistics Support (ILS)
3.1.7	Safety and Security

The draft family specification includes the following readiness material in paragraph 3.1.2.1:

> Readiness is the ability of the family of systems to undertake and sustain a specified set of missions at planned rates in both peacetime and wartime conditions. Therefore the development of design features of the systems which influence readiness shall be a primary objective of the system acquisition process. The design of the systems shall be such that throughout the life of the system they are capable of being maintained at the specified readiness levels by the operating personnel.
>
> The ability of a system to achieve a specified level of readiness is predicated upon relevant factors such as:
>
> Personnel manning and training.
>
> Reliability and maintainability.
>
> The characteristics and performance of subsystems and support systems.
>
> The quantity and location of support services.
>
> Platform operations, i.e., the requirement for the platform to provide the required support and to be restricted, as appropriate, in motions, speed, and other dimensions when necessary to support the response to launch.
>
> Target achievability.
>
> Examples of system readiness measures are combat sortie rate over time, peacetime mission capability rate, operational availability (A_o), percent coverage per day, and asset ready rate.

As the LSA (supportability) input flows to the "family" new needs decisions and the LSA module of the family logistics data base, an important by-product will be creation of a solid basis for refining and improving program LSA policy/guidance.

A "logistics reliability" parameter requirement should be added to the system specification because logistics reliability drives support requirements and frequently differs significantly from mission reliability values. Software R&M is covered below under "Computer Resource Support."

Design-to-maintenance concept. To facilitate commonality among the different subsystems in the "family," the following design-to-maintenance concept is contained in the draft specification:

> A "design to" maintenance concept with maximum feasible service commonality shall be specified in all systems. Deviations from the baseline "design to" concept shall be justified. The maintenance concept herein shall be the departure point concept for austere operational environments. Organizational ("O") maintenance shall require an entry-level three-skill technician to perform assembly, checkout, launch, recovery and turnaround, limited maintenance, and disassem-

bly functions. Limited maintenance shall be restricted to replacement of easily accessible expendable items (lamps, fuses), replacement of failed modules, periodic preventive maintenance task, etc., using simple hand tools (wrenches, pliers, screwdriver). No special or common test equipment shall be required. Diagnostics shall be accomplished by BIT to isolate faults to a specific module. Any adjustment shall be within the capability of the entry-level technician. The technical manual for "O" level shall include maintenance and operator tasks in a single manual. Depot ("D" level) maintenance shall be accomplished by an integration depot to be determined. Depot maintenance shall perform full repair of modules as required. Development of organic capability shall be pursued in accordance with normal economic procedures. Results of Level of Repair Analysis (LORA) shall be used to make depot repair decisions and provide feedback to engineering for possible design improvements. Special and/or common test equipment shall not be authorized below depot level. Any repairs capability, including otherwise available skills and equipment that exists at intermediate or other levels of maintenance, may make repairs to modules but no unique tools or support equipment shall be authorized.

Computer resource support.   The general system specification contains the following software requirement:

> The amount and type of computer resource support needed shall include software (SW) design parameters such as meantime between mission critical software failures (MTBMCF SW), average time to restore a software intensive system to its previous level prior to a critical software fault (MTTR SW), and meantime to reboot (MTTRS). Design principles for software to ensure supportability shall include modularity, adaptability, redundancy, reconfigurability, design for fault tolerance and ease of debugging, design to reduce operator errors, menu, and help capabilities.

2.3.7   Example system specification LSA implications

The readiness requirements are consistent with program LSA policy. Readiness analysis is implied in LSA subtasks 303.2.4 and 402.2.3. It also can be updated as part of tasks 402 and 501.

The inclusion of logistics reliability parameters and software R&M parameters will result in measurement of these parameters in Test and Evaluation (T&E) and thus provide feedback. These requirements should be considered in such LSA tasks as 201, 203, 205, 302, 303, 401, 402, and 501.

The "design to" maintenance concept cited is a target, not a mandatory, requirement. Its inclusion in the system specification is a forcing mechanism to increase ILS commonality.

The ILS section of the system specification identifies those design requirements which heavily affect supportability and readiness. This provides a convenient informal but explicit definition of supportability to design personnel. Design personnel can use this definition as a basis for discussing design supportability aspects in both internal contractor and joint contractor/government program and design reviews. RFP language in this regard is discussed in Appendix 8.

2.4 The ILS, HARDMAN/MANPRINT, Human Systems Integration (HSI), and Maintenance Planning Analysis Processes

2.4.1 ILS

Integrated logistics support (ILS) is a systems management approach adopted as management realized (1) that the cost of support was increasing and was a major part of system life cycle cost and (2) that

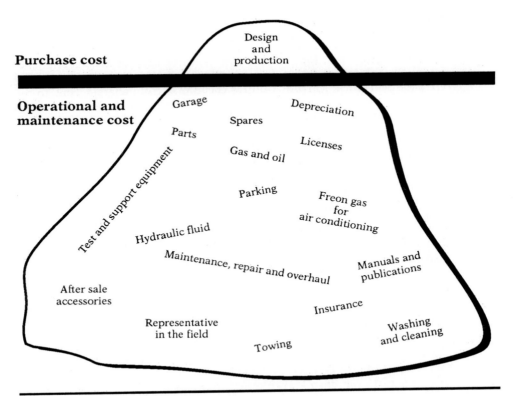

Figure 2.5 Life cycle costs of a car.

many aspects of support were interrelated. The purchase cost aspect and major elements of support are shown for an automobile in Fig. 2.5.

A good statement of a fundamental precept of ILS is to relate support to design and to use an engineering analytical approach, rather than rules of thumb, for designing the logistics support subsystems for acquisitions.

A more formal definition of ILS used by the U.S. DOD in recent years is

> A disciplined, unified and iterative approach to the management and technical activities necessary to:
>
> Integrate support considerations into system and equipment design;
>
> Develop support requirements that are related consistently to readiness objectives, to design, and to each other;
>
> Acquire the required support; and
>
> Provide the required support during the operational phase at minimum cost.

Note that the first two topics above are LSA-related. Obviously, ILS concepts can be applied or adapted in various degrees commercially as well as to defense systems and equipment. There are variations on what support elements are "integrated" by ILS management. The superseded U.S. DOD policy directive, DOD 5000.39, "Acquisition and Management of ILS for Systems and Equipment," listed the following "principal" elements:

- Maintenance planning
- Manpower and personnel
- Supply support (including initial provisioning)
- Support equipment
- Training and training support
- Technical data
- Computer resource support
- Packaging, handling, storage, and transportation

The various U.S. services enhanced the above list, as shown in Table 2.12.

2.4.2 HARDMAN/MANPRINT/HSI

As manpower and personnel costs and skill shortages increase, management may decide to (1) use automation and technology as a substitute and/or (2) manage the "people-related" aspects of a system in a

TABLE 2.12   Variations In ILS Elements

Subject	USMC	Army	AF	Navy
Standardization	X	X	X	
Interoperability		X		
Design Influence				X
R&M			X	
Survivability		X	X	
Energy	X		X	
Configuration Management	X			
Quality Assurance				
Material Fielding Plan	X	X		
Postproduction Support				

"systems" manner. There are numerous examples of the first approach both commercially and militarily. Because "people" are such a major aspect of defense systems, the U.S. DOD has decided to adapt both the first and second approaches. The second approach was initially adopted by the U.S. Navy in its HARDMAN initiative. HARDMAN focused primarily on manpower, personnel, and training. An analysis process was developed to address these concerns. This process is summarized in Appendix 3. Since this process relies heavily on comparative-analysis techniques, it has an obvious interface with the ILS and LSA processes, in particular LSA task 203. The U.S. Army expanded the HARDMAN concept to MANPRINT by including human engineering and safety. This created interfaces with both ILS and SE. The MANPRINT concept is now embedded in the new DODD 5000.1. Many Army organizations require contractors to appoint a common ILS and MANPRINT manager to integrate the two concepts at the working level. DOD has recently renamed this process as *Human Systems Integration* (HSI).

The HARDMAN process is usually composed of five or six steps. The U.S. Navy five-step version and related LSA tasks are as follows:

Step 1: Collect preliminary data and conduct system analysis (LSA tasks 103, 203, 204, and 402).

Step 2: Conduct comparability analysis (LSA tasks 203, 205, and 301).

Step 3: Develop manpower, personnel, and training (MP&T) concept (LSA tasks 101, 302, 303, and 401).

Step 4: Develop MP&T resource requirements (LSA tasks 102, 401, and 501).

Step 5: Develop program documentation input.

Appendix 3 contains a more complete discussion of these steps to aid the ILS manager in relating HARDMAN and ILS/LSA, derived from Ref. 16.

Because of the obvious interfaces, the integration of the SE specialties, ILS, LSA, and HARDMAN/MANPRINT/HSI is a significant challenge to management. Early HARDMAN guidance[16] recommended its application prior to concept exploration/development. More recently, the need to update HARDMAN in later phases has been recognized.

2.4.3 The Maintenance Planning Analysis (MPA) Process

The MPA process is a four-step process depicted in Fig. 2.8. The first step in the process is the identification and analysis of system/equipment failures. This is normally done through a failure modes and effects analysis (FMEA). The FMEA identifies the failures to be prevented or repaired. If a Criticality Analysis (CA) is performed with the FMEA, the resulting analysis is a failure modes, effects, and criticality analysis (FMECA). The FMEA and/or FMECA is normally performed by the reliability and maintainability group.

The critical failures identified by the FMECA are then analyzed using a Reliability-Centered Maintenance (RCM) logic to identify possible preventive maintenance (PM) tasks. Both the FMECA and RCM may result in a request for design change to eliminate or lessen the consequences of failures. RCM analysis may result in a preventive maintenance task or a decision to permit the item to fail, thus generating a corrective maintenance task.

The next step in the MPA process is task analysis of the CM and PM tasks identified by the FMECA and RCM analyses. Additional tasks, such as servicing and lubrication, are also identified and analyzed. Maintenance task analysis defines each task, the required skills and support resources, and the organizational level at which the task will be performed.

Since task analysis includes the maintenance level of repair, a

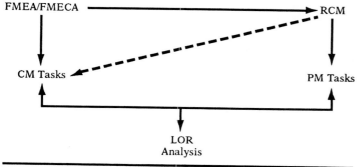

Figure 2.6 MPA process. (*From Draft SPAWAR MPA-LSA Desk Guide.*)

level-of-repair (LOR) analysis is normally conducted. The LOR determines the level at which repair will be performed based on both economic and noneconomic factors.

The MPA process is a subset of the LSA process. This is depicted conceptually in Figs. 2.6 and 2.7. The LSA process can identify the maintenance concept through performance of the various LSA tasks listed in the early phases of a development program. A LOR also can be used to aid in the determination of the maintenance concept.

Subtask 301.2.4 of LSA task 301, Functional Requirements Identification (FRI), records the results of the FMECA and RCM analyses in the LSAR. The FMECA and RCM results are used to identify needed corrective maintenance and preventive maintenance tasks respectively in subtasks 301.2.4.1 and 301.2.4.2. Operator tasks and other support tasks, such as servicing and lubricating, are also identified as part of subtask 301.2.4.3.

The LOR is called out in LSA subtask 303.2.7. The government should specify the LOR model to be used. LOR results are recorded in the LSAR.

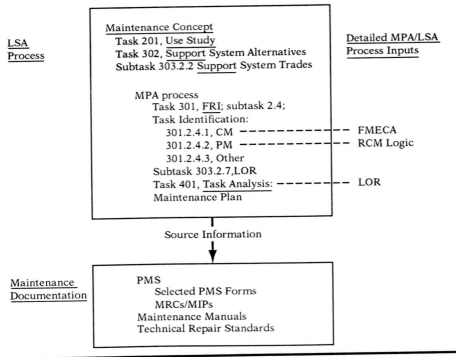

Figure 2.7   LSA inputs to MPA and PMS. (*From Draft SPAWAR MPA-LSA Desk Guide.*)

LSA task 401, Task Analysis, requires a detailed analysis of each task identified in task 301. The analysis recorded in the LSAR includes the procedural steps required and such data as task frequency, task interval, elapsed-time manhours, skill level, logistics resources, and the maintenance level to perform the task. Based on these data, the LSAR system can produce input to the maintenance plan using the LSA-024 Output Report.

2.5 Related Management Initiatives/Trends

2.5.1 General

There are various major management initiatives/trends that also have an impact on LSA/availability engineering. This section discusses

- Total Quality Management (TQM)   (2.5.2)
- Information Engineering/Computer-aided Acquisition and Logistics Support (CALS)   (2.5.3)
- Concurrent Engineering   (2.5.4)
- Acquisition streamlining   (2.5.5)
- Program software intensity   (2.5.6)
- AI/expert systems   (2.5.7)

2.5.2 TQM

General. As global industrialization and competition increase, there is growing recognition of the need for higher quality and user satisfaction. This is resulting in increased management attention to TQM both commercially and militarily. A representative definition taken from the final draft of a DOD Publication dated February 15, 1991, *TQM Guide*, Vol. 1, is as follows:

> TQM is both a philosophy and a set of guiding principles that represent the foundation of a continuously improving organization. TQM is the application of quantitative methods and human resources to improve the material and services supplied to an organization, all the processes within an organization, and the degree to which the needs of the customer are met, now and in the future. TQM integrates fundamental management techniques, existing improvement efforts, and technical tools under a disciplined approach focused on continuous improvement.

It is obvious from this definition that TQM is a management process that can involve everyone in the organization, from the highest to the lowest levels. In fact, the TQM literature emphasizes the active involvement of top management. A typical TQM model for continuous improvement is shown in Fig. 2.8.

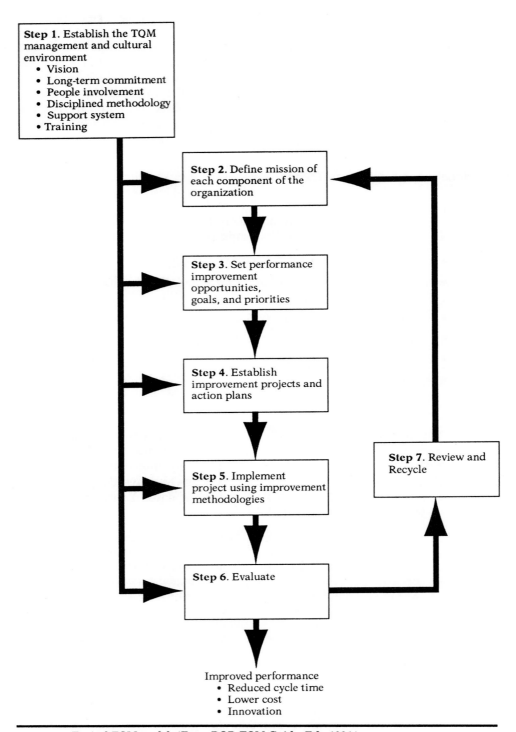

Figure 2.8 Typical TQM model. (*From DOD TQM Guide, Feb. 1991.*)

TABLE 2.13   Some of the Common Tools Used in Implementing TQM

Tools and techniques (listed without regard to priority)	Problem-solving activities					
	1 Bound and prioritize problems	2 Compile information	3 Analysis	4 General alternatives	5 Evaluate	6 Plan and implement
Benchmarking	✔	✔	✔		✔	
Cause and effect diagrams	✔		✔			
Nominal group technique	✔	✔		✔	✔	✔
Quality function deployment	✔		✔		✔	
Pareto charts	✔	✔	✔			
Statistical process control		✔	✔			
Histograms		✔	✔			
Check sheets		✔	✔			
Input/output analysis			✔			
Scatter diagrams			✔			
Concurrent engineering			✔	✔		
Design of experiments			✔	✔	✔	
Cost of quality			✔		✔	
Control charts			✔		✔	
Work flow analysis			✔			✔
Team building						✔
Time management	✔					✔
Shewart cycle	✔	✔	✔	✔	✔	✔

SOURCE: DOD TQM Guide, Feb. 1991.

Table 2.13 contains a list of commonly used tools and techniques in TQM. The list is representative rather than comprehensive, as most readers will readily recognize. Readers may be interested in making judgments about TQM status and progress in their own organizations. Table 2.14 contains a benchmarking matrix useful in this regard.

TQM ILS/LSA impact. Since TQM is "all pervasive," it is difficult to pinpoint its precise impact on ILS/LSA. However, there can be no doubt that TQM, with its emphasis on customer satisfaction and continuous improvement, creates an environment more favorable to support design influence and thus LSA because satisfactory product support can be a major factor in customer satisfaction.

2.5.3   Information Engineering/CALS

The changing environment.   The 1990s will be years of continuous transition featuring

- Movement from paper to a digital environment.
- Migration from islands of automation to interfaced systems.
- Structure of corporate enterprise information models.

TABLE 2.14 Example Guide to Assess Status of TQM Implementation

TQM category	Top-management commitment	Obsession with excellence	Organization is customer-driven	Customer satisfaction	Training	Employee involvement	Use of incentives	Use of tools
5	Continuous improvement is a natural behavior even during routine tasks	Constant, relative improvement in quality, cost, and productivity	Customer satisfaction is the primary goal	More customer state intention to maintain long-term business relationship	Training in TQM tools common among all employees	People involvement: self-directing work groups	Gainsharing (cross-functional teams)	Statistics is a common language among all employees
4	Focus is on improving the system	Use of cross-functional improvement teams	Customer feedback used in decision making	Striving to improve value to customers is a routine behavior	Top management understands and applies TQM philosophy	Manager defines limits, asks group to make decision	More team than individual incentives and rewards	Design and other departments use SPC techniques
3	Adequate money and time allocated to continuous improvement and training	TQM support system set up and in use	Tools used to include wants and needs in design	Positive customer feedback, complaints used to improve	Ongoing training programs	Manager presents problem, gets suggestions, makes decision	Quality-related employee selection and promotion criteria	SPC* used for variation reduction
2	Balance of long-term goals with short-term objectives	Executive steering committee set up	Customer needs and wants are known	Customer rating of company is known	Training plan developed	Manager presents ideas and invites questions, makes decision	Effective employee suggestion program used	SPC* used in manufacturing
1	Traditional approach to quality control Inspection is primary tool (control of defects, not prevention) Better quality = higher cost Significant scrap and rework activity Quality control found only in manufacturing departments MBO improperly used for all departments							

Desired direction ← Standing

*SPC used as an example.
SOURCE: DOD TQM Guide, Feb. 1991.

- Electronic interfacing with customers, partners, and suppliers.

- Evolution of integrated information management processes and standards.

The DOD Computer-aided Acquisition and Logistic Support (CALS) initiative launched in 1985 is one example of management response to general trends in automation of paper and advances in telecommunications. The DOD motivating factors include the current defense restructuring, shrinking budgets, and better quality, shorter lead times, and lower costs. The industry motivation is to meet global competition through better quality, shorter lead times, and lower costs by exploitation of integrated information technologies in terms of enterprise integration and trading partners. CALS is also under way in Europe, Canada, and Australia.

Figure 2.9 shows the planned transition from islands of information to interface systems to the ultimate objective of an integrated weapon system shared data base. The major focus of CALS is on technical information automation, integration, interactive/on-line access, and digital delivery.

CALS/LSA interface/impact. CALS is heavily affecting ILS/LSA. Figure 2.10 summarizes the ILS benefits projected on a major system acquisition. CALS also has heavily affected the new LSAR standard. The development of MIL-STD-1388-2B occurred under CALS auspices/sponsorship. The CALS initiative will make the LSAR much more useful. Examples include the following:

- The change from the flat files in 2A to the relational data tables approach in 2B makes the preparation of ad hoc reports a possible common practice. Conversion of 2A data bases to 2B also will significantly increase the usefulness of these data bases.

- CALS will make possible increases in automated LSAR linkages with other data bases. For example, data in the LSAR with regard to the depot maintenance report formerly required hand extraction of LSAR data from some 10 or more reports, and all the LSAR data only partially provided the data needed for the maintenance report. In the future, computer-aided design (CAD) drawing numbers may be linked directly to the LSAR.

- The CALS initiative is currently expected to result in many cases in the continued update of the program LSAR for use in project support after production of the item has ceased. This, in turn, will enhance use of the LSAR on new but similar programs.

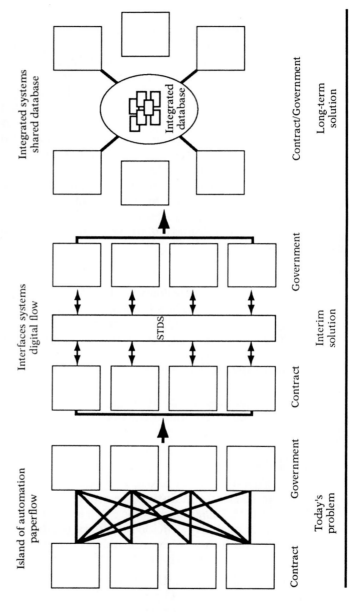

Figure 2.9 CALS transition. *(From DOD Handbook, DOD CALS Program Implementation Guide, Sept. 1990.)*

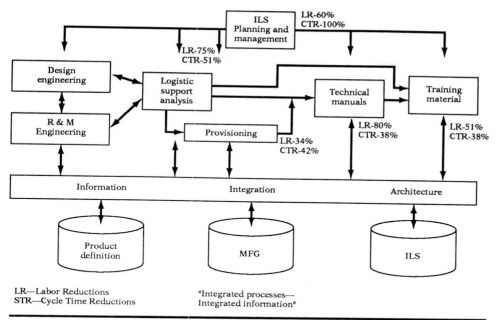

Figure 2.10 Projected CALS acquisition ILS benefits. (*From "Application of CALS to the UAV Family," April 1990.*)

The impact of CALS on R&M and the LSA tasks not heavily LSAR-oriented is discussed in the next subsection.

2.5.4   Concurrent Engineering (CE)

General.   Concurrent Engineering (CE) is a concept gaining wider usage both commercially and militarily in the United States, Europe, and Japan. The industry support group (ISG) to the DOD CALS initiative now calls itself the CALS(CE). Concurrent Engineering is intended to reduce cost, shorten development time, and improve quality. Reference 21 defines CE as

> ...a systematic approach to the integrated, concurrent design of products and their related processes, including manufacture and support. This approach is intended to cause the developers, from the outset, to consider all elements of the product life cycle from conception through disposal, including quality, cost, schedule, and user requirements.

The philosophy of CE is not new. The terms *system engineering, simultaneous engineering,* and *producibility engineering* have been used to describe similar approaches. Various individual specialties, including R&M and ILS/LSA, have long advocated greater and more timely

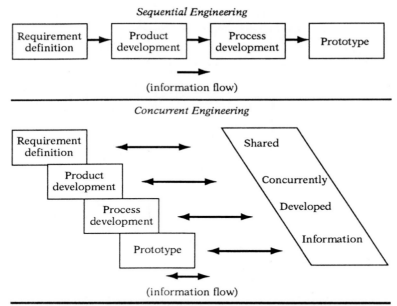

Figure 2.11   A comparison of sequential and concurrent engineering. (*From IDA Report R = 338, "The Role of Concurrent Engineering in Weapon System Acquisition," Dec. 1988.*)

consideration of their concerns in the design process. CE is beginning to make this happen. A comparison of sequential and concurrent engineering is shown in Fig. 2.11, taken from the IDA report on the relationship of CALS and CE. Like CALS, concurrent engineering concepts are being adopted commercially as well as militarily.

CE impact on R&M and LSA.   The CALS and CE initiatives are currently affecting R&M. The R&M community has already automated many of the R&M analysis techniques required. A list of the types of automated analysis tools available from one U.S. vendor includes

- System availability for systems with complex redundancy
- Reliability predictions (electronics)
- Parameter information for microelectronics
- FMECA
- Fault-tree diagrams
- MTTR calculations
- Parts-count predictions

- Drawing of reliability block diagrams
- Calculation of spares cost given stock-out risk

While CALS standards for technical reports, such as the R&M and LSA task reports, do not yet exist, the delivery of such reports by disks using mutually agreeable software is now possible. A number of U.S. DOD programs are planning to do this. More cost-effective means for digital delivery/on-line access will be available in the near future.

However, CALS offers even more opportunities for gain. The CALS Handbook MIL-HDBK-59A suggests that contractors indicate in their proposals their use of R&M automated and, where available, integrated-design R&M tools. Five specific areas of focus are as follows:

Automated R&M analysis procedures tightly coupled to parts libraries and to material characteristics data bases.

Automated R&M synthesis based on design rules incorporating lessons learned from prior design experience and field use.

Fully characterized (tested and validated) component performance and R&M characteristics data bases.

Consistent data management procedures that link major design decisions affecting the R&M characteristics of the end item to the CAE software and data bases used to develop decision criteria and otherwise support the evolving configuration of the product.

A structure of hardware, software, and computer networks to support the procedures and processes of the above and to closely couple R&M specific resources (including personnel) with the rest of the design team.

The development of automated design tools is also fostering the integration of supportability considerations with automated design tools. An example of design tools that permit earlier support consideration in design simulation tools is computer mock-up (COMOCK), shown in Table 2.15. Similar computer mock-up capabilities are now in use in shipbuilding and automobile, aircraft, and commercial building design.

In the longer term, we expect to see many of the LSA task considerations, including performance/cost measures, models such as LORA, LCC, and system readiness, and design/support parametrics integrated with CE tools and techniques in a manner similar to that now beginning to occur in R&M.

TABLE 2.15   Computer Mock-Up Example

Goals	Impact
Metal mock-up sequential design	Computer mock-up
Supportability requirements not fully addressed until after aircraft design	Parallel design
Metal mock-up fabricated after design is completed	Supportability requirements addressed early in design process through zone coordination meetings
Design data are reentered manually to create analysis and manufacturing models	Computer mock-up is used for conceptual and detail design studies as it evolves iteratively as part of the design process
Retrofit design changes uncertain: no engineering data exist for precise tube/wire harness routes through aircraft	COMOK provides central design data base for automated transfer of design data to downstream analysis and manufacturing functions
	Quality of retrofit changes improved L-precise engineering data exist for tube/wire harness routes

2.5.5 Acquisition streamlining

Acquisition streamlining is a common-sense approach to making the U.S. DOD acquisition programs more efficient and effective. Acquisition streamlining concepts and policies are embedded in the new U.S. DOD acquisition directive and instruction. Handbook MIL-HDBK-248, "Acquisition Streamlining," addresses this subject. While acquisition streamlining is a formal defense initiative in the United States, its principles are equally applicable to the commercial sector. Table 2.16 summarizes some major acquisition techniques.

TABLE 2.16   Acquisition Streamlining Techniques

Area of emphasis	Streamlining technique
Performance requirements	Requirements discipline Market analysis Independent feasibility studies Postaward contract requirements review
Technical package	Specification tailoring Computer-assisted document preparation Technical data application and tailoring Tiger teams
Contracting	Streamlining clauses RFP techniques Streamlined source selection Best-value contracting Contracting to reduce risk

Acquisition streamlining policy in DOD 5000.2 states that "acquisitions shall...contain only those provisions which are essential and cost-effective." Requirements are to be stated in terms of performance rather than "how to" design requirements.

Acquisition streamlining policy supports use of nondevelopment items (NDIs) because it states that NDIs will be used to meet acquisition requirements wherever possible. This policy should provide increased defense opportunities for commercial products. Use of the "for comment" RFPs to solicit contractor-suggested improvements recognizes the need for early industry involvement and can be of major benefit to a program.

One of acquisition streamlining's more significant aspects is the greater latitude given contractor design efforts by limiting the tiers of specification references that needlessly complicate requirements. The specification "how to" in LSA requirements is consistent with the "what" but not the "how" advocated in acquisition streamlining. Such "how to" does not specify design solutions, but rather provides a basis for more objective evaluation of proposals using a common set of ground rules.

2.5.6 Program software intensity

The growing use of software in acquisition programs is causing increased management concern both commercially and in defense. The software failures of the New York telephone system and their impact constitute one commercial example.

Militarily, the picture is much the same. The numbers below show the growth in the number of lines of code per system:

F-111B aircraft	100
B-1 aircraft	1 million
AWACS aircraft	4 million
Space station	80 million

It is not unusual for software to account for half the cost and most of the risk in some programs. With the growth in use of software, it is no surprise that the ILS element of computer resource support is growing in terms of magnitude and importance. A number of avenues are being taken to address this growth in software. R&M requirements are being applied to software as well as to hardware (in some cases, the major reliability issue is software, not hardware). The system-specification example earlier in this chapter contains an example of such requirements. The application of ILS and LSA concepts is also evolving.

TABLE 2.17   Hardware/Software Process Comparison

System requirements analysis	Hardware					System test
	Requirements analysis	Preliminary design	Detailed design	Fabrication	HWCI test	
	Software					
	Requirements analysis	Preliminary design	Detailed design	Coding	SWCI test	

It is therefore obvious that ILS and LSA must adapt to these facts. This issue is the subject of more and more speeches and papers as the use of software grows. Fortunately, it is now becoming clear that both ILS and LSA can in fact apply to software as well as to hardware.

While there are many commonalities between the design and use of hardware and software, there are also differences. Table 2.17 contains a comparison of the two processes. Reference 22 cites the distribution of software development costs as follows:

Task	Development cost (%)
Requirements analysis	5
Design	25
Code/unit test	10
Integration/testing	50
Validation/documentation	10

The production of new copies of software coding is much simpler than the production of hardware. Errors in software coding, once fixed, do not normally recur as they do in hardware.

References 22 and 24 state that 70 percent of software life cycle cost (LCC) is associated with logistic support. However, the term *software maintenance,* as commonly used by computer personnel, is defined differently than hardware maintenance because it includes enhancement changes as well as the corrections of deficiencies. Reference 24 states that software maintenance is distributed as follows:

Activity	Distribution of effort (%)
Corrective	17
New requirements	16
Effectiveness improvements	60
Other	5

A brief summary of ILS support of software by ILS element follows.

Maintenance Concept	A key issue: Normally done at "O" level and depot, although three-level maintenance is used.
M&P	A major problem due to skill shortages.
Facilities	May be needed at both "I" level and depot.
ST&E	A major consideration: Includes hardware such as workstations and software tools such as text editors, compilers, assemblers, and other program analysis aids.
Technical Manuals	Operator manuals may combine hardware and software; usually there are also software users and maintenance manuals.
Training	Computer training programs may be embedded in the operational system for on the job training (OJT); there may be training courses for operators, operator "managers," and maintenance.
Packaging, Handling, Storage, and Transportation (PHS&T)	Security may be a concern.
Supply Support	May need to store and stock masks, tapes, disks, etc.
Design Influence	Can be a major consideration (see example at end of Sec. 2.3.6).

Since the system design process allocates functions to software as well as to hardware and personnel, it is obvious that software and hardware design must be integrated. The software requirements and design can have a major impact on the future operating and support (O&S) costs of software maintenance and on operational availability.

Software R&M may be more important to control than hardware R&M. The hardware meantime between failures (MTBF) requirement for one new system is 500 hours, while the software meantime to critical program stops (MTBPCS) is 30 minutes. Software R&M parameters may differ slightly from hardware R&M parameters. Software reliability is frequently expressed as meantime between program stops (MTTCPS) and software maintainability as meantime to restart the system (MTTRS), where *restart* includes software loading, bootstrap, initialization, and restoration of operational data but no logistics time.

Examples of software design requirements are modularity, adaptability, redundancy, and reconfigurability. Software design can lower skill levels and the number of people needed, as well as reduce susceptibility to operator error. Extensive use of menu and help facilities and design for fault tolerance and ease of debugging are additional examples.

Many of the LSA tasks mention software as a consideration or as a

TABLE 2.18   LSA Task-Software Relationship

LSA task	Comment
Use Study	Integrate software considerations
Standardization	Applicable to software as well as hardware
Comparative Analysis	Useful but probably use different data bases
Technological Opportunities	A must since software technology is rapidly changing
Design Factors	Should also apply to software
Functional Requirements	Done by design but probably uses different techniques than on hardware
Support Alternatives	Integrate with system study

Tradeoffs	Particularly in these subtasks:
	    Support system trades
	    Readiness sensitivities
	    Manpower and training trades

Task Analysis	Integrate operator and maintenance tasks in LSAR; link LSAR and software configuration
Early Fielding Analysis	Particularly manpower sources and possible shortfall impact
T&E	Integrate software support-related considerations with other testing to the degree possible

major subject area. The linkage of the LSAR and the software config-uration ensures that the impact of changes in one will be identified on the other. Table 2.18 summarizes potential applications. Software LSA will be discussed further in Chaps. 4 and 6. A good discussion of LSA applied to software is presented in Ref. 23.

2.5.7   Artificial intelligence/expert systems

General.   Another trend of interest is the growing use of expert sys-tems in logistics. Expert systems is one subset of the field of artificial intelligence (AI), a growing facet of computer science. One definition of AI is that application of computer science interested in developing computer systems which emulate the cognitive aspects of human be-havior. Since intelligent human beings have many characteristics, the AI investigations are profound and numerous. In the specific applica-tion of AI to logistics, expert systems have had numerous applications.

Expert systems incorporate the knowledge of a recognized expert in a particular field into a software program. Personnel using the soft-ware can then make decisions based on the captured knowledge and experience of the expert. The inexperienced person will be able to function as a much more experienced and knowledgeable worker by using computers as intelligent assistants that provide advice and make judgments in areas of logistics expertise.

Expert systems can advise, analyze, categorize, communicate, con-sult, design, diagnose, explain, explore, forecast, form concepts, iden-tify, interpret, justify, manage, monitor, plan, present, retrieve, schedule, test, and tutor. They address problems normally thought to require human specialists for their solutions.

Expert systems categories include

- Analysis and interpreting systems
- Predicting systems
- Diagnosis and debugging systems
- Monitoring and control systems
- Planning systems
- Design systems
- Instructional systems

From the preceding list it is not surprising that expert systems concepts would be applied to logistics. The July 1990 issue of *SOLELETTER* reported that there were already over 100 applications to logistics.

An example of a specific logistics application is the Pulsar Radar Intelligent Diagnostic Environment (PRIDE), the first diagnostic expert system within the U.S. Army that is 386 laptop-based. This system was built in 6 months. At its heart are 17 symptoms, 10 failure classes, 214 failures, 192 tests, 3 test procedures, 4 questions, 97 repairs, 4 rules, 16 decision points, and 60 graphics. The intent of the system is to reduce repair costs by better diagnosis of trouble and less reliance on massive parts replacement. A variety of digital pictures and other graphics are used to provide immediate repair diagnosis and repair procedure information.

Expert systems function well in a decision-tree type of environment, where characteristics of the problem can be narrowly defined and variables limited.

LSA implications. It is inevitable that expert systems will be used on both the LSA and the LSAR. One such use already planned is in review of LSAR data.

2.6   Summary

This chapter has described the LSA process and various key management processes and trends to which it relates. A major challenge to the ILS manager is the task of efficiently integrating LSA with these major interfaces.

3

LSA/LSAR Management

3.0 Introduction

While the concepts of LSA/LSAR are simple, the standards and interfaces are complex. Many different skills are needed to prepare requirements and execute the tasks cost effectively. Management of LSA is therefore an involved process.

Both requiring (e.g., the government) and executing (e.g., usually the contractor), ILS/LSA project managers are busy people. The primary objective of this chapter is to save a logistics manager (LM) time by providing readily available information on necessary management activities and detailed management tools (such as LSA team charters and LSA guidance conference checklists).

This chapter has six major sections whose focus is as follows:

General material   (3.1)

Program preaward activities: Preparing LSA requirements   (3.2)

Contractor proposal preparation   (3.3)

Postaward contract activities   (3.4)

Scheduling LSA/LSAR   (3.5)

Costing   (3.6)

Summary   (3.7)

3.1 General

3.1.1 Section content

This section discusses the following general management aspects of LSA/LSAR:

Principles of LSA management

LSA interfaces

Enterprise considerations

3.1.2 Principles of LSA management

This section identifies a number of general principles that should be understood and generally applied to both the preparation of LSA task requirements and their execution to make LSA useful and effective. Most of these principles are based on current guidance and lessons learned in LSA management.

Conceptually, LSA, whether conducted formally or informally, is the foundation of the ILS effort in acquisition. In the early phases, LSA may be the major part of the ILS effort. Historically, ILS prior to engineering manufacturing development (EMD) [formerly full-scale development (FSD)] has not been a very organized effort. This is changing. In the new DOD 5000.1 the LSA program is to be started prior to Milestone II. At a minimum, it is suggested that a comparison with a contemporary system be presented. More "early" LSA is now being performed in concept exploration and demonstration validation (DVAL). Many ILS element managers have now decided that the element must be considered prior to EMD to properly influence design concepts. Lead time for construction of facilities, for example, can be as long as 5 years. Many advocates of integrated diagnostics insist that this subject must be considered in concept exploration. The manpower and training communities advocate analysis much earlier in the acquisition process than in the past. The DOD concurrent engineering effort is attempting to achieve earlier integration of manufacturing and support considerations with design. LSA has the potential to satisfy many of these needs. *LSA can take advantage of reliability and maintainability (R&M) and/or ILS element analyses to reduce the cost of LSA and improve quality.*

The LSA standards allow wide latitude in requirements development and execution. Language can be modified to fit the specific situation. This is discussed in Chap. 4. With the possible exception of task 401, Task Analysis, most LSA tasks invite more detailed specifics in preparing requirements. This is also true to some degree in execution. Examples of this will be given in Chaps. 4 and 5 and some of the appendixes. The recommended team approach to tailoring is discussed in Sec. 3.2.1. Tailoring methods are briefly discussed in Sec. 3.2.3. Chapter 4 contains detailed advice on tailoring.

The timing of the LSA in both requirements and execution is a major concern in managing LSA. The key to LSA usefulness is the timely availability of results for input to the higher-level events and documents they are intended to influence. Less analysis delivered on time is better than a detailed analysis that arrives too late to fully affect the higher-level decisions/documentation they were intended to influence.

A team approach should be used whenever possible. Few individuals know all the things necessary to prepare LSA requirements or to perform all the LSA subtasks. Effective LSA is therefore a team proposition. This is true between the requirer and the executor, as well as between communities within each of these organizations.

Data availability is frequently a major issue, both between the requirer and the executor and between the communities in both organizations. LSA requires much more requirer and executor cooperation than many program management standards (such as R&M). It is not infrequent to have much of the data needed split between the two organizations. As the DOD CALS initiative matures, the necessary data exchange between requirer and executor and between communities should improve significantly, thus expediting and improving the quality of the analysis performed. Presence or absence of data frequently places major limitations on the amount and type of analysis possible. Different organizations have failure data systems of varying capability. Perfect data are seldom, if ever, available. The challenge of adapting to imperfect data to allow pertinent and useful analysis is usually a key management concern in most projects.

The level of detail and accuracy in the LSA analyses performed obviously increases as a project proceeds through the successive acquisition phases. This frequently affects the type of data needed and the models used.

The analysis approach selected should fit the level of resources available, the timing considerations involved, and the depth of detail needed.

Iteration is a normal part of the LSA process. As new information or more detailed data become available, analysis must be updated. Task interactions frequently occur in LSA and cause additional need for iteration. Iteration is useful when a significant factor has been changed and when the results of the analyses can affect the decision process.

Use of quantitative methods and/or models significantly enhances the usefulness and credibility of the LSA process.

These management principles are generally applicable, to some degree, to requiring and executing LSA on any specific program.

3.1.3 Interfaces

The numerous interfaces of ILS/LSA are one of the major challenges to effective LSA management in both the requiring and performing organizations. Interfaces occur at both the management process level, as discussed in Chap. 2, and the discipline and functional level. Most requirements for an LSA plan demand a discussion of discipline interfaces. Jones has an excellent chapter on this subject.[3]

Interfaces within the government, particularly within the government program office, are usually clearly defined. The protocol of where responsibility and authority are vested has been established by organization and precedent. Contractors rarely have such a useful authority structure to guide their responses. A commercial entity has a president, or CEO. He or she is rarely the Program Manager. The Program Manager may be selected from the ranks of management, but his or her responsibility and authority *within the existing entity structure* may not be compatible with the needs of the "Program Manager." This situation is duplicated for every "Manager" within the newly created "Program Office." As an example, within the normal entity structure, dictated by its business profile, the Manager of Technical Publications, with a significant staff, may be at level 7 (with 10 as the CEO). The new proposal may have a relatively small technical publication requirement that is subordinate to larger LSA/ILS interfaces. In contrast, the LSA/ILS program manager may within the entity be at level 5, and the training manager may be at level 6. These entity interfaces need evaluation in terms of the program reality. In the world of the contractor organization, a real level 5 manager is at a disadvantage assigning schedule and budget to a real level 7 and requiring performance. The management of these interfaces is a challenge that must be addressed at the beginning of program planning. There have been attempts to solve this problem by proliferating the manager line in the program structure, where the organization has a series of managers all controlling a function. This usually extracts a penalty, first in cost and second, and more significantly, in control of the work product. In reviewing this tangle of competing areas, a realistic solution is to examine the reality of the customer program office structure. Where does the customer ILS/LSA manager go for an answer? If there are multiple contacts and the final response for the program's many functional areas can come only from the program manager, the structure needs revising.

A useful tool for the performing organization in assessing these complexities and in ensuring inclusion of the various disciplines essential to successful ILS/LSA efforts is the creation of a responsibility matrix. Most contractors scrutinize Request For Proposals (RFPs) and assign responsibility to "lead" functional groups. There was a period, hopefully past, when the portion of the RFP called ILS was detached and sent to the ILS group, never to be seen again until the response was written. The more successful bidders have evolved a matrix chart in which each volume of the RFP is parsed and the columns in the matrix are labeled with the functional disciplines. The rows of the matrix are specific response requirements. An example would be a row designated "Maintenance Training Manual." The functional group tasked as lead may be "Technical Document Production," and in that row/column intersection would be an "L." The R&M group, however, would be a required contributor, and under that row/column interface would be a "c." Progressing through each volume, the lead groups would probably be obvious, but this approach requires a review of contributors as well and tasks the lead group with the responsibility for the inclusion. So too with the LSA tasks; the lead may be the LSA/ILS manager, but the existence of the published matrix alerts contributors. When first published, it may elicit input from groups who have an interest but have been omitted by oversight.

3.1.4 Enterprise considerations

Personnel on a specific project in both the requiring and performing organizations are usually parts of larger organizations. These organizations usually prepare and issue general LSA policy and guidance for specific projects to follow. In the U.S. defense establishment, projects are generally reviewed at various levels depending on the magnitude and importance of the project. Numerous examples of requiring (government) general guidance are referenced and/or discussed in this book.

Many contractors, particularly those with more than one separate, but related, program, issue similar guidance. One document in particular that most contractors should have because of the cross-project benefit, is an LSA/LSAR Procedures Guide. Such a document ensures cross-project "pollenization" and greater comparability of LSA products for use on future programs and provides a data source for corporate memory lessons learned. A table of contents for an applications guide covering a family of related projects (and which bears some relation to this book) is presented in Table 3.1. A contractor's guide could combine material similar to this outline with more specific detail. Appendix 24 contains examples of various conventions in recording LSAR data where options are available in the LSAR standard. Many organizations have adopted different conventions, as illustrated in Appendix 24. This results in different approaches to LSAR data bases between different organizations.

A second possible major activity in both the requiring and performing organizations is research and development (R&D) of new tech-

TABLE 3.1   Suggested Outline: LSA/LSAR Procedures Guide

1. Introduction
    1.1. Purpose
    1.2. Scope
    1.3. Use of This Guide
    1.4. Updates
2. LSA Management
    2.1. General Description
    2.2. LSA Management Responsibilities
    2.3. LSA Objectives
    2.4. General Interface Responsibilities
    2.5. LSA Proposal Preparation
    2.6. LSA Requirement Flow-Down
    2.7. General Schedule Integration
    2.8. Government Reviews
3. LSA Task Execution
    3.1. Task Execution Approaches
    3.2. Task and Subtask Organizational Responsibilities
    3.3. Computer LSA-Related Models Available
    3.4. Internal Task QC/QA
    3.5. Internal Use of Task Reports
4. LSAR Management
    4.1. LSAR Objectives
    4.2. Data Control Input and Management
        Drawing Interface
        R&M Interface
        Task Documentation Conventions
        Support and Test Equipment
        Provisioning and Supply Support
    4.3. LSAR QC/QA
    4.4. Use of the LSAR
        Standard LSAR Reports
        Ad Hoc Queries
Appendixes
    A. LSA Candidate List and LCN Structure
    B. DD-Form 1949-1
    C. Approved Common Tool/Test Equipment List
    D. Abbreviations/Acronym List
    E. DED Cross-Reference List

niques, processes, and automated tools. An example of such a project on the government side is the development of automated programs which, given the desired ILS Contract Data Requirements Lists (CDRLs) and LSAR output reports, automatically prepare the form 1949-1s, which identifies the needed data elements to produce the desired reports. Several such systems, which also could contain a schedule compatibility check feature, now exist in the United States. Examples on the contractor side could include automated linkages between the LSAR data base and related data requirements such as R&M, failure modes and criticality analysis (FMECA), reliability-centered maintenance (RCM), and ILS element data programs. Some progressive U.S. contractors have done this under independent R&D (IR&D) efforts. The DOD CALS initiative is increasing pressure for such linkages.

LSA/LSAR training is another needed activity by both the requirer and the performing organizations. There are also a number of U.S. trade organizations that coordinate with the U.S. DOD in the development of guidance/standards/documentation. Many contractors place representatives on such groups as a means of influencing and staying abreast of upcoming requirements. A recommended LSA library was discussed in Chap. 1, and such a library should be an entity resource.

3.2 Program Preaward Activities: Preparing LSA Requirements

This section discusses:

Responsibilities

Steps in preparing LSA requirements

Tailoring methods

Source selection, sections L&M, and contract incentives

3.2.1 Responsibilities

Many talents and many people contribute to the total LSA effort. The logistics manager is normally the chief coordinator. Because the majority of the people who will contribute the most to the program are not in the direct supervisory chain of the logistics manager, an extraordinary management effort is required.

Requiring organization.   The role of the requiring organization is (1) to determine what tasks and subtasks have to be performed (this includes further tailoring and focusing, as discussed in Chap. 4), (2) to establish when and by whom the tasks will be done, (3) to provide task inputs to the performing organizations, (4) to review the performing organization's LSA program and LSA products, and (5) to use the LSA products produced.

The requiring organization exercises control authority over the LSA program and controls the disbursement of funds to pay for the accomplishment of the LSA program.

Performing organization.   The performing organization executes the tasks and subtasks and provides the LSA data products required by the requiring organization. The performing organization can be either the government [ILS, logistics element managers (LEMs), field activities, or service contractors] or the prime contractor/subcontractor. When no hardware contractor exists prior to EMD, as is the case in many projects, the government or a service contractor must, of necessity, perform any LSA required. Once a hardware contractor(s) exists, it will normally perform the majority of the tasks and subtasks and compile the majority of the LSA data base. The contractor also should use the LSA products internally. The prime contractor will prepare a proposed LSA plan in response to the RFP that describes the contractor's approach to execution of the LSA tasks and LSAR requirements specified in the statement of work (SOW). The contractor details how the program will be managed and the management controls to be employed to ensure proper execution of the program. The prime contractor also will institute and execute the LSA program, as specified, upon contract award.

Responsibilities of the requiring logistics manager. The program LM in the requiring organization is responsible for the overall management and coordination of the LSA program. This includes inputs to program management documents such as the acquisition strategy and solicitation packages. Since management of the LSA process involves the consolidation and coordination of a wide range of disciplines, the proper vehicle for LSA management is a team of specialists under the overall management direction of the ILS manager. It is the LM's responsibility to establish the ILS/LSA management team, select its membership, schedule meetings and activities, cochair meetings with contractors (or performing organizations), coordinate team activities, and ensure that action items are executed.

Responsibilities of the performing logistics manager. Although the government has a clearly defined career category of logistics, this is not universal in industry. It is not an unusual circumstance to have a logistics manager in the performing organization chosen from a variety of functional groups. The responsibilities of the performing manager may be clear to the requiring program manager (PM) but may need to be defined within the performing organization. The performing organization has an existing organizational structure, with associated manpower and budget responsibilities. As has been noted earlier, these realities must be examined for consistency with the logistics tasks of the contract. This is not a cosmetic exercise but is essential to performance and budget. The program manager and logistics, engineering, software, and other associated managers must have a coordinated approach. They must agree on the costs and responsibilities associated with the work breakdown structure (WBS) and what the reporting chain will be.

Reporting responsibilities may be complicated by the real entity structure and the "program management" structure, which may be unique to that program. The performing LM should review, for example, the various plans required by the RFP/contract. These may include the System Engineering Management Plan (SEMP), the System Test Plan, and the LSA plan, to name a few. Using the responsibility-matrix approach, the LM, recognizing the interrelated disciplines, can assist the PM in establishing a structure that can respond to the requirements effectively. The multidiscipline requirements of LSA tasks exact a heavy penalty for organizations operating in a "stove-pipe" structure.

A common requirement illustrating the hidden interrelationships is the Maintainability Demonstration. The schedule must recognize that Technical Manuals and trained personnel must be available and that a systems test (including integrated software) is complete. The technical manual alone presupposes a close relationship with the R&M community, since training on fault isolation and removal procedures, unless the reliability of the items is known, is difficult. These obvious relationships are compelling reasons for an internal game plan. This may require an adjustment in responsibilities not clear in the entity standard organization chart. LSA tasks and the matrix describing "who does what" are an excellent template for analyzing information flow. They also serve to identify communication and information roadblocks.

The performing LM must respond to the requiring LM's LSA tasks and to the schedule of meetings. A further caution exists where there are significant contributing subcontractors. If participation is required from the subcontractors, a line of communication must exist from the performing LM to these organizations. Ideally, the capabilities of the subcontractors and the information and documentation to be received have all been part of the selection process. In addition, the representations required should be identified in the contracts to the subcontractors. Many major subcontracts usually require participation at engineering design reviews; participation also may be required at ILS/LSA meetings. These meetings are necessary, but they are time and manpower consumers. This is an incentive to consider holding these meetings back to back to reduce travel and manpower costs. The performing LM, as a minimum, should identify all the inputs required from the subcontractors, the timing of these data, and their use and impact on the LSA tasks.

LSA team. The ILS/LSA management team in both the requiring and performing organizations does more than review LSA products for acceptability. Its activity begins with the development of the LSA strategy and the preparation of contractual documents for the solicitation package. Accordingly, the team should be created at the start of the preaward period and should exist as long as there are LSA decisions to be made or products to be reviewed. The team concept is the best practical management technique because it integrates the combined expertise of the ILS element managers and other specialists into a single working group under the management control of the LM, as shown in Fig. 3.1. See Appendix 5 for a typical LSA team charter.

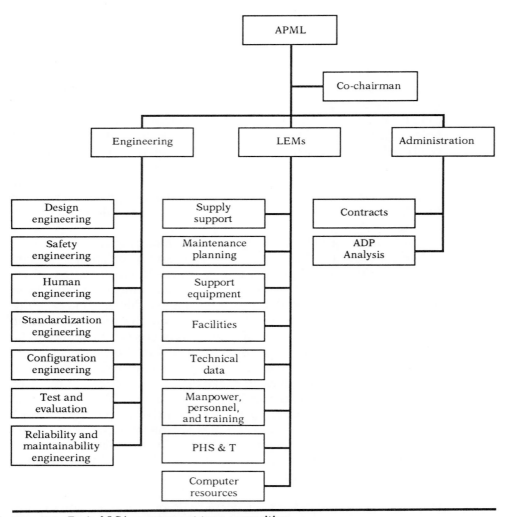

Figure 3.1 Typical LSA management team composition.

TABLE 3.2   Team Input to LSA Requirements Preparation

Activity	LM responsibility	LEM participation	System engineering
Preparation of solicitation document	X		
Establishment of the LSA management team	X	X	X
LSA strategy: tailoring LSA	X	X	X
Identification of performing organization	X	X	X
Tailoring LSAR	X		
Scheduling LSA and LSAR	X	X	X
Preparation of RFP and SOW	X	X	X
SOW	X	X	X
GFI	X		
Data Item Descriptions (DIDs)	X	X	X
Contract Data Requirements Lists (CDRLs)	X	X	X
LSA data record A	X	X	X
SSP Evaluation Criteria	X		
Instructions to Offerers	X		
Award Criteria	X		
Source Selection			
Proposal Review	X		
Grading proposal	X		
After contract award			
Conduct the LSA/LSAR guidance conference	X		
Prepare additional task inputs	X	X	X
Conducts/participates in reviews	X	X	X
Use of LSA Products	X	X	X

SOURCE: Navy LSA Applications Guide, May 1986.

Logistic element managers (LEMs)

Requiring organization LEMs. Table 3.2 cites the minimum activities a requiring LM should aim for in terms of LEM participation. The LEM's participation could be much broader than is shown, on a selective basis, in virtually all aspects of the LSA/RFP requirements preparations. Table 3.3 summarizes suggested areas where the LEM can participate in the tailoring, review, and use of each MIL-STD-1388-1A task, as well as in preparation of the many task inputs. LEM participation is on a selective basis and takes a variety of forms. In tailoring, the LM should ask the LEMs about their requirements for tasks and subtasks. Some tasks, such as task 201, Use Study, and task 302, Support System Alternatives, may cover all ILS elements. Other tasks, such as task 203, Comparative Analysis, and task 204, Technological Opportunities, may only require selected LEM participation.

The requiring LEMs and system engineering specialists can assist the LM by preparing task inputs when these inputs apply to an ILS element area. This is covered in more detail in Sec. 3.4.3. Finally, the requiring LEMs should participate selectively in the review and use of LSA products at the task or subtask level as appropriate. This could

TABLE 3.3   LEM/LSA Task Participation

No.	Title	Tailoring	Inputs	Review	Use
	LSA tasks	LEM participation*			
101	Development of an Early Logistic Support Analysis Strategy	X			
102	LSA Plan				
103	Program and Design Reviews			X	X
201	Use Study	W		X	X
202	Mission Hardware, Software, and Support System Standardization	W	X	X	X
203	Comparative Analysis				
204	Technological Opportunities	W		X	X
205	Supportability and Supportability-Related Design Factors	W		X	X
301	Functional Requirements Identification	W	X	X	X
302	Support System Alternatives	W		X	X
303	Evaluation of Alternatives and Tradeoff Analysis	W		X	X
401	Task Analysis	W		X	X
402	Early Fielding Analysis	W	X	X	X
403	Postproduction Support Analysis	W	X	X	X
501	Supportability, Test, Evaluation, and Verification	W		X	X

*W = LEM general advisor; X = LEM selective participation.
SOURCE: Navy LSA Applications Guide, May 1986.

include general-interest tasks for all LEMs, such as task 201, Use Study, and task 302, Support System Alternatives, and selective subtasks of interest for specific LEMs, such as the manpower, personnel, and training (MP&T) design trades in task 303, Evaluation of Alternatives and Tradeoff Analysis (see Appendix 15 for task-review checklists).

Performing organization LEMs. The logistic manager in the performing organization may not and very probably will not have the range of LEMs that are available to the requiring organization. The services, because of the size and volume of logistic activities, have specialized organizations and personnel in these disciplines. As an example, the Training Command in the service would have a team of experts who serve as required on various programs. The technical data element may have supply support personnel with years of experience in one segment of that discipline alone, provisioning documentation. The title *Logistics Element Manager* (LEM) in the contractor world describes a function rather than a job. While there are notable exceptions in the contractor community, it is not often that the business base justifies the range of LEMs equivalent to that resident in the services. The minimum safety net for contractors must be an experienced logistics manager. This manager brings a knowledge not only of the ILS/LSA processes but also of how these conform or are related to internal processes, perhaps by other names, normally performed within the organization. The recognition of similarities between internal processes and LSA is not unreasonable, since LSA is an orderly progression of analyses from concept through production.

The design-development process and associated analyses are similar in both commercial and military programs. Recognizing this similarity and adapting techniques common to both make transition between the two less burdensome. If, for example, there is no direct equivalent to the government LEMs, the ILS manager may choose to augment his or her resources by using element specialists on a consulting basis. The one absolute essential is timing; if specialists are needed, the expertise must be applied in time for it to influence decisions and meet delivery schedules.

3.2.2   Steps in preparing LSA requirements

Table 3.2 identifies the steps in preparing LSA RFPs and Source Selection Plan requirements for selecting hardware contractor(s). GFI (government-furnished information) in the table refers to any requiring organization data to be made available to prospective bidders. While the same steps in theory would apply to the selection of a government organization or support contractor if a hardware contractor does not and will not exist in the present phase, this process, in practice, may be as simple as preparation of a short task statement for an existing government organization or support contractor. Table 3.3 also cites the LEM/system engineering involvement in the requirements preparation process.

MIL-STD-1388-1A differs from other program-type standards in that there are differences of opinion on which organization, the government or the hardware contractor, should perform some of the tasks and subtasks. In some cases, a strong joint government-contractor team effort is required.

The requiring LM and the LSA team must decide who will perform the tasks (see Appendix 5 for guidance). The criteria for deciding who performs each task include

Who has the information necessary to perform the task/subtask?

Who has the expertise to perform the task/subtask?

When is the task/subtask required?

Who can perform the task/subtask most economically?

Table 3.4 summarizes the selected performing organizations identified for MIL-STD-1388-1A tasks and subtasks in Appendix 5. As can be seen from this table, there can be more than one performing organization within a task; ie., there is no one ideal performing organization.

TABLE 3.4   Summary of Performing Organizations Classifications

Task no.	Task name	G	OEM	"E"	"J"
101	LSA Strategy	1		1	
102	LSA Plan		2		
103	Reviews			1	3
201	Use Studies	2		3	
202	Standardization		4		
203	Comparative Analysis			7	1
204	Technology Opportunities				3
205	Support Design Factors	1		3	1
301	Functional Requirements Identification		6		
302	Support Alternatives			5	
303	Tradeoffs		2	10	
401	Task Analysis		4	5	2
402	Early Fielding Analysis	4		1	
403	Postproduction Analysis				1
501	T&E	2		3	
Totals	(78)	10	18	39	11

SOURCE: Navy LSA Applications Guide, May 1986.

The performing organization classifications in the table are coded as follows:

G = Government (system command, field activity, or service contractor).

OEM = Original equipment manufacturer (design or prime contractor).

E = Either organization could perform the analysis.

J = Either organization is capable of performing, but information from the OEM is required for the government information; it is assumed that this information would be supplied as government-furnished information (GFI) and performance would thus be classified as E.

The totals at the bottom of Table 3.4 and Appendix 5 clearly indicate that the government should or could perform many of the subtasks in MIL-STD-1388-1A because over 55 of the more than 78 subtasks are coded as government or either.

3.2.3   Tailoring methods

Current tailoring approaches can be classified into six broad categories:

1. The boundary or departure-point approach (e.g., see Table III in the 1A standard)

2. The question approach (e.g., see figure 5 in LSA 1A standard)

3. The sample statement of work (SOW) approach

4. The "do them all" approach

5. Any of the above, plus advice

6. Specially tailored readiness and supportability design study requirements (see Appendix 6 for an actual example)

The boundary approach is illustrated by Table 111 on page 98 of the standard. This table suggests the selection of the LSA subtasks to be performed based on program complexity. It in essence suggests (at least as a departure point) weapon system application of all the 1A subtasks at some point in the classical (or full) program development cycle. A major weapon system is one that exceeds certain R&D or production dollar amounts or one so designated by top management. The table suggests reduced application of MIL-STD-1388-1A subtasks to equipments. A number of experts in the field, after issuance of the standard, decided that many of the subtasks suggested for equipment application are not necessary on simple equipments and developed extensions to Table III of the standard that suggest further reductions in subtask applications to "simple" equipments. Chapter 4 and Appendix 7 contain suggested tailoring using this extended approach, since they contain suggested subtask applications for weapon systems, complex equipment, and simple equipments for classical development programs. The advantage of the boundary approach is that it provides a limitation on the tasks and subtasks that need to be considered. The disadvantage is that the boundary may be too high or too low.

The question approach presents the LM with a series of questions, whose answers guide the selection of tasks and subtasks. Figure 5 on page 64 of LSA 1A standard illustrates the approach, although it does not tailor to subtask level (a must) and does not contain all the factors that LSA 1A standard says to consider in tailoring. An advantage of this approach, particularly if not by phase, is its simplicity; a disadvantage is that simple questions can have complex answers. However, a question approach, used by an expert, will probably give the best tailoring. Unfortunately, most people preparing such requirements do not generally have in-depth expertise simply because they do not perform the task very frequently.

The sample Statement of Work (SOW) approach generally provides the LM with a sample SOW for use in each phase with applicable subtasks. The LM can then add or subtract as necessary. The approach is simple but depends to a great extent on the user's knowledge. It is best used where the programs to which the sample SOW is applied are very similar and fit most, if not all, of the assumptions made in developing the sample. If the user is not knowledgeable, a boundary strategy may be better.

More recently, some experienced LSA personnel have suggested a different approach from that in the standard. This approach generally advocates doing most, if not all, of the available tasks suggested for application prior to full-scale development (FSD), without regard to program complexity, with some tailoring occurring at the subtask level. The advocates of this approach do propose the elimination of tasks if they serve no end-objective LSA purpose (such as no design influence possible on NDI cases).

A refinement to all of these methods is discussed in paragraph 40.3.2, "Focusing," in MIL-STD-1388-1A. This paragraph advises the requirer that after the subtasks have been selected, further focusing is needed to concentrate effort in high-leverage areas and to specify any other requirements that seem desirable. Focusing usually should involve modification or restriction of the task or subtask to significant areas, although occasionally requirements may be added. Unfortunately, use of focusing to further tailor the LSA subtasks is seldom seen in practice. A number of possible additions to improve or clarify subtask requirements are in the task discussions in Chap. 4 and in Appendix 8.

All these strategies can and should be augmented by advice to further assist an inexperienced LM in making tailoring selections. The advice can be of general use or apply to specific tasks and subtasks. An example of general advice would be the application of the G (generally apply) and S (selectively apply) codes used in table III of the standard. A possible rule of thumb for G-coded subtasks is to make them a requirement if there is no obvious reason to delete them. However, S-coded subtasks would only be invoked if a good reason to do so is known.

The sixth approach is an extreme example of focusing. It involves the development of highly tailored requirements for a specific system. It includes "what if" supportability-related design studies. Appendix 6 contains an actual example of this approach. The appendix also can be used as a source of possible focusing ideas. This approach can be expensive and thus requires expert help to develop. It can be useful on high-cost or critical programs. The targets of the design studies must be selected based on the contribution of that aspect to the overall program concerns. In suggesting "iteration" or "what if" studies, we are not recommending "fishing expeditions." Nothing is more destructive and wasteful than to have any segment of the design team lose sight of overall program goals, including budget.

Most LSA SOWs today stop after specifying the required tasks and subtasks. Another common practice is for people preparing LSA requirements to delete some subtasks that serve as a needed input to other subtasks. The theory behind this is that since the subtask deleted must be done in order to accomplish the related task, the performing organization will have to do it anyway—hopefully at its own cost. In actuality, the performing organization will probably (1) not do it (and use a "swag" approach), (2) do it minimally (and not very well), or (3) include the cost in the specified subtask. In competitive-bidding situations, the pressure is great to take the first alternative. If it is not required, "don't do it" and "don't bid it" are all too common management edicts.

As one can see from the preceding discussion, tailoring LSA requirements can be complicated. There are several ways to reduce the complexity. One way is good locally developed guidance. DOD-wide or global guidance must, of necessity, be broad and general in nature. Assumptions that may be untrue in a particular situation are implicitly an inherent part of such general guidance. General guidance frequently contains a lot of information not applicable to the local scenario and may ignore or touch lightly on areas of major local concern. Development of good local guidance can reduce the number of variables and best address peculiar local needs. A second way is to develop expert system LSA tailoring systems. There are several such systems in existence in the United States today. The various tailoring options discussed in this book could be combined into an automated program designed to adjust to both knowledgeable and inexperienced personnel. The results of all tailoring schemes should be reviewed carefully to ensure that some hidden assumptions in the scheme have not caused an unrealistic requirement to be included in the final product. This book uses a combination of the preceding tailoring approaches in Chap. 4 and also discusses various tailoring tools (worksheets, etc.) in the appendixes.

Most of these approaches assume a specific acquisition phase. This assumption can create problems in practice because of such conditions as the following:

Actual acquisition phases do not always follow the classic definition of the phases.

Not all systems comprising an acquisition will be in the same phase at the same time.

Other factors peculiar to the program may override the normal tailoring for a phase.

A detailed discussion of tailoring and focusing (further refining, see paragraph 40.3.2.2 of LSA 1A standard) LSA tasks and subtasks occurs in Chap. 4. Specific examples of tasks for which it may be advantageous to engage in further tailoring and focusing include tasks 202,

204, 205, 302, 303, 402, and 403. Tailoring and focusing the LSA tasks to the specific program are major keys to cost-effective LSA.

3.2.4 Source Selection, Sections L&M, and Contract Incentives

Source selection. The SOW, the Contract Data Requirements Lists (CDRLs); section L, "Instructions to Offerers"; source-selection planning and criteria; award factors in Sec. M; and contract incentives are usually interrelated. Source-selection criteria and award criteria are particularly related, because the award criteria are based on the more detailed source-selection criteria developed. Source selection involves the selection of criteria and associated factors and subfactors, the weighting (prioritization) of these criteria and factors, the development of standards by which to evaluate contractor proposals, the possible use of weighting objectives to assist evaluation, and rating and scoring methods. Because a full discussion of these subjects is complex, refer to Appendix 11 for a discussion of source-selection planning and award criteria. The process described is that used by the government, but the concept also may be employed to some degree in prime and subcontractor relationships.

Use of section L. There are a number of possible ways that Sec. L, "Instructions to Offerers," of the RFP can be used to strengthen the supportability/LSA effort. These include

Use to enhance the SOW.

Use to request bidders to specifically address in detail subjects usually treated lightly in contractor bids (e.g., contractor use of the LSA products, specifics on what requirements are to be passed on to major subcontractors, etc.).

Employment to collect information useful in distinguishing between contractors, such as:

1. What major supportability and LSA work has already been done, and what are the results?

2. What does the contractor consider the major readiness and O&S cost drivers to be?

3. What tradeoffs and studies does the contractor plan to do as a result of the first two questions above?

4. What is the association of supportability and LSA to the concurrent engineering effort?

5. What automated linkages to LSA/LSAR will the CALS effort provide?

6. What are the specific technical skills and approaches to subtask execution?

7. What specific LSA/LSAR requirements will be placed on major subcontractors?

Use of Sec. L is particularly advantageous when there is competition or if several awards are planned. It may be necessary to ask contractors to propose an LSA activity they see as useful. Section L also may be the appropriate place in the RFP to inform bidders that the project is willing to accept LSA task reports in alternate, more cost-effective, formats/approaches possible due to contractor CALS(CE) capabilities.

Supportability-related contract incentives. Contract incentives can be used to enhance the stature of the supportability/LSA effort. While use and/or development of such incentives is not a minor chore, this can have measurable benefits in terms of contractor motivation. Possible incentives are discussed in detail in Appendix 12.

3.3 Contractor Proposal Preparation

This section discusses

General considerations

Writing a good LSA proposal/plan

3.3.1 General considerations

The contractor (performing organization) must prepare a proposal in response to the RFP. Jones has a good discussion of the detailed activity involved.[3] There are many contractor counterparts to the government activity in preparing the RFP. Examples include

- The contractor must appoint an individual to prepare/coordinate the LSA proposal effort as part of the initial team.
- The contractor should designate various specialty personnel, comparable with the requirer LSA team, to support the LSA proposal effort to ensure high quality with proper interfacing.
- The contractor LSA focal point should review the entire RFP for support-related design requirements as well as the ILS/LSA material listed in Table 3.3. If there is any concern that all disciplines

may not be sufficiently involved, consider use of the matrix mentioned in Sec. 3.1.3.

- The contractor should employ task tailoring and focusing principles in greater depth in the proposal effort.

3.3.2 Making a good LSA proposal/plan

We have prepared and reviewed many LSA proposals/plans. As a generality, many of the plans reviewed have been disappointingly vague in nature. Such plans can be summarized as stating, almost in the words of the request, that the contractor will do what the RFP says with regard to LSA. The reader is invited to compare the R&M section to the LSA section of the same proposal. The R&M section is usually much more specific and, in addition, presents any analyses already done.

Good LSA proposals/plans can be developed using the same basic approach that is implicit in good R&M sections. Specific, disciplined LSA plans can be a discriminator in contract awards. Some general rules for preparing materials are

- Answer the detail in task 102, LSA plan.
- Identify supportability design requirements (RFP plus your own).
- Identify work already done.
- Be specific where possible, but if the phase is too early for specifics, state why.
- State general approach. When and if specifics may vary (e.g., new information will generate new analysis needs), explain the circumstances.
- State how you plan to use the LSA/LSAR products internally.
- State the analysis models you plan to use.
- State what you do not plan to do and why. This tailoring and focusing is the opportunity to identify cost savings or cost containment.
- State what you expect the government to do. Make this as specific as possible, since the omission of these details may have a significant cost impact. For further guidance, see Sec. 3.4.3.
- Be as specific as possible about the subcontractor approach. Identify the benefits expected from the flow-through of these requirements.

While most of this is obvious, a few points are worthy of more discussion.

Work already done. It may appear at first glance that nothing has been done, since no formal assignments have previously been made. A review of past activity and discussions with key people usually reveal that many supportability discussions occurred informally, particularly by hardware and software design personnel. These discussions may not have been labeled "supportability," but the results may have significant influence. Summarizing this activity improves the specificity and quality of your LSA proposal/plan. Review the points listed above.

Planned internal use of the LSA products.   There is disagreement about whether the LSA task description in MIL-STD-1388-1A and the LSA data item description (DID) require a contractor to use the LSA products internally. Many contractor managements take the position that nothing will be done that is not specifically spelled out in the RFP. In some cases, this is very short sighted. In the instance of using LSA products internally, when LSA products contain specifics, many system and design engineers will make use of the LSA products. Designers, for example, are interested in use studies, maintenance concepts, and specific problems identified on similar systems/equipments. In some cases, all that is required is internal distribution. A review of the individual task tailoring and focusing advice in Chap. 4 (particularly tasks 102 and 103) can provide numerous possibilities in this regard. Stating proposed internal use tells the requirer that you take LSA seriously.

Be specific.   Doing so emphasizes to the requirer that you are serious. Ideas for being specific include tailoring and focusing individual tasks. The advice for tailoring and focusing in Chap. 4 can be equally useful in tailoring the contractor's proposal/plan. Appendix 6, "Readiness Design Definition," is another possible source of ideas for studies, and Chap. 5 contains possible specific approaches for task execution. Being specific has another benefit after contract award when the LSA plan becomes a part of the contract. Specifics decrease the chance that the customer and the contractor will be involved in wrangles over products that were insufficiently described.

Use of models.   Most people associate the use of models primarily with task 303, evaluation of alternatives and tradeoff analysis. The fact is that models can be employed on many of the other LSA tasks and subtasks. This is discussed in more detail in Chaps. 4, 5, and 7. AMC-P 700-4, "LSA Techniques Guide,"[6] has a table discussing models useful for various tasks and subtasks. Discussion of your proposed use of models by subtask demonstrates your commitment to a quality LSA effort.

TABLE 3.5 Major Subcontractor LSA Task Pass-Through

Task	Pass-through
201 Use Study	No; possible team member on field visit
202 Standardization	No
ILS	Possible selection/tradeoff criteria
HW	Possible selection/tradeoff criteria
High-level	Check for other standardization requirements
Low-level (parts)	Maybe
SW	
203 Comparative Analysis	
Setting goals	Possibly no; R&M goals, etc. usually set; could identify comparative items informally
Identifying drivers	Yes if prime can't and subcontractor has unique data
204 Technological Opportunities	
ILS	Probably no; prime can do
HW/SW	Possibly
205 Goals/Thresholds	Possibly subtask 205.2.3
301 Functional Requirements/Task Identification	Possibly yes; subcontractor best source of LSAR data
302 Support Concepts	At most concept alternatives; 302.2.1
303 Tradeoffs/Alternatives	Possible tradeoff criteria; selection subsystem tradeoffs
401 Task Analysis	Strong candidate but tailor requirements
402 Early Fielding Analysis	At least tailor at subtask level and below; consider 402.2.1 and 402.2.4
403 Postproduction Support	At most a highly tailored requirement
501 T&E	At most a highly tailored requirement for 501.2.2

*Major subcontractors, not vendors; results are heavily dependent on nature of actual program.

Subcontractor provisions. We have seen few proposals/plans that discuss this subject in terms of specific assignments or data contributions. In actuality, depending on the magnitude of major subcontractor involvement, considerable pass-through can and should occur. Table 3.5 illustrates the possibilities in this regard. A review of subcontractor LSA-related requirements and schedules may preclude late generation of specific requirements (both data and documentation) and numerous schedule incompatibilities.

Explicit approaches to subcontractor LSA requirement pass-through frequently can result in superior rating in this regard compared with competitors.

An example. Table 3.6 contains an actual case illustrating some of the preceding points. The actions proposed were presented in a Best and Final (B&F) proposal after the contractor was advised that its original LSA proposal was rated last competitively. While the new approach in this table was not the only factor in the B&F, the proposer was awarded the contract.

TABLE 3.6   Supportability and Readiness Integration Program

Management

1. Equal status of ILS with technical.
2. Appointment of an LSA-R&M coordinator under the technical manager to expedite appropriate interaction and teamwork.
3. Appropriate colocation of technical, R&M, LSA, and ILS element personnel to foster and expedite information exchange.
4. Review of ILS-R&M and LSA/LSAR status at program and design reviews.
5. Conduct of a joint LSA seminar for engineering, software, and ILS personnel shortly after contract award.
6. Addition of a highly qualified service contractor to assist in ILS, LSA, and EMI support and to provide review services in the R&M, CM, and QA areas.

Assessment

1. Participation of R&M, QA, and LSA personnel in in-house design reviews.
2. Discussion of supportability design requirements by design personnel in design reviews.
3. ILS signoff on drawings and specifications.
4. Analysis of the LSAR for design issues.
5. Inclusion of supportability considerations in test and evaluations.

LSA tasks

1. Identification and dissemination of support-related design requirements and information, such as the LSA Use Study, as an integrated package to appropriate technical and ILS personnel.
2. Collaboration of design and ILS personnel in performance of appropriate LSA tasks, such as the Comparative Analysis and task 303, Tradeoffs.
3. Identification and analysis of low-reliability hardware for improvement opportunities:
   a. Low-cost units—reliability and A_o improvement; and
   b. High-cost units—operating and support cost reduction.
4. Input of supportability considerations into tradeoff studies.
5. Performance of a TIGER simulation analysis as part of the LSA effort to relate design, R&M, and ILS resources to operational availability.
6. Support system optimization through the LSA tasks.

LSAR

1. Use of a contractor-developed ADP system to more efficiently record the LSAR and provide management flexibility to develop unique LSA output reports as the situation dictates.
2. Automated transfer of FMECA analysis to the LSAR B sheets.
3. Integration of ILS element information, such as task descriptions for technical manuals and provisioning information, with the LSA Record D and H sheets.
4. Documentation of software in the LSA Record.
5. Location of a read-only computer terminal in the Navy project office for continuing computerized LSAR review.

3.4   Postaward Contract Activities

3.4.1   General

This section discusses:

LSA guidance conference

Task inputs

Task execution

LSA/LSAR reviews

Use of the LSA products

3.4.2 LSA guidance conference

One of the requirements normally included in the SOW is that the contractor host and participate in an LSA/LSAR Guidance Conference. This conference, normally held within 30 to 90 days after contract award, provides a forum for establishing a greater mutual understanding of the government LSA contract requirements and the contractor's plan for execution. During this conference there should be a thorough discussion of the LSA and LSAR requirements. The discussion, at a minimum, should cover key areas, such as approaches to execution of the LSA tasks and subtasks; the scheduling of the LSA tasks, subtasks, and data records; and the specific role various contractor communities will play in executing specific LSA/LSAR requirements. LSAR requirements should be covered to the degree necessary considering the phase of the program. This could include the list of LSA candidates for documentation in the LSAR, the proposed LSAR Logistics Control Number (LCN) structure, and any data-element definitions at issue. The LSAR data record A or the data required to accomplish the data record A should be provided by the requiring authority no later than, and preferably prior to, the LSA/LSAR guidance conference. A formal review of the LSA data record A should be accomplished by the ILS/LSA management team during the LSA/LSAR guidance conference. Since LSA tasks and LSAR scheduling are frequently problems, the LM may request a contractor review of these events.

The requiring LM should provide an agenda and chair the guidance conference (see Appendix 13 for suggested agenda items). Government representatives should include the LM, assistant LMs, selected LEMs, project engineers, PMs, and the systems command (SYSCOM) ILS/LSA office. The contractor should have its major ILS players present, host the conference, and take minutes. The requiring LM must review the minutes, make sure that unresolved issues are recorded and assigned to specific individuals for resolution, and follow up on action items.

The guidance conference is a key step, since it sets the tone for the entire future LSA effort.

3.4.3 Government task inputs

LSA tasks and subtasks require specific input information necessary to their performance. Since the quality of these inputs significantly affects the quality of the LSA task reports, the requirer should place special emphasis on such government inputs. The inputs for each LSA task accompany the individual task write-up in MIL-STD-1388-1A. Over 100 inputs are identified in the standard. Some inputs are marked by an asterisk to indicate that they are to be provided by the government. The requiring LM is responsible for ensuring that government inputs are provided to the performing organization. Over 78 of MIL-STD-1388-1A task inputs can or should be provided by the government to the performing organization. Over 20 of these task inputs are administrative in nature. Thirty-five or more of the 78 have the potential to require direct LEM participation (see Table 3.4). There are also over 30 additional task inputs that can be provided by the government or the contractor at the discretion of the program manager and LM. See Appendix 5 for a table of MIL-STD-1388-1A tasks along with suggested skills, performing organization, and rational for selection. Since the government occasionally asks the contractor to collect the requirer inputs, the contractor may need to enlist its LSA team members' help in this regard.

3.4.4 Approaches to task execution

Approach alternatives. There are three general approaches to task execution:

Off-line (usually by an LSA analyst)

Interdisciplinary teams

Division of task responsibilities between ILS and engineering

The off-line approach was often used in initial MIL-STD-1388-1A applications. However, as the wide variation in different skills needed for the various subtasks was increasingly recognized, the use of interdisciplinary teams has grown. Table 3.7 illustrates this approach. More recently, as LSA has been recognized as a part of system engineering and more managers realize that many of the LSA tasks and subtasks are engineering tasks (Appendix 5 identifies general skills by subtask), the third approach is becoming more common. Table 3.8 illustrates this approach. One U.S. DOD acquisition organization actually places LSA tasks and subtasks designated as pertaining to engineering in the engineering part of the SOW, and their engineering organization is responsible for review and use of the results.

The three approaches can be combined or mixed. If a specific approach is favored, for whatever reason, a properly phrased question in the instructions to offerers (ITO) could encourage the bidder to use the preferred approaches. In fact, this issue could be a source-selection factor.

TABLE 3.7   Interdisciplinary Approach

Task/subtask	Team members
LSA Plan	LSA team
Design, Program Reviews	R&M: LSA
Use Study	LSA: HW/SW design: HE
Standardization	HW/SW design: R&M: LSA
Comparative Analysis	HW/SW design: R&M: LSA: cost
Functional Requirements	R&M: safety: ILS: HE
Support System Alternatives	ILS elements: LSA
Tradeoffs	
Support system	LSA; ILS elements
Readiness	R&M: LSA
Manpower	Maintainability: LSA
Training	Training: LSA
Level of repair	R&M: LSA
Diagnostics	Maintainability: LSA
Task analysis	ILS elements: LSA
Supportability T&E	HW/SW design: R&M: HE: safety: HE

TABLE 3.8   Performing Organization: One Approach

Lead responsibility	Task
Logistics	101
Logistics	102
Engineering	103
Engineering	201
Logistics	202
Engineering	203
Engineering	204
Engineering	205
Engineering	301
Engineering	302
Engineering	303
Logistics	401
Logistics	402
Logistics	403
Engineering	501

The same issue applies to support organizations or support contractors doing LSA prior to EMD if no hardware contractor exists. However, if a number of different support organizations are involved, use of the interdisciplinary team method may be difficult.

Detailed task-execution variations.   There are significant differences in practice in the "how to" aspect of task execution. This is particularly true of LSA prior to EMD. This problem is exacerbated in some cases by the fact that few support contractors or government organizations have in-depth experience executing the LSA tasks. Recently, this issue and possible solutions have begun to appear in the literature and in meetings of the CALS(CE) Industry Support Group (ISG) LSA subcommittee. The Material Readiness Support Agency (MRSA) has actually contracted for study of the "how to" front-end issue. This book discusses possible task-execution approaches in detail in Chap. 5, which includes sample forms for documenting non-LSAR-related task results. The material in this chapter provides support and hardware contractor departure points for developing LSA task execution approaches as a means of improving task-report quality.

3.4.5 Review of LSA products

Review of LSA products will be conducted by the performing and requiring ILS/LSA management teams under the direction of the requiring LM, who is responsible for establishing the review procedures, selecting reviewers, scheduling reviews, chairing the reviews, and ensuring that action items from each review are completed. The ILS/LSA reviews should be convened by formal correspondence to the contractor and outside activities, with distribution to all the internal team members.

Appendix 15 contains a checklist for contractor management and government reviews of draft LSA task reports (not LSAR reports). These checklists have been used successfully on other programs to quickly identify the quality of contractor-submitted tasks. Appendix 16 contains a checklist for the LSAR tables. Chapter 6 discusses the review of the LSAR in considerable detail.

3.4.6 Program use of LSA results

LSA task reports are used in two ways: "as is" and as an information base for development of input to various program documents. "As is" use means distribution of the reports to interested requiring and performing organization parties who need the information to perform their work. This obviously includes the ILS manager and the ILS elements managers. However, most members of the design community are interested in:

Item use (e.g., task 201)

Support and maintenance concepts (e.g., task 302)

Specifics on past problems (e.g., task 203)

Tradeoff results (e.g., task 303)

Logistics goals/constraints affecting design (e.g., task 205)

Table 3.9 identifies general uses of the LSA task reports as an input source to various program documents.

TABLE 3.9 LSA Task Usages

Documents	Tasks														
	101	102	103	201	202	203	204	205	301	302	303	401	402	403	501
Solicitation documents	X														
Define contractor program		X													
Program initiation documents				X	X										
Program requirements documents					X					X					
Program decision documents					X		X	X	X	X					
Development specifications						X	X	X	X						
Contracts		X	X	X	X	X	X	X	X	X	X	X			
Internal contractor use										X	X				
ILSP															X
T&E master plan															X
Status evaluation				X											
ILS resource determination												X	X	X	

SOURCE: Navy LSA Applications Guide, May 1986.

3.5 Scheduling LSA/LSAR

LSA/LSAR scheduling must occur within the structure of the overall program schedule. Execution of the tasks and subtasks will have considerable overlap, as illustrated in Fig. 3.2. This figure shows the scheduling at task level on a current program.

The timing of completion of the task and subtasks concerned with design influence is a major consideration. For maximum impact, each task must be scheduled to match major engineering review and decision points. Incorporation in the initial design is more cost-effective and increases the probability of acceptance. Figure 3.3 shows the major design reviews and their major functions.

Properly scheduling the LSAR and derivative products is critical to high-quality, timely LSAR data. LSA schedule compatibility with related data is a common problem in practice. The scheduling must take into account interim product deliveries, final product deliveries, and updates (see Table 3.10). The delivery of data must be available in time for the development of its related data item description (DID) product. For example, the provisioning data should be available in sufficient time prior to production delivery date to enable the purchase of spares concurrent with production assets; the LSAR data detailed task description must be available in advance of the technical manual delivery date; and so on. Key considerations in scheduling the LSAR are discussed in Chaps. 4, 5, and 6.

While detailed LSA/LSAR scheduling is usually developed by the contractor within the overall schedule specified by the government, the requiring LM should review CDRL delivery dates to ensure that the schedule fully supports the requiring organization's need date and

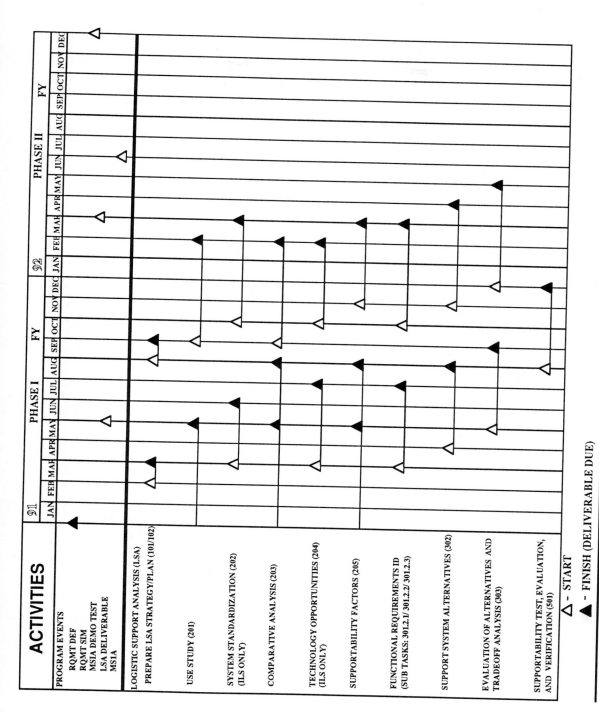

Figure 3.2 Example of overlap and iteration in task scheduling.

Figure 3.3 Overview design review process. (From Navy LSA Applications Guide, May 1986.)

Preconcept	Concept exploration	Demonstration and validation	Full scale development	Production deployment
0	0 / I	I / II	II / III	III

PROGRAM REVIEWS

Triangles: FRR/SRR — ARR/SDR — PDR — CDR — IPR — FDR

FRR/SRR (Concept exploration)

Functional requirements review/ system requirements review

Objective:
Ascertain system definition

When accomplished:
After functional analysis and preliminary requirements allocation

Elements:
Resolve differences between CD and preliminary FCI
Establish basis for functional baseline
Initiate LSA planning for system life cycle

ARR/SDR (Demonstration and validation)

Functional requirements review/ system requirements review

Objective:
Document/evaluate preliminary ACI

When accomplished:
After charateristics defined on items identified and ACI established

Elements:
Update FCI
Establish ACI
Provide basis for allocated baseline
FCI and ACI continues cycle application of configuration management

PDR — Preliminary design review

Objective:
validate selected design approach

When accomplished:
After preliminary design prior to start of detail design

Elements:
Interface control established prior to PDR
Initiate detail design for FSD
Initiate PCI
Initiate LDA Records

IPR — Integration pretest review

Objective:
Ascertain system readiness for test

When accomplished:
After fabrication of prototypes
After system documentation updated
After configuration prior to release for testing

Elements:
IPR demands traceability of specifications LSA and related data to functional and physical characteristics
Normally conducted by contractor serves to minimize risks of test

CDR — Critical design review

Objective:
Establish system elements
Design Compatibility assess
System productibility

When accomplished:
When detail design essentially Complete prior to production release

Elements:
Most detail design activity PCI
LSA records completed
PCI and PCI update
LSA EMD production operational and support requirements finalized

FDR — Formal qualification review

Objective:
Verify system performance in operational environment meets performance specifications

When accomplished:
After PCI finalized and ready for establishment of PBI

Elements:
Review configuration audit minutes
Evaluate changes resulting from audits and qualification tests
Serves to certify readiness for production delivery and logistic supportability

TABLE 3.10 LSAR Scheduling

CDRL delivery date	Data records Data/item description	| 1388-2A Data sheets | A	B	B1	B2	C	D	D1	E	E1	F	G	H	H1	J	X	| 1388-2B Tables | A	B	C	E	U	F	G	H	J	Data tables Data/item description
31 Jul 89	DI-R-7085		X	X	X												X	X	X							DI-ILSS-81163
15 Sep 90	DI-L-2085A	X	X	X	X	X											X	X	X						X	DI-ILSS-80646
31 Dec 91	DI-H-7091					X	X				X	X									X		X			DI-ILSS-81165
01 Feb 92	DI-H-7090					X	X	X	X	X	X	X									X		X			DI-ILSS-81150
	DI-S-6177B						X	X	X	X																DI-ILSS-81167
	DI-V-6180												X	X										X		DI-ILSS-80292A
	DI-V-6186A												X	X						X				X		DI-ILSS-81143

SOURCE: Draft UAV LSA Applications Guide, Block 1, July 1991.

the timing of events supports the intermediate program milestones. The requiring LM also could require the contractor to perform a schedule-compatibility analysis and present the results at the Guidance Conference. Contractors interested in an on-time, quality LSA product should perform this schedule review for their own benefit.

3.6   Costing

At some point, the government must make an estimate of the cost of the LSA/LSAR effort for budgeting and planning purposes. The contractor must make similar estimates for bidding purposes. One source of information for such estimates is the Army material readiness support activity pamphlet, "Cost Estimating Methodology for LSA Guide (CELSA)," MRSAP 700.11, which can be used as a departure point for making such estimates.

Table 3.11 shows the tasks that account for most of the LSA costs in each phase. The percentages will vary some with the type of system and the number of LCNs. The table is based on the more extensive data in Appendix 17. As a generality, mechanical systems for the same number of LCNs are more manhour-intensive than electrical systems, primarily because of higher task-analysis costs.

The cost-estimating methodology for LSA (CELSA) is a departure point, since documented cost figures are difficult to secure and costs vary with the depth of experience and the degree of LSA automation. The LSA task cost estimates should be reviewed for two possible pitfalls. First, is every activity covered in the WBS? Are the contributing disciplines funded? Second, has each LSA task been reviewed to ensure that it is fully integrated? The products of each discipline should reinforce the product, not produce duplicate answers.

TABLE 3.11   LSA Cost Drivers by Phase

Phase	Task
Preconcept	Task 203, 50%
CE/V	Task 103, 20% Task 303, 20%
DEMVAL	Task 103, 14% Task 303, 10% Task 401, 50%
EMD	Task 103, 10% Task 303, 8% Task 401, 64%
PROD	Task 103, 51% Task 501, 23%

3.7 Summary

This chapter identified key management considerations for both the requiring and performing organizations in implementing LSA/LSAR projects and specific appendices that provide useful detail concerning actions the LMs in the requiring and performing organizations must undertake.

Preparing LSA Contract Requirements

4.0 Introduction

The main thrust of this chapter is how to prepare cost-effective LSA requirements. The discussion expands on the general discussion on preparing LSA requirements in Chap. 3. The first part of the chapter discusses each task in MIL-STD-1388-1A in terms of its purpose, benefits, what each task says, the general considerations the logistics manager (LM) should take into account in tailoring the task, and the use of the task results. These discussions assume a full development program (one that proceeds through all the normal phases cited in DOD acquisition directives). Many programs (probably more than half) do not follow the full development acquisition process. Chapter 2 mentioned such cases (PIPs, NDIs, etc.). Adjustments for such cases are discussed in the second part of this chapter. The specific order of discussion is

Tailoring approaches (4.1)

Task discussion format (4.2)

The 200 series tasks (4.3)

The 300 series tasks (4.4)

The 400 series tasks (4.5)

The 500 series tasks (4.6)

The 100 series tasks (4.7)

Tailoring for various types of acquisitions (4.8)

Summary (4.9)

4.1 Tailoring

4.1.1 MIL-STD-1388-1A task format

Since tailoring of the LSA tasks is fundamental to cost-effective LSA, the format of the tasks in the standard was designed to encourage tailoring. Each task and subtask is presented as a discrete entity to be selected or rejected by the LM as part of the tailoring process. Each task is presented in the following format:

1. Purpose of the task

2. Description of the task (usually as a series of subtasks)

3. Task inputs (information needed to perform the task)

4. Task outputs (the products that result from performance of the task)

4.1.2 Tailoring approach in this chapter

Six different approaches to tailoring MIL-STD-1388-1A and their advantages and disadvantages were discussed in Chap. 3. The reader who is preparing to develop an LSA request for proposals (RFP) may wish to review this material. The tailoring advice in this chapter is a combination of the question approach, the boundary approach, and advice on further tailoring and focusing. Remember, *focusing* is further definition of the subtasks in the LSA standard. For example, the government could specify the alternate concepts to be evaluated. Focusing helps both the requirer and the executor to concentrate the limited resources available on the key issues rather than executing a "shotgun" approach. Focusing is seldom seen in practice, but we are strong advocates of its use. Our task write-ups suggest many places where focusing may be useful.

The tailoring and focusing approach in this chapter first uses a series of questions to reduce the number of tasks and subtasks that need to be considered. The first search question examines what LSA end objective (mission-system design influence, support-system design influence, logistic resource requirement identification) the task serves. If the objectives served are not applicable to the program, the task or subtask should be eliminated from consideration.

The second question involves a search by phase (see Table III, "LSA Task Application and Documentation Matrix," in MIIL-STD-1388-1A). If the task is not appropriate for the program phase, it can again be eliminated from consideration. A third search question is by program complexity. Since this type of search is questioned in some quarters, the step is labeled optional. If it is applied, the figures in the task discussions in this chapter, in Table III in the LSA standard, and/or in Appendix 7, including one for simple equipment, can be used.

Further advice on selecting, tailoring, and focusing is then used to refine the requirement. The approach is summarized in Table 4.1.

The next step involves advice on what to do with the remaining subtasks, which may include tailoring and focusing by subtask in some cases. Frequently, this requires consulting with ILS element managers and system engineering specialists about the desirability of specific tasks or subtasks.

Chapter 5 discusses subtask execution from the "do it all" approach. The reasons for using this approach are twofold. First, this approach

TABLE 4.1 **Suggested Tailoring Approach**

1. Does the program follow the classic acquisition process?
   Yes: Continue.
   No: You may wish to review the material in the second part of this chapter for tailoring advice on your type of acquisition program.
2. Is the top-level LSA objective (see Table I in standard 1A) served by this task applicable to your program?
   Yes: Continue.
   No: Go to the next task.
3. Is the task appropriate for your program phase?
   Yes: Continue.
   No: Go to the next task.
4. (Optional) Is the task or subtask appropriate for the complexity of your program (see task descriptions in book or Appendix 7)?
   Yes: Continue.
   No: Go to the next task.
5. Read the task benefits task description and the tables with LM guidance on subtask selection and focusing.
6. Consult the appropriate expert (system engineering specialty, ILS element manager) about tailoring and/or focusing of the task or subtask as appropriate.
7. Consider task/subtask refinement suggested in the individual task discussions.
8. Will there be competitive awards?
   Yes: See Sec. 4.8.7.
   No: Continue.
9. Would an "ask versus tell" approach be useful (see Sec. 3.2.5)?
10. Using the generic SOW in Appendix 10, prepare the appropriate LSA requirements.

shows the requirer what is not being done when a task or subtask is deleted in tailoring requirements. Second, this approach provides possible focusing in more cost-effectively bidding and executing the subtasks required. Too many LSA proposals are too general. The use of focusing in bid preparation can improve the quality of proposals.

4.1.3 Tailoring aids in the appendixes

Overview. Chapter 3 discussed the general tailoring approaches to the LSA. Section 4.1.2 discussed the tailoring approach in this chapter. In addition to this guidance, the following aids to tailoring and focusing LSA requirements appear in the appendixes:

Appendix 7: LSA Tailoring Worksheets

Appendix 8: 1A Task-Enhancement Sample SOW Language

Appendix 9: CALS-Related LSAR SOW Language

Appendix 10: Sample Common Core SOW

Common core SOW. LSA statements of work (SOWs) contain general requirements, with little variation from SOW to SOW, and specific requirements, which may vary considerably. The tailoring advice for different types of acquisitions in the later parts of this chapter illustrate this variation. A typical outline of an LSA SOW is as follows:

1. Management
   a. LSA planning
   b. LSA guidance conference
   c. Reviews
2. LSA documentation
   a. General
   b. Review and approval
   c. Delivery
3. LSA requirements (LSA tasks and subtasks)
4. LSA Record (LSAR)
   a. General
   b. Candidate list
   c. Logistics Control Number (LCN) Structure
   d. LSAR updates
   e. Validation
5. Forms 1949 series
6. Contract Data Requirements Lists (CDRLs)

While such outlines may vary, most address the subjects listed above. The tasks and subtasks selected for the program appear in section 3 of the outline. MIL-STD-1388-2A/2B tailoring can appear in sections 3, 4, 5, and 6. The remainder of the outline is a relatively stable or "common core" requirement. Appendix 10 contains a sample *common core* LSA SOW.

A common core SOW has several uses. Stored on a word processor, the LM's LSA tasks and subtasks can be inserted directly into the document for use in the SOW. LSA management also can combine the common core SOW with the other tailoring advice in this book to construct a variety of sample SOWs for the major types of organization acquisition situations.

Tailoring worksheets. Tailoring worksheets for major systems and complex equipments appear in Appendix 7. This appendix also contains worksheets for simple equipments, since the advice for equipments in the LSA standard is "heavy" in the opinion of many practitioners. A distinction has therefore been made between *complex* and *simple* equipments. Although we have seen no clear-cut definition of the difference between the two, this is clear at the extremes. An aircraft fire-control system is obviously complex. A basic ohmmeter is obviously simple. The worksheets in Appendix 7 expand on the suggested phasing and application guidance in the LSA standard when the standard does not tailor to the subtask level. Perhaps most important, the worksheets are a means to record the reasons for task and subtask selection. Such decisions are not always clear-cut in practice. We have seen a number of cases where lack of written backup on why tailoring choices were made was a later source of embarrassment. It is wise to record the reasons for selections for justification to management and review groups.

Suggested LSA task-enhancement language. The importance of focusing and enhancing the task and subtask statements in SOWs was emphasized in Sec. 4.1.2. Specific suggestions for this are contained in the task discussions in this chapter, particularly in the discussions on tasks 102 and 103. Appendix 8 contains suggested SOW language for many of these suggestions and the rationale behind them.

CALS-related LSAR requirements. Chapter 2 discussed the impact that the DOD Computer-Aided Acquisition and Logistics (CALS) initiative has had on the LSAR and its expected impact on the LSA tasks. The use of section L, "Instructions to Offerers," to apply new LSA requirement strategies, including CALS considerations, was discussed in Chap. 3. Appendix 9 contains suggested language concerning the LSA, LSAR, and related subjects such as Reliability and Maintainability (R&M), Reliability-Centered Maintenance (RCM), Level of Repair Analysis (LORA), and possible training requirements. This appendix specifically includes suggested language for data automation, data system integration, on-line access, digitization of LSA and LORA plans and reports, and various CDRL examples.

4.2 Format for Task and Subtask Discussions

This part of the chapter discusses each task in terms of its purpose, the benefits to be derived from performing the task, what the task requires and phasing, specific guidance to the LM about the task, and use of the task results. Special emphasis is placed on tailoring, focusing, and enhancing the words in the 1A standard. The tasks are purposefully not presented in numerical order. The 100 series, which deals with LSA management, is presented last because the tasks are easier to comprehend after one has an understanding of the analysis-oriented tasks in the 200 through 500 series.

Most of the task descriptions have summary charts, descriptions, inputs, outputs, 1A suggested phasing, and special information where appropriate. The layout of this information is shown in Fig. 4.1. The information presented in the task figures is taken from MIL-STD-1388-1A. In some cases, the information in the 1A standard is ex-

Input	Subtasks	Output documentation
3.1 Numbers to the upper left are the last two digits of the inputs for the task. Administrative-type inputs are not listed.	2.4 Numbers represent the last two digits of the subtasks that are required to be performed under that task.	4.4 Numbers represent the last two digits of the task outputs. This is the documentation which results from performance of the task.

Phasing	Subtask	Concept	DVAL	EMD	PROD	
Provides tailoring guidance to indicate which acquisition phase the subtask is performed in. Distinguishes between major systems and equipments.						Special information

Figure 4.1   Layout of task summary figures and explanation of content. (From Navy LSA Application Guide, May 1986.)

panded to be more comprehensive. Subtask titles in the task discussions are those used in Table I in the standard. They summarize the subtask information in the figures.

4.3   The 200 Series Tasks: Mission- and Support-System Definitions

The 200 series tasks establish supportability objectives and supportability-related design goals, thresholds, and constraints through comparison with existing systems and analyses of supportability, cost, and readiness drivers.

4.3.1   Task 201: Use Study

Purpose. This task answers questions concerning system deployment, its use, and its intended environment, as well as the design and support implications of the planned use.

Benefits. The Use Study is a relatively cheap (see Appendix 17) but key task. It assembles and develops the information available on usage and the environment of use in one document. It thus provides a starting point for formulating supportability design and support requirements by providing the basis for the LSA tasks, such as tasks 202, 203, 205, and 301, and for future ILS planning. Information on expected mission and functional requirements for the new system is needed to determine in general terms whether there are any major changes in support resources from existing similar systems and their support structure. When an explicit and detailed mission scenario is developed, it also can serve as the basis for readiness modeling in subtask 303.2.4.

Task description. The Use Study should be initiated in the preconcept phase, when possible, because it is a useful input to task 101, Development of an Early LSA Strategy. Figure 4.2 summarizes the four subtasks in the Use Study.

Subtask 201.2.1, Supportability Factors, identifies the support fac-

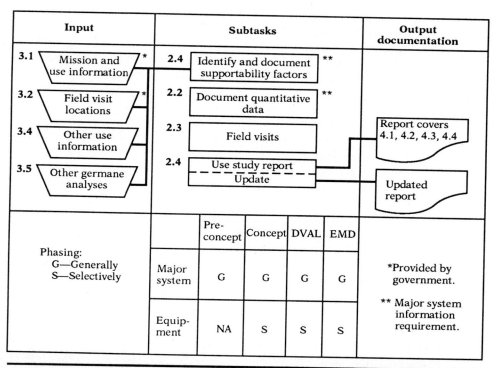

Figure 4.2 Task 201, Use Study. (From Navy LSA Applications Guide, May 1986.)

TABLE 4.2   **Examples of Supportability Factors**

Mobility requirements	Number of systems
Deployment scenarios	Planned usage and employment (peace, war)
Mission frequency and duration	Allowed maintenance periods
Basing concepts	Environmental requirements
Anticipated service life	Operational availability
Interfaces (other systems, items)	Logistics delay time/turnaround time
Operational environment	Preplanned product improvement
Human capabilities and limitations	ILS elements factors and constraints
Existing support available for the new system	

tors that affect the intended use (peacetime and wartime) of the system. Examples of such factors are listed in Table 4.2.

Subtask 201.2.2, Quantitative Factors, develops specific values for the supportability factors identified from inputs and data in subtask 201.2.1, Supportability Factors.

Subtask 201.2.3, Field Visits, involves field visits to operational units and support activities to ensure that the executor has a realistic picture of the operating environment and conditions.

Subtask 201.2.4, Use Study Report and Updates, results in a report on planned usage that summarizes and analyzes the data collected in the previous three subtasks, and updates are prepared as necessary. Note that this subtask is really two subtasks, a report and an update.

Most of the input information normally comes from requirer sources. Other sources of information for the Use Study include mission area analysis, Justification for Major System New Starts (JMSNS), mission effectiveness model outputs, field visits, war plans, and previous system studies. This task is therefore a candidate for performance by the requirer. However, if very detailed information is required, performance by the prime contractor may be desirable. If subtasks 203.2.1, Identify Comparative Systems, and 203.2.3, Comparative System Characteristics, are done, they are substantial inputs for task 201.2.2, Quantitative Factors. A system-level Functional Requirements Identification is also a useful input for this task, particularly if any quantitative analysis is to be done.

Although software is not listed as one of the Use Study considerations, this should obviously be considered, since the nature of use could affect various software design characteristics, such as response times, restart times, software reliability, and operator ease of use.

See Table 4.3 for selection and focusing advice.

Use of results.   The report should be used by both the requirer and the executor LMs and the LEMs as a basis for ILS planning. Other requirer and executor communities, such as system and design engineering and R&M, are also usually interested in the Use Study as the

TABLE 4.3   Task 201 Subtask Selection and Focusing

LM guidance
1. *LSA end objectives:* Influence of mission and support design.
2. *Phasing:* See Fig. 4.2 (Appendix 7 for simple equipments). Phasing is generally the same regardless of program complexity. The task complexity is generally the result of program complexity. The task is generally desirable for DEMVAL or EMD starts, at least in some minimal form.
3. *Complexity:* Figure 4.2 lists the task as selective for equipments. Most complex equipments warrant a Use Study. It is suggested that at least a minimal effort be expended on simple equipments since the types of data in the Use Study drive the conduct of the other LSA tasks and future ILS planning. Some minimal effort, whether formal or informal, is necessary to derive an A record.
4. *Tailoring and focusing:*
a. Field visits are not necessary in CE if the requirer and the executor are the same. Field visits are particularly desirable when the contractors involved are more commercial than defense oriented (e.g., fire trucks).
b. If a minimal effort is all that is needed, the effort could be accomplished by the requirer or a support organization or contractor, particularly in CED or DEV.
c. The LM must provide the information necessary for the contractor(s) to perform this task.
d. The LM should determine appropriate locations for any on-site visits and arrange for such visits. Field visits present an opportunity for detailed identification of ILS resources (i.e., test equipment, manpower quantities and skills, facilities, etc.) available at operating and maintenance organizations.
e. The LM should request the LEMS and system engineering to review the Use Study report for comment and use.

SOURCE: From Draft UAV LSA Applications Guide, Block One, July 1991.

basis for planning. The study should therefore be widely circulated in all these communities.

The task also can be an input to task 101, Early Development of an LSA Strategy, in preconcept exploration (see Chap. 5). Appropriate Use Study results are also documented on LSAR A Data Tables. This includes system-level information in the AA, AB, and AG tables and interoperability and transportability information in the AH and AJ tables, as well as peculiar inputs in the AK table, with little or no additional analysis required.

4.3.2   Task 202: Mission Hardware, Software, and Support-System Standardization

Purpose.   This task determines what supportability benefits the use of standardization can provide to the program.

Benefits.   Use of standards has benefits to design and production as well as to logistics. A study made in the 1960s concluded that standardization benefits accrued as follows: design, 25 percent; production, 35 percent; and logistics, 40 percent. Benefits to design include savings in design time and testing. Production benefits are lower costs through economies of scale. Logistics savings include technical manual and training course development costs, costs to enter an item in the inventory, decreased range and depth of stock, and use of existing test and support equipment.

Examples of standardization include standard electronic modules, Navy standard computers, and standard test equipments. By standardizing, the DOD reduced the number of makes and models of mobile power generators used from over 2000 in the Vietnam conflict to approximately 36 presently. Standardization within a program also can be beneficial (i.e., fasteners, cables, connectors, etc.).

Standardization should therefore be an important program consideration. However, whether or not a particular standardization proposal is good for a program must be evaluated in the context of that program.

Standardization can be imposed at any level, from major subsystems [aircraft engines are frequently government-furnished equipment (GFE)] to the piece-part level (nuts, bolts, fasteners). Generic standardization of fasteners on one major aircraft program resulted in significant savings. Standardization decisions are made at various levels in both the requirer and executor organizations.

Task description and phasing. The LSA standard is careful to note that it is not the primary implementing requiring document for standardization. Figure 4.3 shows the inputs, subtasks, outputs, and application by phase for a typical program. Beginning in concept exploration (CE), government inputs may be a key part of this task. These include specific program constraints and requirements as well as DOD information on existing or planned logistic resources. Specific program constraints may be a hard contract requirement, or they may be cited as a desirable constraint. Some examples of such constraints are use of existing automatic test equipment (ATE), restriction of repair tools to mechanics' standard tool kits, and so on. The other subtask products represent input to decisions; their function is to help both Navy and contractor decision makers achieve the best balance between performance, cost, schedule, and supportability.

The first subtask, 202.2.1, Supportability Constraints, is a contractor extension of the government inputs cited above. It is oriented toward ILS elements.

Subtask 202.2.2, Supportability Characteristics, requires contractor logistic engineering input for company-generated standardization proposals to evaluate their impact from a supportability viewpoint. It is oriented toward mission hardware and software as well as to support.

The third subtask, 202.2.3, Recommended Approaches, represents suggestions for mission hardware and software standardization.

Subtask 202.2.4, Risks, analyzes the possible risks involved with

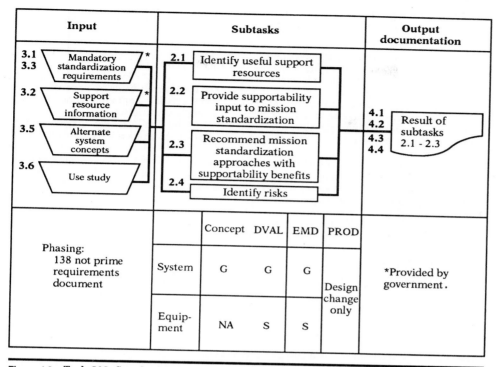

Input	Subtasks	Output documentation

Input	Subtasks	Output documentation
3.1 3.3 Mandatory standardization requirements *	2.1 Identify useful support resources	
3.2 Support resource information *	2.2 Provide supportability input to mission standardization	4.1 4.2 4.3 4.4 Result of subtasks 2.1 - 2.3
3.5 Alternate system concepts	2.3 Recommend mission standardization approaches with supportability benefits	
3.6 Use study	2.4 Identify risks	

Phasing: 138 not prime requirements document		Concept	DVAL	EMD	PROD	
	System	G	G	G	Design change only	*Provided by government.
	Equip-ment	NA	S	S		

Figure 4.3   Task 202, Standardization. (From Navy LSA Applications Guide, May 1986.)

the constraints proposed in the first three subtasks, although the language can be interpreted to apply only to the first subtask owing to use of the word *constraints*.

This task explicitly mentions software in the second and third subtasks, and it is implied in the other subtasks. Examples of possible software standardization include standard development software, standard high-order languages, and other developed software/documentation. The U.S. DOD is currently heavily emphasizing reuse of existing software. (See Table 4.4 for selection and focusing advice.)

Use of results.   Specific results should be communicated to the applicable requirer and executor designers and LEMs to ensure that subsequent acquisition of support equipment, spare parts, etc. benefit from the task. The supportability design criteria/constraints developed also serve as possible inputs to the comparison candidates in task 203 and to standardization and interoperability data in the LSAR A tables and task 205, Supportability Design Factors, particularly subtask 205.2.3, Specification Requirements.

TABLE 4.4 Task 202 Subtask Selection and Focusing

LM guidance
1. *LSA end objective:* The subtasks appear to support all three top-level LSA objectives, although this differs from the classifications in Table I, "Index of LSA Tasks," in standard 1A.
2. *Phasing:* May be applicable to all phases prior to production.
3. *Complexity:* Figure 4.3 suggests selective application of the task to complex equipments. Appendix 7 is a tailoring breakout for simple equipments developed from several existing efforts in this regard. Appendix 7 suggests selective application of the first subtask for simple equipments.
4. Most contractors use standards where possible, due to the development benefits, even if they have no formal program. If LSA funds are tight, this task or some of the subtasks are candidates for elimination.
5. Figure 4.3 does not tailor by subtask, but this is a good task to tailor by subtask, with possible further focusing at the subtask level. Although CELSA data (see Appendix 17) does not indicate it, the second and third subtasks could involve a very large and expensive work load in EMD, since they could in theory require a review of each standard piece part proposed. These subtasks should be restricted to major items and generic areas. Section 5.3.5 has other suggestions for such restriction.
6. The third subtask could overlap with other possible standardization requirements (MIL-STD-965 and MIL-STD-680) invoked elsewhere in the contract. The APML should check with engineering on the desirability of these subtasks.
7. The LEMs system engineering and the SYSCOM standardization office, if one exists, can help identify sources of information. Additional systems are in Chap. 5.
8. The LM should require the contractor to maintain a list of its LSA standardization proposals and actions taken on these proposals.
9. The requirer should include any desired constraints/requirements in the task input (see task inputs 202.3.1 and 202.3.3). These should have been developed in the previous work to minimize possible adverse impacts on the work of the performing organization.

SOURCE: From Draft UAV LSA Applications Guide, Block One, July 1991.

4.3.3 Task 203: Comparative Analysis

Purpose. This task has two main objectives. The first is to provide data to serve as a basis for selecting supportability objectives in subtask 205.2.2, Supportability Objectives and Associated Risks, and to judge the feasibility of the objectives and goals resulting from the tradeoffs in task 303, Evaluation of Alternatives and Tradeoff Analysis.

The second objective is to examine the history of similar systems to identify the supportability drivers and qualitative problem areas on such systems. In essence, this is a "corporate memory" effort.

Benefits. A rational basis is needed for the LM and program management to select or evaluate the feasibility of the new system supportability objectives and goals selected in task 205, Support-Related Design Factors, and task 303, Evaluation of Alternatives and Tradeoff Analysis. The determination of such goals involves the R&M as well as the design communities and program management. Such objectives and goals are program decisions to which the LM contributes.

The first three subtasks, 203.2.1, Identify Comparative Systems, 203.2.2, Baseline Comparison System (BCS), and 203.2.3, Comparative System Characteristics, provide the LM with a sound basis on which to evaluate such goals and to make recommendations. The comparative system is based on data collected from known systems and as such is more meaningful than hypothetical data. The results of these subtasks are also inputs to tasks 205 and 501.

The fourth, fifth, and sixth subtasks, 203.2.4, Qualitative Supportability Problems, 203.2.5, Supportability, Cost, and Readiness Drivers, and 203.2.6, Unique System Drivers, explicitly identify problem areas in the comparative or BCS system to highlight points where design improvements will be most cost-effective. They are key inputs to task 204, Technological Opportunities, in which the contractor and the LM use the results of the corporate memory subtasks to focus the design effort in task 204 on the areas of highest payoff. The results of these subtasks also feed the second and third subtasks in task 301.

If done prior to concept formulation, this task provides excellent inputs to task 101, Development of an Early LSA Strategy, and the development of supportability constraints in program initiation documents. Subtask 203.2.3, Comparative Supportability Characteristics, can be an input or update to subtask 201.2.2, Supportability Characteristics, depending on the time frame of execution of the two subtasks.

Except for preconcept, when Cost-Estimating Methodology for LSA (CELSA) estimates (see Appendix 17) it to be 50 percent of the cost of LSA, this task is not very costly in terms of the LSA effort in most phases.

Subtask description.   The eight subtasks are summarized in Fig. 4.4. Risks should be analyzed each time the task is done. Therefore, risk analysis should precede the first update.

Subtask 203.2.1, Identify Comparative System, can be combined directly with subtask 203.2.3, Comparative System Characteristics, for simple, high-level comparisons. Figure 4.4 illustrates such a comparison on Air Force cargo aircraft.

The C-130E, C-141, and C-5 are existing aircraft. The numbers for the C-X, which is the new aircraft, are added to the existing data in subtask 303.2.9, Comparative Evaluation (see Table 4.5). Other factors that might be compared include mean time between failures (MTBF), A_0 mean flight hours between unscheduled maintenance actions, mean time between removals (MTBR), annual preventive maintenance manhours, and annual support costs.

The second subtask, 203.2.2, Baseline Comparison System (BCS), can be developed for complex system/equipment by creating an artifi-

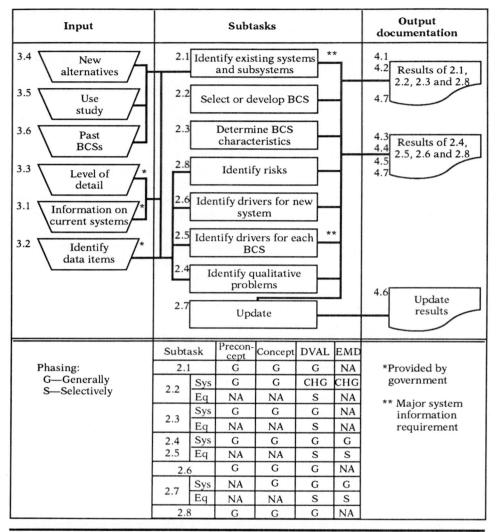

Figure 4.4 Task 203, Comparative Analysis. (From Navy LSA Applications Guide, May 1986.)

TABLE 4.5   Simple Comparative Analysis (C-X)

	C-130E	C-17	C-141	C-5
Maintenance manpower per flying hours	17.0	19.0	19.2	54.9
Mean flying hours between maintenance action	0.46	0.65	0.42	0.17

cial system through the selection of components from different existing systems. The JVX aircraft example in Appendix 18 uses 57 subsystems selected from 6 different systems to form a BCS. A system-level functional requirements identification can provide a basis for BCS component selection as soon as it is available.

The third subtask, 203.2.3, Comparative System Characteristics, identifies operating and support (O&S) costs, readiness and R&M values, and logistic support resource requirements of the systems/subsystems selected for comparison. Other design features that drive support requirements, such as weight, size, etc., are also useful for comparison. Such key support-related design parameters could be identified in task 201, although the 1A standard is ambiguous in this regard. An example of the support-related design characteristics for a software-intensive system is presented in Table 4.25.

The qualitative support problems in subtask 203.2.4, Qualitative Supportability Problems, are factors subordinate to top-level readiness, R&M, manpower, and cost concerns. Examples of such problems are human factors (e.g., visibility, accessibility, comfort, freedom of movement, task complexity), skill levels, modularization, accessibility, and environmental constraints.

The supportability, cost, and readiness drivers in subtask 203.2.5, Supportability, Cost, and Readiness Drivers, include major problem areas such as low-reliability, high-maintenance items in terms of manhours and skill levels, resupply and recycle times, and expensive spare parts.

The sixth subtask, 203.2.6, Unique System Drivers, must, of necessity, rely heavily on engineering input. Estimating by analogy is frequently employed in this subtask. Inputs to this subtask and subtask 203.3 can include data from maintenance data systems such as the Naval Aviation Maintenance and Material Management Program (3-M data), visibility and management of operating and support costs (VAMOSC), casualty reports (CASREPs), O&S cost estimates and specification values for items currently in development, task 201, Use Study, and task 202, Standardization. If the LSAR is maintained for the life of the system due in part to the CALS initiative, these data bases also can be a source of data in the future.

Although application of this task to software is not mentioned explicitly in the standard, it could be interpreted to be implicit. Certainly the concepts of reviewing existing comparable software systems in terms of design characteristics and approaches and the examination of costs and failure reports for problem areas can be applied to software. In some cases, software problem-reporting systems do exist. This should be increasingly true with the growing use of software.

Comparative analysis on both the mission system and ILS elements is a source of possible standardization candidates. The BCS can be done by ILS element as well as by key supportability design and cost parameters. For example, parts of the HARDMAN/MANPRINT analysis could be an integral part of the BCS. Any comparative analysis performed on ILS elements should be an input to Task 202, Standardization, and to task 302, Support-System Alternatives. If an ILS element BCS is required, data from manpower authorization documents, training requirements, support equipment, Tables of Allowance (TAs), and the Standardization Handbook on Ground Support Equipment (MlL-HDBK-300) are possible useful inputs. Any LSAR data bases on the BCS items also would be useful. The application of comparative analysis to ILS elements in this task should be done on a selective basis. A summary of the JVX Comparative Analysis is contained in Appendix 18 for illustrative purposes.

A key to these tasks is data availability. Without data, the task reduces to subtask 203.2.1, Identify Comparative Systems, subtask 203.2.4, Qualitative Problems, subtask 203.2.7, Updates, and subtask 203.2.8, Risks and Assumptions, unless expert opinion, engineering estimates, and specification values for items currently in development are used as data sources. See Table 4.6 for selection and focusing advice.

Use of results.   This task generally provides data of interest to both the design and ILS communities, so the report should be widely circulated. The drivers identified should result in various design actions. This task is a key step in the LSA drive to influence design.

4.3.4   Task 204: Technological Opportunities

Purpose.   This task identifies and evaluates applications of new and existing technology to improve readiness and reduce the operating and support costs of the new system.

Benefits.   This is a relatively inexpensive task (see Appendix 17). It is an obvious follow-on to task 203, Comparative Analysis, that identified drivers and problems with existing systems or the BCS. It enlists the design community to help eliminate specific past problems to improve readiness and supportability. Opportunities developed that are not immediately usable can become part of a Preplanned Product Improvement (P^3I) package. Figure 4.5 illustrates the notion of technology insertion, such as P^3I, over time in a program. This task can help the LM and the program manager determine a planned approach to adopt specific new technology to improve supportability as technology evolves.

TABLE 4.6   Task 203 Subtask Selection Focusing

LM guidance
1. *Top-level objectives:* This task can influence both mission-system and support-system design.
2. *Phasing:* See Fig. 4.4 and Appendix 7 (for simple equipment). Table III in standard 1A suggests that only the first, sixth, and eighth subtasks be performed for equipments prior to DEMVAL. There are other subtasks that can be usefully performed in CE on complex equipments. Appendix 7 on simple equipments suggests selective application of the first, third, fourth, and fifth subtasks in CE. Experience suggests the first, third, sixth, and eighth subtasks as a minimum for simple equipments.
3. *Complexity:* See Fig. 4.4 and Appendix 7. Table III in standard 1A classifies the first, sixth, and eighth subtasks as mandatory for equipment and the other subtasks as selective. Appendix 7 classifies all subtasks as selective or NA for simple equipments. The most complex equipments warrant the subtasks standard 1A classifies as selective, and most simple equipments warrant at least the first, third, sixth, and eighth subtasks.
4. The LM may wish to do the first subtask internally and provide the results to the executor as a departure point for this task.
5. The LM should coordinate the need for any ILS element BCS analysis with the individual ILS element managers. MP&T, S&TE, and facilities are particularly good BCS candidates.
6. The LM should not specify the subtasks requiring a data base if none exists or he should specify that use of expert opinion or engineering estimates as a basis is acceptable.
7. The LM may wish to specify the levels of indenture for this analysis or ask the contractor to state its approach in its proposal. In terms of top-level goal setting, beyond subsystem level, each additional level adds costs while the amount of increased precision decreases. Analysis below subsystem level should focus selectively on drivers. This was done on a major aircraft program to the fourth and fifth levels.
8. The LM should require government approval of the support parameters to be used in subtask 203.3. A logistics-related reliability parameters, such as mean time between maintenance actions, should be required, if available, because logistics reliability can differ significantly from mission reliability.
9. This task generally requires contractors to use existing data bases, e.g., 3-M, 66-1, VAMOSC, etc. An alternate or additional approach potentially useful in DEMVAL and EMD is to use the R&M analysis of the proposed system to identify readiness and O&S cost drivers through identification and study of *a.* Low-reliability, high-unit-cost items—O&S cost improvement *b.* Low-reliability, low-unit-cost items—readiness improvement *c.* Low-reliability, high-MTTR items—manpower improvement This approach has the added advantage of a higher end-objective orientation. Sample SOW language to require this approach is in Appendix 8.
10. The task is generally considered an excellent candidate for performance by a requirer service organization, since the data required resides in the government.

SOURCE: From Draft UAV LSA Applications Guide, Block One, July 1991.

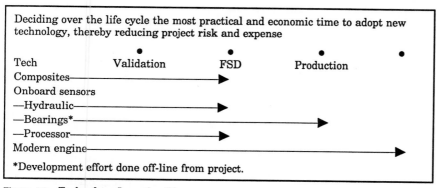

Deciding over the life cycle the most practical and economic time to adopt new technology, thereby reducing project risk and expense

Figure 4.5   Technology Insertion Plan.

The proposed improvements are an input into subtask 205.2.4, Supportability Objectives and Associated Risks, and subtask 205.2.5, Specification Requirements. Updates of task 203, Comparative Analysis, to lower indenture levels create new targets of opportunity for task 204, Technological Opportunities, consideration.

Task description.   Figure 4.6 summarizes the task and typical phasing. The three subtasks involve the following:

1. Establishing design technology approaches to improve supportability of the new system by (a) identifying technological advancements and design improvements that have the potential for reducing support costs or improving readiness, (b) estimating the potential impact of these improvements in terms of costs, risks, and readiness of the system, and (c) identifying design improvements to ILS elements associated with the system to increase supportability.

2. Updating the design objectives as a result of improvements planned in the first subtasks.

3. Identifying the risks associated with the updated design objectives and the costs and schedule impacts of the proposed improvements.

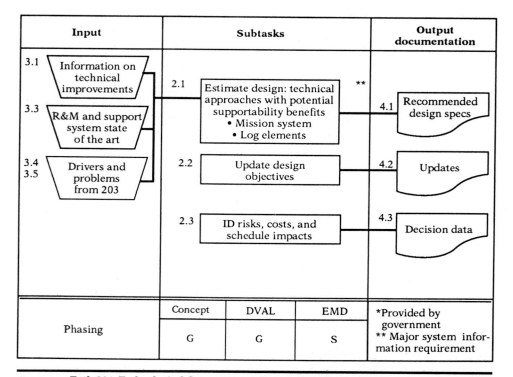

Figure 4.6   Task 204, Technological Opportunities. (From Navy LSA Applications Guide, May 1986.)

TABLE 4.7 Technology Areas

Electronic packaging and interconnection	Diagnostics
Very high speed integrated circuits	Power supplies
Cabling and connectors	Fiber optics
Computer-aided design/computer-aided technology	Testing manufacturing
Nondestructive evaluation	Structural composites
Mechanical systems conditioning monitoring	

Some examples of major technological advancements for consideration in this task are listed in Table 4.7. Inquiries about the possible use of such technologies can be made to both the government and contractor design communities.

Software is also not mentioned explicitly in this task in the standard. However, the current advances in software technology make its application in software-intensive programs very desirable. The LM may wish to include such a software study in the SOW.

Task 201, Use Study, subtask 203.2.4, Qualitative Supportability Problems, subtask 203.2.5, Supportability, Cost, and Readiness Drivers, and subtask 203.2.6, Unique System Drivers, are key inputs to this task. See Table 4.8 for selection and focusing advice.

TABLE 4.8 Task 204 Subtask Selections and Focusing

LM guidance
1. *LSA end objectives:* Influence the design of the mission system and/or the support system.
2. *Phasing:* See Fig. 4.6 and Appendix 7 (for simple equipment). This task should begin in the concept exploration phase, while design is still flexible. It normally tapers off somewhat in EMD, since design becomes fixed in that time frame. It could continue into the production and deployment phase if a P^3I program is planned. This could be particularly appropriate for accelerated programs which do not have the development time to include new technology currently evolving. This task should have heavy input from the design.
3. *Complexity:* The 1A standard makes no distinction concerning program complexity in its tailoring guidance. However, Appendix 7 on simple equipments calls for selective application in CE and DEMVAL.
4. Inputs to this task could come from both the government and contractor. For simple equipments, the government could perform this task; for complex equipments, the prime contractor is generally needed. Since many data sources could be researched, this task, in a sense, could constitute a "hunting license," so the requirer may wish to place some limits on the work done, particularly if there is no competition. One way to do this is to instruct the bidder to be very explicit on how it proposes to do this task. A second approach is to direct the executor to confine activity to the problems and drivers developed in task 203.
5. This task could be tailored by separating it into mission-system technology issues and ILS-element technology issues. If both areas are required, the two segments could be done by different organizations, so the LM should require the contractor to identify which organization contributes to which segment.
6. The second subtask could be accomplished as part of subtask 205.2.4, Supportability Objectives and Associated Risks, and subtask 205.2.3, Specification Requirements.

SOURCE: From Draft UAV LSA Applications Guide, Block One, July 1991.

Use of results. The results of this task are a series of proposals for design improvements to the mission system and ILS elements such as support equipment and training devices. The proposals may result in new design criteria/requirements or guidelines. Since new technologies may bring new support burdens, the support impact must be evaluated and decisions made to either implement or reject the proposals. The proposals may vary in terms of implementing authority. Some will be in the purview of the contractor, but many will require government approval. For those proposals requiring government approval, the LM should see that the proposals enter the proper evaluation channels and follow up to see that decisions are made and appropriate implementation actions are taken.

4.3.5 Task 205: Supportability and Supportability-Related Factors

Purpose. This task assists in the determination of reasonable support and support-related design objectives, goals, and constraints and identification of associated risks for the new system.

Benefits. The main thrust of this task is to transform the results from previous 200 series analyses into specific requirements for both design and support. It is therefore a relatively inexpensive task (see Appendix 17). This is a key design influence task because it captures known supportability design needs and translates them into design requirements, if not previously done. This enlists the design organization's help in addressing the LM's supportability concerns.

The requirements developed can be both quantitative and qualitative. They can be system-level or lower. Quantitative parameters are objectives subject to tradeoff until engineering manufacturing development (EMD), when they become firm goals or thresholds. Examples of possible goals include hardware (HW) and software (SW) reliability, repair turnaround time, mean time to repair (MTTR), mean time for removal, mean time between maintenance (MTBM) actions, support equipment utilization rates, etc. A logistics-related reliability parameter, as well as a mission reliability parameter, should be required, since these numbers can vary by as much as 10 to 1. As these objectives become firm goals or thresholds, they become the basis for supportability for the remainder of the program. Who takes action depends on the nature of the goal or constraint. Some actions can be implemented by the contractor without government approval. Top-level requirements are usually program management decisions that require consultation within the program office. The goals and constraints become design criteria/requirement inputs to various requirements documents, program management documents, specifications, and contracts, as well as program ILS requirements.

Task description. Figure 4.7 summarizes this task and its phasing. Subtask 205.2.1, Supportability Characteristics, identifies quantitative support characteristics for alternative equipment design and operational concepts. The support characteristics should be derived from those identified in task 201, Use Study, and task 203, Comparative Analysis.

Subtask 205.2.2, Sensitivities, requires sensitivity analyses on equipment variables connected with cost, readiness, and supportability. Sensitivity analysis allows the manager to measure the impact of possible changes, as well as to answer "what if" questions. It can show

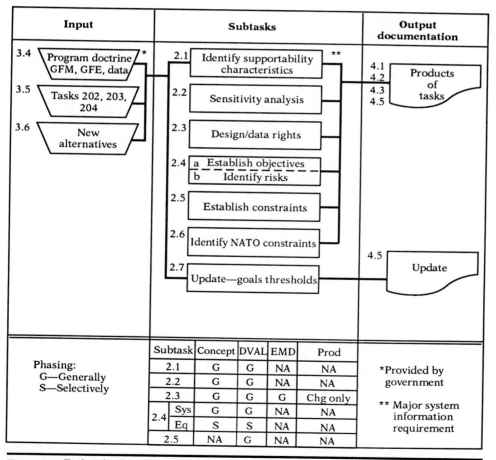

Figure 4.7 Task 205, Supportability and Supportability-Related Design Factors. (From Navy LSA Applications Guide, May 1986.)

where a slight change in one parameter can cause a major change in another. Figure A21.2 in Appendix 21 illustrates this phenomenon. At another range shown in this figure, further improvement in reliability has no significant impact on sortie rates. Sensitivity analysis can be useful in selecting quantitative goals.

Subtask 205.2.3, Design/Data Rights, addresses identification of hardware/software on which the government will not have full design rights. These rights are important because they can significantly affect future support costs. However, such rights are probably addressed elsewhere in the RFP/contract.

Subtask 205.2.4, Supportability Objectives and Associated Risks, can be considered two subtasks. The first part requires establishment of support, cost, and readiness objectives based on the first subtask. The second part requires a risk or uncertainty analysis of the objectives recommended and of any new technology planned for use. This part of the task can be related to the second subtask on sensitivities. The objectives in this subtask are normally established at the end of concept formulation.

Subtask 205.2.5, Specification Requirements, establishes quantitative and qualitative design constraints to be included in the requirements documents, specifications, and contracts. These can be design requirements that flow down from the top-level supportability requirements (e.g., inherent reliability versus operational reliability) or constraints developed as a result of task 202, Standardization, or subtask 203.2.4, Qualitative Supportability Problems.

Subtask 205.2.6, NATO Constraints, seeks to identify any constraints that would prevent the use of a NATO system/equipment to meet the mission need. It is not recommended by Table III in the 1A standard on less than major systems.

Subtask 205.2.7, Supportability Goals and Thresholds, requires the refinement of support, cost, and readiness objectives and the establishment of firm goals or thresholds by Engineering Manufacturing Development (EMD). It is based on the update of the fourth subtask and any tradeoff results available from task 303, Evaluation of Alternatives and Tradeoff Analysis.

Although the standard does not mention software explicitly in this task, consideration of software support-related design parameters and principles is an obvious necessity in software-intensive programs. This could include software reliability, since software reliability in some newer systems is the major driver in determining system-readiness goals. See Table 4.9 for subtask selection and focusing.

Use of results. Task 205 provides support-related design goals and a departure point for tradeoffs in task 303, Evaluation of Alternatives and Tradeoff Analysis. The final goals or thresholds and other supportability requirements (subtask 205.2.3, Specification Require-

TABLE 4.9 Task 205 Subtask Selection and Focusing

LM guidance
1. *LSA end objective:* To influence the design of the mission and support system.
2. Initiated during the CE phase, updated as needed, and generally finalized at the end of DEMVAL. However, subtask 205.2.3, Specification Requirements, can be required during EMD.
3. *Complexity:* No distinction is made in Fig. 4.7 in tailoring due to complexity except for the sixth subtask involving NATO. However, the need for the sixth subtask should be made on a program basis without regard to complexity. Appendix 7 on simple equipments suggests that the third subtask be used.
4. The LM should determine which of the first three subtasks are needed. Sensitivity analysis may not be needed. Data-rights analysis is complex and expensive, but it is important because it has a significant impact on future ILS costs. However, it should be covered in other parts of the RFP/contract. If so, do not include it in the LSA SOW.
5. The need for the sixth subtask also should be determined without regard to complexity.
6. The LM should ensure that logistics reliability, as well as mission reliability, is addressed in the HW and SW reliability parameters and goals. In many of today's systems with high hardware reliability, SW R&M is the major reliability concern.
7. If contract incentives on supportability requirements are a consideration, they should be at the system level to prevent optimization at lower indenture levels which may not really represent truly optimal solutions at the system level.
8. The support burden and other effects of the GFE/GFM, administrative and logistic delay time, and other items outside the control of the performing activity must be accounted for. For example, if the overall threshold for manpower is 100 manhours/system/year, then the contract should reflect a threshold of 75 manhours/system/year for performing activity-developed hardware.
9. The translation from supportability goals and thresholds to specification requirements is also important for readiness parameters. When the item under procurement is a complete system, applicable readiness parameters may be suitable for inclusion in the specification. However, if the item is less than a system, then other logistics-related parameters may be more appropriate.

10. Some key review points are
  a. Is the list of supportability characteristics complete?
  b. Do the numbers (objectives, goals, etc.) recommended represent a good balance between performance and supportability?
  c. If a sensitivity analysis was performed, is the analysis credible? Are the underlying assumptions valid and reasonable?
  d. What design requirements has the contractor developed (e.g., subtask 203.2.3, Specification Requirements)? These should be on file and available for requirer review.
  e. Has the prime contractor implemented detailed recommendations within its purview (e.g., subtask 205.2.3, Specification Requirements)?

SOURCE: From Draft UAV LSA Applications Guide, Block One, July 1991.

ments) also become the basis for evaluating supportability in task 501, Supportability Test, Evaluation, and Verification. Appropriate task results are documented on the LSA A and B data tables. This task is a major step in the LM's effort to influence design.

4.4 The 300 Series Tasks: Task Preparation and Evaluation of Alternatives

The 300 series tasks optimize the support system for the new item and identify the new system concept alternatives that achieve the best balance between cost, schedule, performance, and supportability.

4.4.1 Task 301: Functional Requirements Identification

Purpose. This task identifies the functions and tasks that must be performed to operate and support the new system and initiates corrective action on any problems identified as a result.

Benefits. This is a relatively inexpensive task. Identification of the support functions provides the basic foundation for development of support-system alternatives in task 302, Support System Alternatives. The concept of identification of functions is not unique to LSA. System engineering and hardware and software design and some R&M analyses use functional block diagraming techniques as a basis for identifying functions that the designed system/equipment must perform. The first half of task 301, Functional Requirements Identification (FRI), is a comparable analysis of operation and support functions. A system-level functional requirements identification is a useful input to task 201, Use Study, and task 203, Comparative Analysis. The functional requirements identification also helps focus on functions that may require special attention, since new functions may create new support problems. For example, use of composites in airframes requires special repair capabilities. Task identification (TI) sets the stage for conduct of the detailed analysis in task 401, Task Analysis. On a large system there may be several thousand maintenance tasks. Both the FRI and TI provide the basis for possible design feedback. Functional requirements and redesign/supportability problems are placed in the LSAR B data tables. Reliability-centered maintenance (RCM) data are recorded in the BF and BG data tables. Failure modes and criticality analysis (FMECA) results are recorded in the LSAR BF through BL data tables.

Task description. The subtasks in task 301 are shown in Fig. 4.8. Subtask 301.2.1, Functional Requirements, identifies the operating and support functions the new system and its support system must perform. Subtask 301.2.2, Unique Functional Requirements, analyzes the functions identified in the first subtask to identify those which are either unique or new to design technology and those which are supportability or readiness drivers. Subtask 301.2.3, Risks, analyzes the risks associated with satisfying the functions identified in the first subtask.

Subtask 301.2.4, Operations and Maintenance Tasks, requires the identification of required preventive and corrective maintenance tasks based on the failure modes and criticality analysis (FMECA), reliability-centered maintenance (RCM) analysis, and all other tasks required to meet the functions identified. Entries in B and C LSAR data tables are made as a result of this subtask.

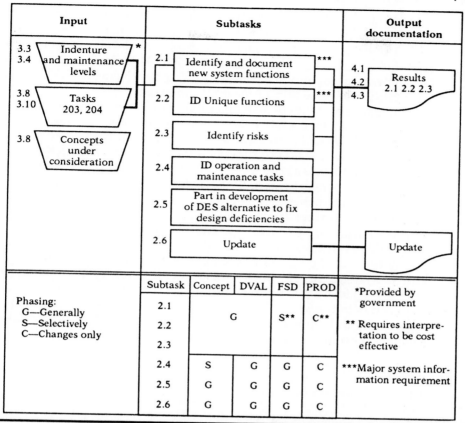

Input	Subtasks	Output documentation

Figure 4.8 Task 301, Functional Requirements Identification. (From Navy LSA Applications Guide, May 1986.)

Subtask 301.2.5, Design Alternatives, involves use of the first four subtasks as a basis for design feedback. Subtask 301.2.6, Updates, provides for updates to the FRI and TI as design progresses to succeeding levels of indenture and can result in new data in the LSAR B and C data tables. Table 4.10 presents a more detailed picture of the inputs to the 301 subtasks.

TABLE 4.10 Task 301 Inputs

301 subtasks	Inputs
2.1	201.2.4, 203.2.1/2
2.2	203.2.3/4/5/6
2.3	203.2.3/4/5/6
2.4	205, 301/2.1, RCM, FMECA

The draft DOD RCM standard[27] now in development is discussed in Chap. 7. The standard will have two tasks that merit consideration here. The first, task 100, Influence Design, could be employed in concept exploration (CE). The second, task 500, AGE Exploration, could be required in engineering manufacturing development (EMD) and production. See Appendix 19 for draft language.

Although the standard only mentions software explicitly in the fourth subtask, since system functions are allocated to software as well as hardware and personnel, software should be considered in this task. Examples of software functions include data base backup, online software diagnostics, and crash dumps to record software failures. Software failure modes can be both single- and multiple-cause. Single-fault failures include exception, stop, crash, hang, incorrect data response, and incorrect timing. Multiple-fault failure modes include desynchronization, inconsistency, and partitioning. The functions of comparable systems identified in subtask 203.2.1, Identity Comparative Systems, or the BCS developed in subtask 203.2.2, Baseline Comparison System, can aid in the analysis of the new system functions and can help identify any driver or unique functions of the new system alternatives in the first subtask. Functions related to subtasks 203.2.3 can be used to help evaluate the risks (subtask 301.2.3) involved in achieving new system functions. The functions related to the fourth,

TABLE 4.11   Task 301 Subtask Selection and Focusing

LM guidance
1. *LSA end objective:* The first three subtasks assist support-system design, the fourth aids support-system design and detailed resource identification, and the fifth supports design influence.
2. *Phasing and complexity:* Table III in 1A standard codes the FRI as applicable until EMD, when it becomes selective (see Fig. 4.8). Appendix 7 for simple equipments suggests the FRI not be performed. The TI is generally applicable beginning with DEMVAL regardless of complexity.
3. The first three subtasks should not be required if there is no flexibility in terms of support concept alternatives, unless they are needed as a basis for initial task identification not associated with PM or CM.
4. The FMECA is normally an R&M requirement. The FMECA and RCM are frequently done functionally when a drawing structure is not yet available. The TI can be similarly constructed. However, at some point, the LCN structure must be a top-down drawing structure. While a partial crosswalk between a functional and a drawing structure is possible, this frequently leaves gaps which must be filled if the final LSAR is to be properly completed. The LM should coordinate the FMECA with R&M when preparing the SOW to ensure that either reliability or LSA, but not both, require an HW-oriented FMECA. A sample clause requiring the contractor to identify and advise the government of any inconsistencies between the structure of the FMECA, RCM, and LSAR requirements is included in the Acronyms.
5. The LM should require the prime contractor to maintain a log of recommended design improvements and actions taken. Key points of focus could be the use of the FRI and TI for design feedback by the contractor and the compatibility and completeness of the FMECA, RCM, and LCN structure.

SOURCE: From Draft UAV LSA Applications Guide, Block One, July 1991.

fifth, and sixth subtasks in task 203 can be used to help identify driver functions and those functions with potential qualitative support problems to assist in identifying areas for design improvement. The various objectives, goals, and design constraints from task 205 can be used to help identify design deficiencies and to focus on the supportability design improvement effort. See Table 4.11 for task 301 subtask selection and focusing.

Use of results. The task results provide a basis for performing task 302. They are also a possible source of design feedback. The LM should take appropriate steps to ensure that the results are conveyed to the proper organizations for action.

4.4.2   Task 302: Support-System Alternatives

Purpose. This task develops viable support-system alternatives for the new system alternatives if support concept flexibility exists.

Benefits. Different support-system concepts can vary significantly in terms of cost, readiness, and manpower requirements. This is illustrated by the example of alternative support concepts for the Sparrow Missile in Table 4.12. Task 302 identifies viable support concept alternatives for further analysis in subtask 303.2.2, Support System Tradeoffs. It also provides the basis for support-system and mission/support/operating-concept tradeoff studies in subtask 303.2.3, System Tradeoffs. The risk analysis identifies areas of high risk on which the LM may wish to develop alternate backup plans. Portions of the concepts are recorded in the LSAR B data tables for maintenance significant items.

TABLE 4.12   Support Concept Evaluation

USAF AIM-7F/M missile (Sparrow) analysis

● Evaluated three support concepts:

Alternative 1: Partial all-up-round (AUR) handling and extensive periodic testing (current concept)
Alternative 2: Total AUR handling and reduced testing
Alternative 3: Total AUR handling and no periodic testing

● Results

	ALT 1	ALT 2	ALT 3
Manpower (FY 88)	58	7	0
LCC	$56M	$47M	$23M
Availability (FY 90)	89%	84%	66%

Task description. MIL-STD-1388-1A defines a *support concept* as a complete system-level description of a support system in which the ILS element concepts are in harmony with the design and operational concepts. A support-system concept must address all elements of ILS. A *support plan* (the third and fourth subtasks) is defined by the 1A standard as lower-indenture-level support concepts. Support concepts and plans are oriented toward the system and use and are very different, although hopefully related, to the acquisition ILS plan and ILS element plans. A concept/plan must meet task 205 requirements and satisfy the functions identified in task 302. Figure 4.9 summarizes the subtasks in this task.

Concepts/plans should address the major drivers identified in task 203, Comparative Analysis, and task 204, Technological Opportunities, and the unique functional requirements identified in task 301. An example of such emphasis is the LAMPs MK 111 helicopter, which

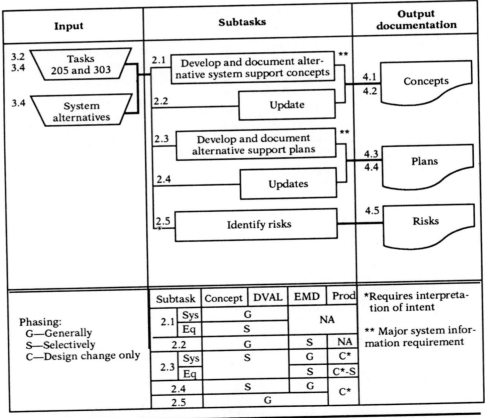

Figure 4.9 Task 302, Support System Alternatives. (From Navy LSA Applications Guide, May 1986.)

TABLE 4.13 Task 302 Subtask Selection and Focusing

LM guidance
1. *LSA end objective:* Table 1 in 1A standard lists support-system design as the major LSA objective of this task. However, most designers are very interested in knowing the support and maintenance concepts so that the results of this task should be disseminated to design organizations.
2. *Phasing and complexity:* See Fig. 4.9 for major systems and complex equipments. Appendix 7 for simple equipments suggest selective application of the system concepts subtask beginning with CE.
3. The first issue for the LM in this task is the amount of flexibility available in support-system alternatives. If no flexibility exists to vary the concept, there is no reason to perform this task. For example, support equipment is frequently repaired at the intermediate level as a matter of policy. In some cases, use of such models as LORA in task 303, Evaluation of Alternatives and Tradeoff Analysis, may determine the support-system alternatives.
4. To help focus this task, the LM should ask the LEMs what alternate support concepts/plans, if any, they wish the contractor to study. Every ILS element is a potential candidate for innovative concepts. The best sources of information in this regard are the prime contractors and LEMs.
5. The LM may wish to limit the study of contractor support alternatives. Interim contractor support is widely used, but total contractor support is the exception rather than the rule.

SOURCE: From Draft UAV LSA Applications Guide, Block One, July 1991.

goes aboard ship with a pack-up kit of weapon replaceable assemblies (WRAs) and a flight and maintenance crew. Pack-up kit selection and crew composition are obvious points of focus.

Note the requirements in this task to investigate innovative concepts and to consider contractor logistics support. In many cases, a decision to obtain interim or lifetime contractor logistics support is made on the basis of necessity, not economics. When this is the case, the study of contractor support should be deleted from the task.

The standard explicitly mentions software in the third subtask. Software maintenance may be at two or three levels. Examples of software support considerations include equipment, facilities, personnel skill levels, documentation, and software tools. Software analysis from this task can be an input to the Computer Resources Life Cycle Management Plan (CRLCMP) or vice versa. See Table 4.13 for subtask selection and focusing.

Use of results. The results of this task should serve as a basis for defining the specific ILS element-related tradeoffs in task 303. The report should be reviewed by the design community and the ILS managers. It also can be used as input to various program documents.

4.4.3 Task 303: Evaluation of Alternatives and Tradeoff Analysis

Purpose. This task is used to determine the preferred support-system concept. Its subtasks answer the following questions for the LM:

1. What are the best performance, support, and operating concepts for the program (subtask 303.2.2, Support System Tradeoffs, and subtask 303.2.3, System Tradeoffs)?

2. What are the best combinations of ILS elements and design features (subtask 303.2.5, Manpower and Personnel Tradeoffs, subtask 303.2.6, Training Tradeoffs, subtask 303.2.8, Diagnostic Tradeoffs, subtask 303.2.10, Energy Tradeoffs, subtask 303.2.12, Transportability Tradeoffs, and subtask 303.13, Facility Tradeoffs)?

3. What is the sensitivity of system readiness to key design parameters [R&M, built-in test (BIT)], ILS elements (spares, manpower), logistic system parameters (mean delay time), and cost (subtask 302.2.4, Readiness Sensitivities)?

4. What are the optimal levels of repair (subtask 303.2.7, Repair Level Analysis)?

5. How do the system's proposed supportability parameters (the results of subtask 303.2.3, System Tradeoffs) compare with the results of task 203, Comparative Analysis, and task 205, Support-Related Design Factors, and what risks are involved in achieving the system readinesss goals (subtask 303.2.9, Comparative Evaluations)?

6. What are the energy and survivability implications of design alternatives (subtask 303.2.10, Energy Tradeoffs, and subtask 303.2.11, Survivability Tradeoffs)?

Benefits. This task is one of the LSA cost-driver tasks (see Appendix 17). The results of these subtasks provide quantitative measures of the readiness, supportability, and cost impact of various design/support alternatives. These results provide the kind of information needed to convince program managers and sponsors to take actions that result in a more supportable system. For example, this information can lead to design changes, a more effective support concept, a more effective operational concept, and/or more support funds. Appendix 21 provides an example of how these tradeoff studies and evaluations benefit the LM's ILS effort. It describes an actual case history illustrating the uses and impact of readiness modeling and sensitivity analysis.

Task 303 is a key study. It provides the basis for

1. Selecting the support concept.

2. Establishing the best balance between performance, support, and operating aspects of the program.

3. Establishing specific supportability goals, including readiness.

4. Establishing the best balance between the individual ILS elements and design.

Task description. Figure 4.10 summarizes the subtasks in this task. The first subtask is a key subtask because it specifies criteria, analytical relationships or models, and parametric and cost-estimating relationships to be used in the remaining subtasks. Models specified should include O&S and life cycle cost (LCC) models to be used. However, this requirement can be used to identify the models and criteria to be used in the other LSA tasks as well. Models can be used in tasks 202, 203, 204, 205, 402, and 501, as well as in task 303. Such possible use is discussed in the task-execution descriptions in Chap. 5. Reference 6 contains tables matching various cost models with the LSA subtasks. Appendix 26 contains a macro LCC model that can be modified for use prior to demonstration validation (DEMVAL) on individual programs.

The second subtask, 303.2.2, Support System Alternatives Analysis, can vary significantly in terms of cost and impact on readiness. Any

Subtask	Phasing			
	Concept	DVAL	EMD	Prod
2.1 Tradeoff Criteria	G	G	G	Design changes only
2.2 Support System Tradeoffs*	G	G	G	
2.3 System Tradeoffs*	G	G	G	
2.4 Readiness Sensitivities	G	G	G	
2.5 M&P Tradeoffs*	G	G	S	NA
2.6 Training Tradeoffs*	G	G	G	Design changes only
2.7 Repair Level Analyses	S(1)	G	G	
2.8 Diagnostic Tradeoffs	G	G	S(1)	NA
2.9 Comparative Evaluations*	SYS-G EQ-NA	SYS-G EQ-S	S(1)	Design changes only
2.10 Energy Tradeoffs	SYS-G EQ-NA	SYS-G EQ-S	S	
2.11 Survivability Tradeoffs	SYS-G EQ-NA	SYS-G EQ-S	SYS-G EQ-S	EQ-S
2.12 Transportability Tradeoffs	SYS-G EQ-NA	SYS-G EQ-S	NA	NA
2.13 Facility Tradeoffs	SYS-G EQ-NA	SYS-G EQ-S	G	NA

Code:
G — Generally
S — Selectively
1 — Requires interpretation to be cost-effective

*Major system information requirement

Figure 4.10   Task 303, Evaluation of Alternatives and Tradeoffs.

HARDMAN/MANPRINT/LSAR data available could help satisfy the requirement in this subtask to identify any new or critical logistics resource needed. Tasks 301 and 302 could provide possible inputs. In fact, all the 303 subtasks except the third subtask could be inputs to this subtask.

The third subtask represents the top-level system engineering tradeoffs and is normally done by the systems engineering group with inputs from the logistics group. System engineering tradeoff weighting methods are discussed in Chap. 7. Appendix 20 contains an example using engineering judgments. All the other subtasks in task 303 can be inputs to this analysis. Since this subtask is intended to optimize the combination of design, support, and operating considerations, it could easily include software aspects. If system engineering tradeoff models and weights are prepared, the LM should attempt to include supportability factors. Examples of such possible factors include preventive and corrective maintenance, standardization, and software maintenance factors.

Subtask 303.2.4, Readiness Sensitivities, is a key subtask. As a minimum, sensitivity analysis must be performed on the pertinent system factors. However, more sophisticated methods may be useful. Different readiness modeling approaches are summarized in Table 7.13. All the U.S. services have large-scale simulation models that are often used to perform this subtask. A case example of a readiness analysis that illustrates its usefulness to the LM is contained in Appendix 21. However, the simpler and less costly techniques in Table 7.13 can be used for many situations.

The fifth, sixth, eighth, twelfth, and thirteenth subtasks all represent various design/ILS element tradeoffs. The eighth subtask, 303.2.8, Alternate Diagnostic Concepts, is a key study because it may be a major driver of other analyses. The diagnostic or testing approach must support the maintenance concept. Appendix 22, excerpted from Ref. 28, contains possible departure-point SOW diagnostic language and parameters.

The seventh subtask is the traditionally performed level-of-repair (LOR) analysis. The ninth subtask compares the new system objectives/goals with the comparative analysis performed in task 203, Comparative Analysis, based on the results of the other task 303 subtasks and examines the risks of goal achievement. The tenth subtask looks at design-energy tradeoffs. The eleventh looks at survivability and battle damage characteristics. An effective battle damage assessment requires (1) combat modeling to determine lethal and repairable equipment casualties, (2) attack-mode and material system modeling to provide estimates of combat damage, and (3) analysis of historical combat damage and repair techniques. The logistic implications of battle damage are (1) repair procedures, (2) added personnel to repair battle damage, (3) supply stockage, and (4) additional transportation needs.

While the standard does not explicitly include software in the task, it is an obvious necessity. Reference 24 includes software in most of the subtasks. Since software models and parametric/cost-estimating relationships usually differ from hardware relationships, this can begin with the first subtask. Support-system tradeoffs, system tradeoffs, and manpower and diagnostic tradeoffs can have software implications. Although we know of no actual inclusion of software in complex simulations used for readiness analysis, as the impact of software on system readiness increases, such inclusion is inevitable.

Example support-related software factors cited in Ref. 24 include maintenance capacity, maintenance task simplicity, and support automation. Software factors in tradeoffs can be weighted in the same manner as hardware (see Chap. 7 and Appendix 20 for a discussion of this technique). See Table 4.14, Task 303 Subtask Selection and Focusing.

Use of results.   The results of these studies are used to make a number of key decisions that affect both design and support and provide data for input to program approval documents. The LM must ensure that the studies are acted on and that the decisions are used as a basis for future design and support planning activity.

The results of task 303 are fed (1) to subtask 205.2.5, Supportability Goals and Thresholds, to assist in final goal/threshold selection, (2) if appropriate, to subtask 205.2.3, Specification Requirements, to identify design constraints, and (3) to task 401, Task Analysis, and form the basis for supportability assessment in task 501, Supportability Test, Evaluation, and Verification.

4.5   The 400 Series Tasks: Determination of Logistics Resource Requirements

The 400 series tasks identify the logistics support resource requirements of the new system in its operational environment and develop plans for postproduction support.

4.5.1   Task 401: Task Analysis

Purpose.   This task analyzes the operation and maintenance tasks required for the new system/equipment and is the major data source for the LSAR. The results of these analyses identify logistics support resources both qualitatively and quantitatively. This task determines the detailed support resources necessary to the program.

TABLE 4.14 Task 303 Subtask Selection and Focusing

LM guidance
1. *LSA end objective:* All the subtasks could affect mission-system and support-system design, although Table 1 in 1A standard does not classify the seventh and ninth subtasks as design influence. This table suggests that the second, third, fourth, and eighth subtasks would not normally affect identification of detailed logistic resources, but this appears to be a matter of definition and judgment.
2. *Phasing and complexity:* See Fig. 4.10. Appendix 7 for simple equipments uses much the same phasing but classified all subtasks as selective.
3. Whereas tailoring normally involves the elimination of selected subtasks, the tailoring of task 303 can involve the addition of subtasks. For example, there were about 300 to 400 major tradeoff studies on the F-16 and F-18 aircraft programs. The tradeoffs listed in this task represent those believed to be the most widely useful. The LM can add any additional tradeoffs known to be of major interest to the specific program. See Appendix 6 for ideas on possible additional tradeoffs. Be sure to be as specific as possible or the contractor may not perform the exact study desired.
4. New tradeoff possibilities may occur to the contractor as the program evolves. The contractor's LSA plan should address how it proposes to conduct such trades and integrate them with planned LSA tradeoffs.
5. The subtasks are excellent candidates for focusing. For example, maintenance may wish to test a two- versus three-level concept or a repair versus design for discard policy. Such focusing will direct attention to specific areas and reduce contractor work load and costs. The driver analysis in task 203 is an excellent source of focusing possibilities. The LM should consult the LEMs and the system engineering community about the need for and the focusing possibilities of some of task 303 tradeoffs:

Community	Subtask
M&P	303.2.5
	303.2.6
Test Equipment (and Maintainability)	303.2.8
System Engineering	303.2.10
Vulnerability/Maintainability	303.2.11
Transportation	303.2.12
Facilities	303.2.13

6. Supportability considerations should be included in the design tradeoff studies and in any system engineering tradeoff model the contractor employs. Government approval of the factors and weights, if any, used in the contractor's system engineering tradeoff model should be required. This information could be requested as part of the contractor's proposal.
7. The ILS/LSA management team should review and approve the models and analysis methods which the contractor uses on this task.

SOURCE: From Draft UAV LSA Applications Guide, Block One, July 1991.

Benefits. Properly used, the LSAR data base serves as a cost-effective integrator among all the ILS elements by eliminating duplicate data requirements and ensuring the use of the same data values for specific data needs. Task analysis provides a very cost-effective means for ensuring accurate identification of detailed ILS resource requirements. It identifies new support equipment and training requirements early, thus allowing greater lead time. It also can be used to document software operation and maintenance tasks and identify detailed software support requirements.

The task is one of the most expensive tasks in DEMVAL and EMD. Appendix 17 contains tables showing the estimated cost of this task in relation to the total estimated LSA cost by phase. Table 20A.5 shows the relative costs of each subtask. If depot maintenance task analysis is delayed until the production phase as a result of interim contractor support, some costs would shift to this phase. LSAR CALS linkages will be a major source of possible cost reduction.

Task analysis helps to determine and provide the following data to document each operation and maintenance task:

1. Task descriptions/procedures, including safety recommendations, frequency of performance, manhours required, and elapsed time to accomplish the work

2. Maintenance levels to be used

3. Personnel required by number, skill specialty, and level

4. Spares, repair parts, and consumables required

5. Tools and support equipment; test, measurement, and diagnostic equipment (TMDE); and test program sets (TPS)

6. Training requirements, including training materials, rationale, and recommended locations

7. Facilities required

8. Transportation/transportability requirements

The task analysis is also a source of information on design problems, operation and/or support cost reductions, logistics resource optimization, and readiness enhancement. Other communities may do task analysis. These include HARDMAN/MANPRINT/HSI, Human Engineering, and Maintainability. HARDMAN/MANPRINT/HSI may require task analysis on the BCS items in task 203 during CE. The LM needs to coordinate with such efforts to reduce duplication.

The output from these subtasks (see Table 4.15 and Figure 6.1 on LSAR output report usage) identifies detailed support resource requirements and provides source data for ILS documents.

Task 401 is generally called out first in the DEMVAL SOW, primarily to identify new and/or critical resources. A running issue in the LSAR world is how much LSAR data are needed in this phase. The LSA standard suggests selective application of this task in the DEMVAL phase. Different parts of the LSA and LSAR standards give conflicting advice on how much LSAR to do, and there are conflicting opinions on this subject among professionals in the field.

Possible uses of the LSAR during DEMVAL include

1. Document long-lead-time item provisioning information.

2. Develop the support-system package (SSP) data needed to support DEMVAL testing.

3. Design feedback source data.

TABLE 4.15   Task 401 Subtask Inputs and Outputs

Subtask table	Applicable requirements	Additional deliverable data
401.2.1/2	C	
401.2.3	E, F, G	DI-S-4057, Scientific and Technical Report, to document critical resources
401.2.4	C, G	
401.2.5		DI-S-3606, Trade Study Reports (alternative design approaches)
401.2.6		DI-S-3606, Trade Study Reports (management action to minimize risk associated with new or critical logistics support resources)
401.2.7	J	
401.2.8	H	
401.2.9		DI-3606, Trade Study Reports (LSAR validation)
401.2.10		DI-L-7146 through DI-L-7191*
401.2.11	Update	DI-L-7146†

*If the LSA reports are to be obtained from the contractor, they must be called out individually in the CDRL. See Chap. 6 for advice concerning report selection.

†DI-ILSS-81173, LSAR Data Table Exchange/Delivery. This DID should be an iterative quarterly submission. The amount of documentation required is indicated on the DD Form 1949-1 contained in the contract. See Chap. 6 for advice in completing this form.

Early LSAR data can provide a potentially useful source of design feedback. However, rapid design change can virtually wipe out the usefulness of premature LSAR data. During this phase, only higher-indenture-level information is normally available. Some of the standard LSAR output reports are structured to provide design feedback. A list of the LSAR output reports classified by LSA top-level objective (design influence, etc.), by primary user (requirer or executor), and by community and the purpose of use is contained in Figure 6.1. This table can thus be used to select LSAR output reports useful for design feedback in both DEMVAL and EMD. Each report has mandatory data elements needed to run the report. There are tables in 2A standard that have this information. Also consider any data needed for long-lead-time provisioning and the source selection plan (SSP) in task 501. If the LSAR can be used economically to generate these data, use it for this purpose. Other uses are possible, depending on the specific circumstances. In the long run, CALS or other automated linkages of the LSAR to, for example, the FMECA data base should make this approach more viable since its automated linkages will make entry of such data very economical.

During EMD, the LSAR obtained from the subtasks will mature as the design hardens. Proper accomplishment of subtask 401.2.8, Provisioning Requirements, during the EMD phase provides the basis for spares procurement during production. During production, the 401 subtasks are normally applied to design changes. An interesting issue facing the LM is whether to maintain the LSAR for the life of the program after production of the end item ceases. Possible uses of the LSAR both during and after production are

1. Use on follow-on procurements.

2. Application to engineering change proposals (ECPs) and modifications and special studies related to these. The use of the LSAR for special ad hoc studies will grow as CALS matures and more LSAR data bases employ relational data bases, thus making rapid ad hoc reports feasible.

3. Input to a postproduction support data base to monitor disappearing manufacturing sources (such problems can occur long before production ceases).

4. Use on other programs employing the same subsystem, assembly, etc. (some numbers such as R&M may change owing to the different environment, but much of the data does not need to be redone).

5. Use on other programs employing the same subsystem, assembly, etc. to develop HARDMAN/MANPRINT analysis.

Task description. Table 4.16 illustrates the subtasks and inputs from other tasks. Subtask 401.2.1, Task Analysis, and subtask 401.2.2, Analysis Documentation, are the basic subtasks to task 401. These should be completed to the lowest reparable assembly (LRA, LRU) level as soon as the initial design can be established. In many cases, this will be during EMD, although some information may be obtainable during DEMVAL.

TABLE 4.16   **Task Analysis Subtasks and Inputs**

No.	Title	Inputs
401.2.1	Task Analysis	Task steps, all of 301.2.4, 303.2.8 Resources Frequency/interval, 301.2.4.1/2 Maintenance level, 303.7
401.2.2	Analysis Documentation	
401.2.3	New/Critical Support Resources	303.2.2
401.2.4	Training Requirements and Recommendations	303.2.6
401.2.5	Design Improvements	205.2.2/3/4
401.2.6	Management Plans	
401.2.7	Transportability Analysis	303.2.12
401.2.8	Provisioning Requirements	
401.2.9	Validation	
401.2.10	ILS Output Products	
401.2.11	LSAR Updates	

Subtask 401.2.1 analyzes each operator and maintenance task previously identified by task 301, Functional Requirements Identification, using a four-step process:

1. Task allocation assigns the task to an appropriate maintenance level based on maintenance characteristics of the system, maintenance level capabilities, and the level-of-repair analysis (LORA). This is documented in the LSAR CA data tables.

2. Task frequency is based on the type of task. Operator and functional alteration task frequencies are dependent on the amount of system use and the type of mission. Preventive maintenance tasks are determined as a result of the RCM logic analysis. Preventive and corrective maintenance task frequencies are determined according to the formulas in MIL-STD-1388-2B and are documented in the LSAR CA data tables.

3. Task design/description is the development of procedures explaining how to perform operator and maintenance tasks. These are documented in the LSAR CC data tables.

4. Resource enumeration is the listing of all resources needed to perform the task. This is documented in the LSAR CD, CG, and CI data tables.

Appendix 23 discusses the R&M LSA supply support dependency aspects of the LSAR. While the LSAR standard contains a great amount of detail, considerable flexibility in task documentation still exists. Establishment of key task documentation conventions is therefore desirable on individual programs. Appendix 24 gives possible departure points in this regard.

The performance of subtask 401.2.3, New/Critical Support Resources, identifies resources that will require either development or special management attention due to schedule constraints, cost implications, or known scarcities. These resources will normally be documented in the LSAR E, F, and/or G data tables.

Subtask 401.2.4, Training Requirements and Recommendations, identifies training requirements based on task procedures and personnel assignments. A method of training will be recommended along with a rationale for the training requirements.

Subtask 401.2.5, Design Improvements, should be performed by both the requirer and the original equipment manufacturer (OEM) to determine which tasks fail to meet established goals. This task verifies the supportability and supportability-related design goals previously established (task 205, Supportability-Related Design Factors).

The use of LSA summary reports is beneficial to the effective accomplishment of this subtask.

Subtask 401.2.6, Management Plans, may be done by the requirer and/or the OEM. The intent of this subtask is to identify any actions that may be taken to lessen the risks associated with new or critical logistics resources. This subtask is required by 1A standard for major system acquisitions.

Subtask 401.2.7, Transportability Analysis, may be performed by either the requirer or the OEM. This is the only 401 subtask that may be done during DEMVAL based on anticipated design and then not called out in later SOWs unless a transportability problem is anticipated. When the general requirements of MIL-STD-1366 ("Material Transportation System Dimensional and Weight Constraints") are exceeded, the transportability characteristics should be documented in the LSAR. These data should be reviewed by the Military Traffic Management Command, and transportability problems should be fed back to the design agency.

Subtask 401.2.8, Provisioning Requirements, is used to satisfy this requirement of DOD Instruction 4151 ("Uniform Technical Documentation for Use in Provisioning of End Items of Material"). This instruction states that all provisioning will be accomplished using the LSAR. Coordination of data requirements with personnel at the inventory control point (ICP) responsible for provisioning the new item is essential.

The validation of LSAR documentation (subtask 401.2.9) should be accomplished by both the OEM and the requirer, and it should be coordinated with other system engineering tests and demonstrations.

Although the standard does not mention software in this task, as stated earlier, the task can be used to document software tasks and needed support resources. The LSAR does provide linkage for maintenance modules and software, since the tables refer to the Unit Under Test (UUT). A similar linkage for operational software and hardware is required.

Software developed in accordance with MIL-STD-2167, Defense System Software Development, to support operational requirements is justified by a functional requirement. At some level, this functional requirement is involved with some hardware function. In one program, a linkage was proposed using the logistics control number (LCN) of the hardware module. This reference was embedded in a matrix that referenced the software module which addressed the same hardware. This is a minimum bookkeeping system that provides a cross-check when changes are made. There may be more sophisticated systems that provide the same information. The ability to flag software changes that need to be assessed in concert with the hardware configuration, and vice versa, is essential.

The selection of LSA Output Reports (subtask 401.2.10) should be coordinated among all ILS elements. If these reports are to be deliverables from the OEM, the appropriate data item description (DID) must be called out, as well as the depth of the data to be reported. Possible uses of the various standard output reports by both the requirer and the performer are contained in Table 6.1.

The LSAR is a product of iterative development, becoming more complete and detailed as the equipment design evolves. In order to make the best use of this data base, it is necessary to use it as it evolves. For this reason, subtask 4.2.11, LSAR Update, is necessary. See Table 4.17, Task 401 Subtask Selection and Focusing.

Use of results.   The task analysis and the LSAR data base identify the detailed support resources needed. The LSAR data base is also a unique source of design feedback at the detailed level. The results of this task are also inputs to task 402, Early Fielding Analysis, and task 501, Supportability Test, Evaluation, and Verification. The LSAR should be distributed to the appropriate ILS element manager; design problems should be forwarded to engineering.

4.5.2   Task 402: Early Fielding Analysis

Purpose.   This task assesses the impact of the new system on operating organizations, the logistics infrastructure, and ILS element resources.

Benefits.   The introduction of new systems into the operating forces can have a major impact on the operating organizations and logistics infrastructure. Examples are guns/rockets that expend ammunition in greater quantities and systems that significantly increase fuel requirements, which, in turn, affect storage and transportation requirements.

This task is one of the more inexpensive tasks (see Appendix 17). It provides the LM with an early assessment of the nature of the support problems likely to be encountered when the system/equipment is deployed into the operating organizations. Early identification of these problems facilitates timely problem resolution and thus improves the readiness and supportability of the item.

Task description.   This task consists of a number of subtasks addressing issues that need to be reviewed prior to the activation of operating

TABLE 4.17   Task 401 Subtask Selection and Focusing

LM guidance
1. *LSA end objectives:* Table I in 1A standard lists detailed resource classification as the focus of most of the subtasks; however, the fifth, seventh, ninth, and eleventh subtasks are also cited as potential mission-system and support-system design influences.
2. *Phasing:* LSAR generation normally begins in DEMVAL, with most of the data being generated in EMD and/or production.
3. *Complexity:* Table III of standard 1A makes slight distinctions between major systems and complex equipments. Appendix 7 for simple equipments is a little more selective.
4. This task requires extensive coordination among the ILS elements and human engineering. The amount of detail to be included in the task description should be coordinated with the tech manual LEM. The needed data elements must be identified.
5. Timing of deliverables requires careful attention. The task cannot be performed effectively prior to receipt of the input information from design. However, it cannot be delayed past a point where the outputs will not be available to develop the required ILS element documentation. The flow from the engineering data and drawings, the FMECA, the RCM, the LSAR, and the ILS element data requirements needs to be checked for scheduling compatibility. An analysis of these and other LSAR-related time linkages should be performed prior to finalization of the RFP. The contractor also should be required to submit a schedule-compatibility analysis as part of its proposal or at the LSA Guidance Conference. Sample SOW language for this is contained in Appendix 8.
6. Depth of performance is also a tailoring consideration. It may be possible to delay depot maintenance until production. If interim contractor support is planned, it may be possible to slip other LSAR data requirements timewise.
7. LSA output reports offer the APML and LEMs an important source of information. Output report requirements should be coordinated with the LEMs. If the APML has the resources available, he or she may consider ordering a copy of the OEM master files on magnetic tape and run the reports in-house. This offers several advantages to the program: (a) the reports would not have to be called out individually in the contract, (b) reports would not have to be run only to fulfill contract requirements when the amount of data does not warrant it, (c) additional reports could be run without a contract modification, and (d) reports could be run to any indenture level or using various sets of selection criteria. With the advent of CALS, interactive access to the contractor's data base is another option rapidly gaining wider use. Sample SOW language to require interactive access by a read-only terminal in the government program offices is in Appendix 9.
8. If the LM wishes to emphasize design feedback, he can specify that this task will begin as soon as preliminary drawings are available.
9. The LM may wish the contractor to relate system hardware LSAR to system software, particularly in software-intensive programs, in order to readily identify when a change to one affects the other.

SOURCE: From Draft UAV LSA Applications Guide, Block One, July 1991.

units. Figure 4.11 summarizes the subtasks in task 402, Early Fielding Analysis.

The first subtask basically addresses the impact of the program on operating organizations and the logistics infrastructure. It particularly invites focusing, since it requires assessment of the impact of fielding the new system on existing systems. The assessment could cut across all ILS elements at each of the operating and support levels, including depot. Accuracy in assessments at I (Intermediate) and D (Depot) levels is difficult because resources are frequently shared. A summary of this subtask is shown in Table 4.18.

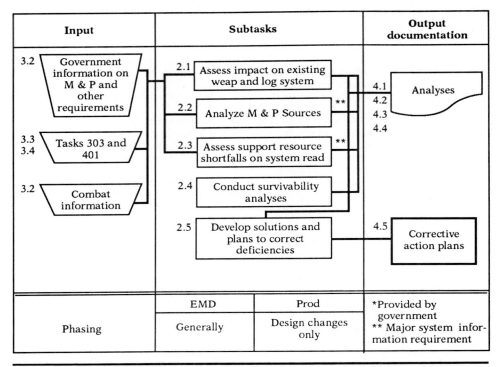

Input	Subtasks	Output documentation

Figure 4.11   Task 402, Early Fielding Analysis. (From Navy LSA Applications Guide, May 1986.)

TABLE 4.18   Subtask 402.2.1 Summary

Operating organization	Infrastructure
ATE	ATE
POL	Depot work load
Transportation	Provisioning
MP&T	Inventory factors

The second subtask checks program manpower and personnel requirements against probable available manpower and personnel assets at the infrastructure level (i.e., fleet-wide projected requirements versus probable availability). The second subtask assesses manpower and personnel impact. An executor can assess this impact from the viewpoint of its system. The executor also can assess the impact at specific service organization levels (e.g., O and I levels). However, in terms of the highest levels and long-range projections, the requirer is best fitted to assess total needs over time and likely accessions into the service manpower pool. Furthermore, the high-level projections are just projections. Many actions to increase accessions (bonuses, higher pay, etc.) are taken by both the military services and private contractors (e.g., drilling for oil in Alaska, off-shore drilling rigs, contract support specialists on ships or in combat areas, etc.). Such impacts are most visible on large programs. The Army's introduction of the XM-1 tank and the Bradley fighting vehicle is having a significant impact on the Army's logistics structure. The tank's greater capability requires more petroleum, oil, and lubricants (POL); more ammunition; and more transportation to support the new systems. A similar, although less obvious, impact can occur from the introduction of a number of equipments. While the introduction of each individual equipment may represent a small incremental impact on manpower, the cumulative effect of many equipments can have a devastating impact. This is particularly true in the mushrooming area of electronics.

Normally the LSAR data base is a major input to the first two subtasks. If so, the first two subtasks should not be initiated until enough of the LSAR is available. There are basically three types of actions to address any problems identified in the first two subtasks: (1) actions by the program office (i.e., improve reliability, design skill-level requirements), (2) actions by the operating organizations (i.e., conduct additional on-the-job training), and (3) actions by the infrastructure (i.e., increase recruiting of personnel with scarce skills or increase reenlistment bonuses). In practice, the most feasible option will vary. The LM's role is to identify the problems, initiate action, and follow up to ensure implementation of the course of action selected.

The third subtask assesses the impact of support-resource shortfalls on system readiness requirements. This can be done for those ILS elements included in the readiness sensitivity analysis in subtask 303.2.4 if a complex simulation model is used. If a sparing to availability analysis (e.g., the ACIM model) was performed, budgets could be compared with this analysis. Other tradeoffs in subtask 303 also may serve as a point for judgment. A last point of information may be the growing practice of relating support-resource budgets to system readiness, if this was done on the program. The third subtask provides the LM and LEMs a basis for adjusting ILS element planning or for requesting design changes to achieve system readiness goals.

The fourth subtask provides the LM and LEMs a basis for planning battle damage wartime logistics needs. If subtask 303.2.11, Survivability Tradeoffs, was done at the general level, the results of this analysis may need further study to develop explicit plans and actions in terms of battle damage planning and wartime versus peacetime support/test equipment and supply inventories. ILS element considerations are shown in Table 4.19.

TABLE 4.19   Battle Damage ILS Element Considerations

ILS element	Battle damage consideration
Maintenance planning	Assess impact and repair needs
Supply support	Compute stockage requirements
Technical manuals	Develop/incorporate instructions for assessment and repair
Training	Develop battle damage training requirements
Transportation	Evaluate intertheater and intratheater transportation requirements due to battle damage
Manpower and personnel	Identify impact on M&P requirements

Subtask 402.2.5 represents the normal management actions necessary to develop solutions and plans to correct the deficiencies identified in the previous subtasks. The readiness analysis in subtask 303.2.4 is an input to the third subtask, and the survivability analysis in subtask 303.2.11 is an input to the fourth subtask.

Some of the 402 subtasks are perfect illustrations of the common 1A standard problem of determining who the performing organization should be, the requirer or the executor. The second subtask is a good example of this issue. The subtasks also illustrate the need for the requirer and the executor to both input data to the subtask if a good product is to be obtained. The amount of work to be performed is also a major consideration, since it can vary tremendously depending on task interpretation. It is therefore important that the requirer and executor fully understand in detail what the cited requirement involves.

Although the standard does not mention software explicitly in this task, it could be included as a consideration in the second, third, and fifth tasks. The executor may need a large amount of input information from the requirer on the systems and support infrastructure affected by the new system. See Table 4.20, Task 402 Subtask Selection and Focusing.

Use of results.   The nature of the actions taken will vary with the specific results of the task findings. The major point to remember is that this task provides a quick look and review of the status and nature of any problems that may surface as deployment occurs.

4.5.3   Task 403: Postproduction Support (PPS) Analysis

Purpose.   This task assesses what needs to be done to ensure that the program is supportable after production ceases.

Benefits.   This task identifies and analyzes probable postproduction problems and alternative solutions and develops a plan for implementing the most effective solutions. There is a growing recognition that today's

TABLE 4.20 Task 402 Subtask Selection and Focusing

LM guidance
1. *LSA objective served:* Detailed logistics resource determination.
2. *Phasing:* EMD only.
3. *Complexity:* No distinction is made in Table III in 1A standard for complexity. Appendix 7 suggests selective application to simple equipments only in unique cases.
4. *Tailoring and focusing:*
a. This task could become very expensive (see CELSA data in Appendix 17); therefore, the LM should restrict this task to areas of true need. A conservative approach is suggested. The need for each subtask should be carefully examined. The LM should consult the LEMs in regard to tailoring the first subtask and MP&T about the second subtask. The first and second subtasks could be performed by a service organization based on the LSAR.
b. The LM should select the performing organizations (requirer or executor) for each subtask. The organization that performed the readiness analysis in subtask 303.2.4 should be considered for the third subtask. The subtasks should not be performed until new data become available. This may not occur until development and operational testing results become available. Consult R&M.
c. The battle damage subtask could be performed by either the prime contractor or a service organization. Consult system engineering.
d. One way to execute the first subtask in terms of the operating organizations would be through further field visits in task 201, Use Study.

SOURCE: From Draft UAV LSA Applications Guide, Block One, July 1991.

PPS problems need greater management attention and new approaches. A number of factors have combined to increase the need for more attention to postproduction support problems. These problems are being exacerbated by interaction of longer platform lives and rapidly changing technology. Some (battleships, aircraft, etc.) are mothballed and later recalled for use. Ships are in the active inventory for up to 40 years. Many B-52s are older than their pilots. The Minuteman missile originally had a design life of 10 years but is now planned to be operational until the year 2010. Over half the original Minuteman manufacturers no longer exist. A system scheduled for removal from the active inventory may suddenly have its planned service life extended significantly.

Although the CELSA data (see Appendix 17) indicates this is a low-cost task, development of a PPS data system can be fairly expensive. See the Chap. 5 discussion for a description of the work that could be involved.

Task description. Task 403 is summarized in Fig. 4.12. The PPS problem has many facets. The following is a list of common problems:

1. The original manufacturers no longer exist or no longer make the item.

2. Technical data deteriorates (no money for updates) or disappears.

3. Support and test equipment is no longer produced or is transferred to another program application.

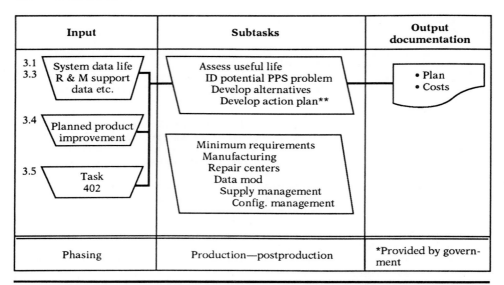

Input	Subtasks	Output documentation

3.1 3.3 System data life R & M support data etc.

Assess useful life
ID potential PPS problem
Develop alternatives
Develop action plan**

• Plan
• Costs

3.4 Planned product improvement

3.5 Task 402

Minimum requirements
Manufacturing
Repair centers
Data mod
Supply management
Config. management

| Phasing | Production—postproduction | *Provided by government |

Figure 4.12 Task 403, Postproduction Support Analysis. (From Navy LSA Applications Guide, May 1986.)

4. Data rights are poor.

5. High-cost spares were purchased in low quantity.

6. There is weak organic engineering capability.

7. "Fire drills"—modifications and service life extension.

8. Differences exist in functional versus system-management priorities.

The following are possible areas on which the postproduction plan should focus:

1. Postproduction system readiness objectives

2. Postproduction organizational structures and responsibilities

3. Assessment of the impact of technological change and obsolescences

4. Alternative strategies to accommodate the loss of production sources (see Chapter 5, Table 5.6)

5. Possibility of service life extensions and the logistic implications

6. Possible foreign military sales (FMS) and the impact on logistics

7. Currency of the technical data base

8. Provisions for handling government and contractor factory test equipment, tools, and dies

PPS problems can begin as soon as a system/equipment moves from the first series/model to another. The F-4s are now up to series J.

TABLE 4.21 Task 403 Subtask Selection and Focusing

LM guidance
1. *LSA objective served:* Support-system design and detailed logistics resources.
2. *Phasing:* Usually during the production phase, although there is some contradictory guidance on this point.
3. *Complexity:* See Fig. 4.12. Appendix 7 for simple equipments suggests selective application. The rapidity of technology changes, rather than complexity, is generally the key factor. Commercial electronics, for example, has a rapid obsolescence rate.
4. The general nature of this task makes it a natural candidate for tailoring and focusing. The LM should consult the LEMs and R&M about any specific needs.
5. Due to the fluidity of guidance on this subject, the LM should review the latest in developing SOW requirements for this task.
6. The problem of diminishing or disappearing manufacturing sources is a key point of focus. Some OEMs have set up a simple management system to track sources likely to disappear. By identifying these ahead of time, the LM can plan and budget ahead, avoid crisis management, and implement more cost-effective and timely solutions. The following are other suggested areas for immediate focus: configuration and technical data currency, readiness and R&M assessment and improvement, P^{3}I, LORA, and LSAR maintenance and update.

SOURCE: From Draft UAV LSA Applications Guide, Block One, July 1991.

Therefore, planning must begin sooner than commonly visualized. See Table 4.21.

Use of results. Task 403 provides the LM with the basis for developing plans and actions to ameliorate PPS problems. It could be part of the ILSP and/or OLSP. The important thing is to establish a management mechanism, begin to identify problems, and take action as early as possible. An integrated approach to PPS management is still evolving. DOD is currently taking the initiative to address PPS problems. PPS is a subject of discussion in the new DOD Instruction 5000.2. This could result in increased attention to PPS management problems.

4.6 The 500 Series Tasks: Supportability Assessment

The 500 series tasks assist the LM in ensuring that specified requirements are achieved and deficiencies corrected.

4.6.1 Task 501: Supportability Test, Evaluation, and Verification

Purpose. This task assesses how well support-related requirements are met, identifies the reasons for any shortfalls, and develops actions to correct deficiencies as soon as possible in the acquisition. The first four subtasks focus on acquisition testing. The last two subtasks collect actual in-service use data to find problems not revealed by earlier testing.

Benefits. The CELSA cost data in Appendix 17 indicate that this may be a relatively inexpensive task until production. Supportability data from test and evaluation (T&E) and field feedback-reporting systems (such as 3M) help the LM assess the support program and take appropriate action. This information allows comparison of actual results against the support goals. Actual results are more credible than predictions and can be used to validate the analyses. They also identify problems and provide a basis for corrective action. Testing replaces paper analysis with actual data and thus reduces uncertainty and enhances the credibility of the issue.

T&E results can have major impacts on production decisions. Major programs have been held in limited production or have been redirected because of problems identified in T&E. Initial readiness assessments can be made based on T&E R&M results, with actual field data (3M, etc.) used later. Early correction is much more cost-effective than action later in the acquisition process.

MIL-STD-1388-1A heavily emphasizes testing in Development Testing/Operational Testing (DT/OT) and tends to test the entire logistics system. This approach delays the availability of data and thus may continue the risk or uncertainty longer than may be necessary. While full system testing in development and operational testing is a major source of support status data, more limited assessment can be made earlier from mockups, test bench setups, prototypes, etc. DT/OT testing is also the most expensive. Lastly, the entire logistics system is generally not in place at the time of DT/OT, so such testing is not completely representative of ultimate actual use. While T&E data are valuable to the LM in assessing the status of the supportability parameters, such assessment has limitations. For example, technical manuals may be only preliminary, or personnel conducting the test may be more highly skilled than fleet personnel. The *Hawthorne effect* (people exceeding normal productivity as the result of special attention) also may skew the results. Because of these factors, additional assessment is required after the system/equipment is fielded and all aspects of the logistics environment are present. This is the objective of the last two subtasks.

Subtask description. The first four subtasks collect and analyze data from contractor testing, development and operational testing, etc. The last two subtasks involve data collected from operational units after initial deployment. Figure 4.13 summarizes the subtasks.

The subtasks should be integrated with other T&E planning wherever possible to conserve time and resources. The subtasks are well defined in MIL-STD-1388-1A and will not be repeated here. The objective is to assess the status of the major supportability parameters.

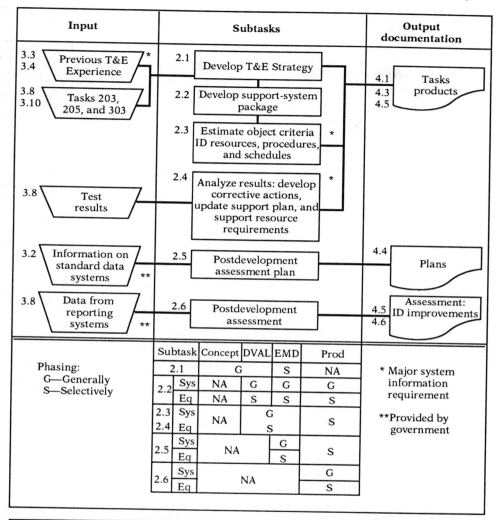

Figure 4.13 Task 501, Supportability Test, Evaluation, and Verification. (From Navy LSA Applications Guide, May 1986.)

Most of these are identified in tasks 201, Use Study, task 203, Comparative Analysis, and task 205, Support-Related Design Factors. The suggested phasing of the subtasks is a major point of interest. LSA T&E planning needs to begin early to ensure that all the necessary details are in place.

The standard explicitly mentions software in the fourth subtask. Including software results in various T&E plans and activities should be

TABLE 4.22 **Task 501 Subtask Selection and Focusing**

LM guidance
1. *LSA end objective:* This task can affect all three major objectives.
2. *Phasing:* See Fig. 4.13. This guidance assumes that testing is done in EMD. As the discussion indicates, data can usually be collected from earlier testing in many cases. The first four subtasks should be phased to do so.
3. *Complexity:* While Table III of the 1A standard lists most of the subtasks as selective for equipments, use the subtasks wherever useful data can be collected.
4. The key to a cost-effective supportability assessment is to incorporate supportability assessments with overall assessments of mock-ups, prototypes, and T&E activities when possible. The APML should tie logistics testing to the overall program testing.
5. Logistics testing should focus on the key design-related parameters identified in tasks 203 and 205, key logistics issues, and whatever ILS element data that can be meaningfully gathered within the constraints of information availability. The LEMS and R&M should be consulted to help focus these subtasks.
6. This task could be performed by a service organization.

SOURCE: From Draft UAV LSA Applications Guide, Block One, July 1991.

a major consideration in software-intensive programs. A special software failure data collection system may be desirable as part of the fifth subtask.

Task 203, Comparative Analysis, task 205, Supportability Design Factors, and task 303, Evaluation of Alternatives and Tradeoff Analysis, are sources of input to task 501, Supportability Test, Evaluation, and Verification. They provide information on supportability drivers, past problem areas, system supportability parameters, and major areas of supportability risk. Task 203 also can provide information on the support system, LOR, and diagnostics. The LSAR from task 401, Task Analysis, can provide data for planning ILS element resources to support the testing. In turn, the various supportability assessments cited above provide opportunities to help validate the LSAR. Where possible, personnel of specified skills should be used in these assessments to evaluate the predicted skill levels and task times. See Table 4.22.

Use of results. The LM should take action to ensure that various subtask results are included in related documents such as Test and Evaluation Master Plan (TEMP). The LM uses the first three and the fifth subtasks to develop the details of supportability to be evaluated prior to and after deployment. The results of the actual analyses (subtask 501.2.4, Updates and Corrective Action, and subtask 501.2.6, Supportability Assessment) provide the basis to initiate corrective action in problem areas. The specific corrective action developed could involve changes to design, support concept, or the amount of ILS resources required.

4.7   The 100 Series Tasks: Program Planning and Control

The 100 Series tasks provides for formal program planning and review actions.

4.7.1   Task 101: Development of an early LSA strategy

Purpose.   This task assesses how much and where to invest in LSA.

Benefits.   This task is used to provide information that serves as the basis for preparing the LSA requirements, such as the SOW and other ILS planning documentation. Its intent is to help the LM focus LSA requirements on those areas of greatest need and benefit. See the tailoring discussion in Chap. 3 and the beginning of this chapter.

Task description.   This is one of the less expensive LSA tasks (see Appendix 17). It is summarized in Fig. 4.14. The task requires identifi-

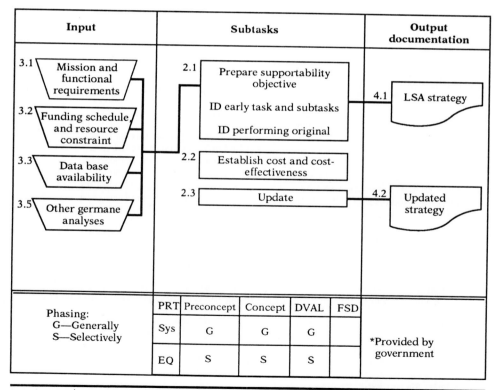

Figure 4.14   Task 101, Development of an Early LSA Strategy. (From Navy LSA Applications Guide, May 1986.)

cation of program supportability objectives, early LSA tasks and subtasks, and suggested performing organizations. It should begin, if possible, before concept exploration (CE). Software-intensive programs should address software-support implications. We suggest the deletion of the requirement in 1A standard to identify the cost-effectiveness of each task and subtask, except in the most general way. While general judgments can be made about which tasks are probably the most productive in terms of supporting the top-level LSA objectives, a mediocre LSA effort could affect a poor program more than a great LSA effort can affect a good program.

In practice, most LSA strategies do not meet many of the requirements for this task cited in 1A standard. In fact, the staff guidance of some organizations states that a strategy consists of identifying the tasks and subtasks to be required. We advocate a middle position. While the total requirement in 1A standard may have its unrealistic aspects, simply selecting tasks and subtasks ignores too many refinements, such as focusing, that can make future LSA requirements more cost-effective. See Table 4.23.

Use of results.   As stated under "Benefits."

TABLE 4.23   **Task 101 Subtask Selection and Focusing**

LM guidance
1. *LSA end objective:* Plan of action to do LSA work.
2. *Phasing:* See Fig. 4.14 and Appendix 7. The strategy should be developed as late as possible in the present phase to take advantage of current LSA task results but must be performed in time to affect the LSA RFP for the next phase. By EMD, the 1A standard suggests selective application.
3. *Complexity:* The 1A standard and Fig. 4.14 suggest selective application for equipments. Appendix 7 for simple equipments does not recommend its application. However, the preparation of an LSA RFP requirement inherently implies some assumed strategy, formal or informal, simple or complex. Our suggestion is the use of at least some minimal formal documentation, if only a simple LM memo for the record summarizing the strategy logic.
4. Complexity is the key to tailoring and focusing. The options range from the entire task as stated in the 1A standard (without undue concern for estimating the cost-effectiveness of each task and subtask) to a memo for the record stating the reasons for the task and subtask selections.
5. If task 101 is done internally, the LM should consult project engineers, R&M personnel, and ILS element managers. The reports could consist of the resulting SOW and some documenting recording reasons for selection. Such documentation is useful for justifying the results to program management and to various review groups.
6. This task should be scheduled for completion in sufficient time to be an input into the preparation of LSA SOW requirements. The potential supportability objectives should be based on tasks 201 and 203, if one or both have already been performed.
7. Existence of data bases can affect subtask selections. If the predecessor item is not in the field (3M, etc.) data base, it is difficult to do subtask 203.2.3, 203.2.5, and readiness drivers unless expert opinion is used as a basis for estimated values.

SOURCE: From Draft UAV LSA Application Guide, Block One, July 1991.

4.7.2   Task 102: LSA Plan (LSAP)

Purpose.   The LSAP details how the contractor will execute the LSA RFP requirements.

Benefits.   The LSAP is a key document. It normally becomes part of the contract and represents the performing organization's road map of what LSA is to be done and how it will be done. The LSAP serves as one of the bases for judging the contractor's LSA efforts. It provides the LM with visibility of the performing organization's approach to LSA. It also serves as a basis of dialogue between the LM and the contractor so that full understanding of the contract LSA requirements is achieved. The LSA plan is one of the lower-cost LSA tasks (see Appendix 17).

The LSAP may be used for the following purposes:

1. To ensure that the LSA program will be timely and effective.
2. To evaluate contractor proposals in source selection if the plan is part of the proposal.
3. To review the contractor's progress.

Task 101, LSA Strategy, if performed, any other LSA task results available, and the LSA RFP requirements are possible inputs to the plan.

Task description.   Figure 4.15 summarizes the task requirements. Tailoring and possible focusing of this task are discussed in the next section. While the standard does not explicitly mention support-related software considerations, the LM may wish to specifically require the plan to address software considerations in such key areas as task execution and the LSAR data base. See Table 4.24, Task 402 Subtask Selection and Focusing.

Use of results.   The LSAP is a key document because it normally becomes part of the contract, thus serving as the basis for monitoring the effort as execution occurs.

4.7.3   Task 103: Program and Design Reviews

Purpose.   This task establishes the requirements for the contractor's support organization participation in internal reviews and government participation in program, design, and LSA reviews.

Benefits.   These reviews provide the LM with communication channels to determine the status of contractor progress in executing the

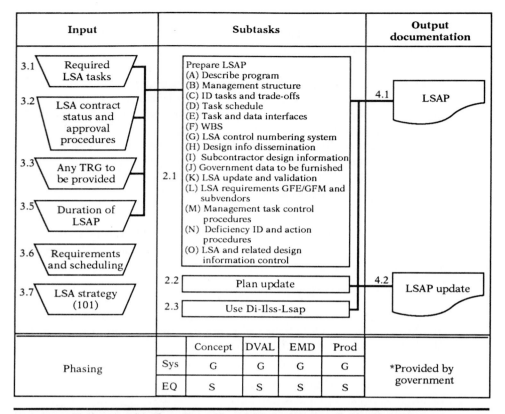

Input	Subtasks	Output documentation

Figure 4.15 Task 102, LSA Plan. (From Navy LSA Applications Guide, May 1986.)

approved LSAP, assess developments on design supportability, identify problem areas and issues, and initiate timely and appropriate actions on those problems and issues. Dialogue between design and logistics personnel at design and LSA reviews represents a significant design influence opportunity for the LM.

The product of these reviews is action on problem areas and issues. The LM should follow up periodically to determine that required actions have been taken and integrated into the appropriate program activity.

Planning for these reviews should receive the LM's full attention, since the CELSA data (see Appendix 17) indicate that they are 10 to 20 percent of the cost of LSA in various phases. Use of on-line access to reduce the cost of such reviews should be considered. Design reviews can include both hardware and software.

TABLE 4.24 Task 102 Subtask Selection and Focusing

LM guidance

1. *LSA end objective:* Provide a plan to do LSA work.
2. *Phasing and complexity:* At least some minimal plan should generally be required, although Fig. 4.15 suggests selective application for equipments in CE, and Appendix 7 suggests selective application.
3. An alternative strategy is to provide the results of the previous phase LSA effort to the bidders and direct them in the instructions to offerers (section L of U.S. RFPs) to state in their proposals what LSA tasks and subtasks they intend to perform.
4. If feasible, require the LSAP as part of the proposal to allow work to begin as soon as possible. If the plan is not extensive (e.g., on smaller programs), it could be included as part of the contractor Integrated Support Plan (ISP). Another option to reduce the size of the contractor's plan is for the contractor to include procedural detail in internal contractor LSA/LSAR instructions or an LSA/LSAR procedures guide (see Table 3.1).
5. The LM should consider the following additional requirements (suggested SOW language is in Appendix 8):

   a. Contractor LSA/ILS sign-off on engineering drawings and specifications. Provides contractor ILS/LSA a chance to comment prior to final design decisions and ensures access to the most current data.

   b. Appointment of an LSA focal point in design engineering. Provides ILS/LSA a designated point to coordinate design input to LSA tasks and the LSAR and data exchange with LSA.

   c. Contractor conduct of an engineering-LSA supportability seminar. Promotes greater understanding between the two contractor communities.

   d. Dissemination of all support-related design requirements and support-system requirements as an integrated package to design management and designers by treating support-related design as an entity rather than the fragmentation in most specifications. Table 4.25 is an example of this in a software-intensive program.

   e. Inclusion of supportability and O&S costs in all contractor tradeoff studies.

   f. Colocation. Intended to promote dialogue and data exchange by causing close location of design, R&M, and LSA personnel.

   g. Contractor LSA/LSAR products should be disseminated to contractor design and ILS personnel. Examples are the Use Study and support-system alternatives. However, the LSA plan requirements in MIL-STD-1388-1A do not address contractor use of LSA/LSAR products. The contractor should be required to describe the use of the LSA products internally in the proposal/plan.

   h. Contractor LSA performing organization and specialists. Designed to have the contractor identify in the LSAP the specific contractor organization(s) and specialists who will perform each subtask and prepare the LSAR records.

   i. Identification of models. Required the contractor's plan to identify the models/formula used in tasks other then 303 (such as tasks 202, 203, 204, 205, 402, and 501).

SOURCE: From Draft UAV LSA Applications Guide, Block One, July 1991.

Task description. Figure 4.16 summarizes the subtasks in task 103. Subtask 103.2.1, Establish Review Procedures, provides contractor personnel with the authority to participate in internal design reviews. The remaining subtasks cover joint government-contractor program, design, and LSA reviews. The names of the reviews in Fig. 4.16 are for EMD. Differently named reviews may be conducted in earlier phases. See Table 4.26, Task 103 Subtask and Selection Focusing.

TABLE 4.25   Example of Supportability-Related Design Requirements

Paragraph no.	Title
3.1.4.2.7	Training Subsystem
3.1.4.2.8	Maintenance
3.1.7	Operational and Organizational Concept
3.2.2.2	Equipment Mounting/Interconnection
3.2.2.4	Equipment Layout
3.2.2.5	Storage and Transportation Packaging
3.2.2.6	Health and Safety
3.2.3	Reliability
3.2.4	Maintainability
3.2.4.1	Maintenance Concept
3.2.4.3	Maintainability Design
3.2.4.4	Maintenance Complexity
3.2.4.5	Lowest Repairable Unit
3.2.4.6	Repair Times
3.2.4.9	Fault Detection/Isolation
3.2.4.10	Standard/Common Elements
3.2.4.11	BIT and Diagnostic Software
3.2.5	Availability
3.2.7	Redundancy
3.2.9	Transportability
3.3.2	Parts
3.3.5	Interchangeability
3.3.6	Human Performance/Human Engineering
3.3.8	Safety
3.5	Logistics
3.5.1.1	Test Equipment
3.5.1.3	Equipment Accessibility

Use of results.   The reviews in this task provide the LM with a major source of information for managing the LSA effort and for interfacing with other activities and personnel involved in the project. Attendance and participation in these reviews are essential.

4.8   Specific Tailoring Advice for Various Types of Acquisitions

4.8.1   Overview

This section discusses LSA tailoring approaches for various types of acquisitions that differ from the full development acquisition but which frequently occur in practice. The cases discussed are

EMD starts

Product improvements

Nondevelopmental items (NDIs)

Technology demonstrations

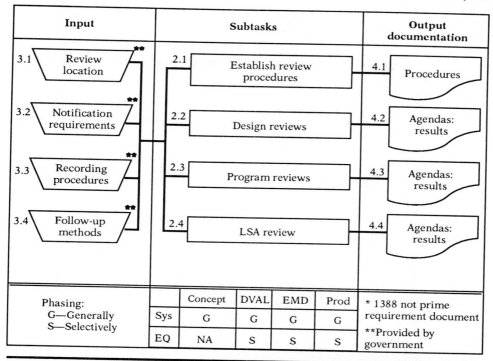

Input		Subtasks					Output documentation	
3.1	Review location ★★	2.1	Establish review procedures				4.1	Procedures
3.2	Notification requirements ★★	2.2	Design reviews				4.2	Agendas: results
3.3	Recording procedures ★★	2.3	Program reviews				4.3	Agendas: results
3.4	Follow-up methods ★★	2.4	LSA review				4.4	Agendas: results

Phasing: G—Generally S—Selectively		Concept	DVAL	EMD	Prod	* 1388 not prime requirement document
	Sys	G	G	G	G	**Provided by government
	EQ	NA	S	S	S	

Figure 4.16   Task 103, Design and Program Reviews. (From Navy LSA Applications Guide, May 1986.)

Accelerated programs

Competition

In practice, two or more of these cases may occur together. The discussions cover the LSA that "may" be appropriate. The management principles cited in Chap. 3 still apply. Do *not* duplicate other work, but do tailor to the individual situation.

4.8.2 EMD starts

Many acquisition programs begin in Engineering Manufacturing Development (EMD). Examples are technology demonstrations that become acquisitions, contractor Independent Research and Development (IR&D) projects that gain service sponsorship, and test and support equipment or training equipment needed to support major systems. The major considerations in tailoring LSA for EMD starts are

1. What LSA has been done?
2. Is the LSA sufficient?

TABLE 4.26 Task 103 Subtask Selection and Focusing

LM guidance
1. *LSA end objective:* This is dependent on the nature of the program, the type of review, and the specific phase. In principle, this task can serve all three top-level LSA objectives.
2. *Phasing:* Applicable to all to some degree.
3. *Complexity:* Figure 4.16 suggests selective application for simple equipment, but it is wise to participate in any reviews which the government requires.
4. LSA reviews in the CE and DEMVAL phases should emphasize progress on the 200 and 300 series tasks. In-process LSA reviews (possibly quarterly for major programs) prior to task completion should be conducted to ensure any necessary early redirection of analysis effort.
5. When possible and appropriate, the LM should schedule these reviews and ILS reviews in conjunction with each other. Contractor design personnel should discuss supportability issues as part of the design reviews (see 8 below). A using organization logistics representative should attend LSA and design reviews.
6. The LM must develop requirer review procedures to assist in LSA reviews (see Chap. 5). The LM must assist in LSA reviews (see Chap. 4). The LM must establish follow-up procedures for LSA to ensure that actions and changes identified in these reviews are implemented. Contractor design personnel and, where appropriate, subcontractor personnel, should attend LSA reviews.
7. The LSAR data packages for government review in DEMVAL and EMD should be made available to the government review team members for preliminary review 30 days in advance of the review. On-site LSA and LSAR reviews provide an opportunity to physically view hardware and assess what contractor communities are actually participating in the work on specific subtasks and LSA records. Some programs are now locating read-only terminals in the requirer offices with on-line access to the contractor's LSAR data base. This allows data to be reviewed at the convenience of the requirer and can reduce the amount of on-site time needed. The LM may wish to have the government review team examine records from subtask 103.2.1, Establish Review Procedures, to evaluate the design influence achieved by the contractor supportability specialists.
8. Possible task enhancement. The following enhancements should be considered by the APML. Actual sample SOW language is in Appendix 9.
a. Discussion of support-related design requirements as an integrated package by contractor design personnel at contractor-government design reviews. Related to clause *d* in task 102: forces designers to be familiar with supportability aspects of their design.
b. Contractor engineering personnel attendance at LSA/LSAR reviews. This clause is intended to ensure that contractor design personnel are available for discussion purposes at government LSA reviews.

SOURCE: From Draft UAV LSA Applications Guide, Block One, July 1991.

3. If LSA has not been done, would it serve a purpose if done now?

4. Are supportability goals/thresholds established? If so, were they done on a solid analytical basis or should the goals be reexamined?

5. How much design influence can be exerted: none, minimal, or substantial?

The first three questions generally apply to the 200 series tasks and most of the tradeoffs in task 303, Evaluation of Alternatives and Tradeoff Analysis. These tasks, if not already performed, should be required in an FSD start only when a good reason to do so exists. In terms of 303 subtasks, subtask 303.2.4, Readiness Sensitivities, and subtask 303.2.7, Repair Level Analyses, should be done unless reliability is very high, the cost of the item is low, or repair will obviously be at a predetermined level.

If a check on supportability goals is desired (question 4), at least two of the first three subtasks in task 203, Comparative Analysis, should be specified. In terms of question 5, possible minimal design influence subtasks are

103.2	Design Reviews
201.2.1	Supportability Factors
201.2.3	Field Visits
202.2.1	Supportability Constraints
202.2.4	Risks
205.2.3	Specification Requirements (possibly combined with subtask 203.2.4, Qualitative Supportability Problems)
301.2.5	Design Improvements (feedback from task 301, Functional Requirements Identification, if required)
401.2.5	Design Improvements (feedback from the LSAR)

If a substantial amount of design influence is possible or the program plans a P^3I effort, consider the following subtasks:

203.2.5	Supportability, Cost, and Readiness Drivers (requires subtask 203.2.1 or subtask 203.2.2)
203.2.6	Unique System Drivers (if unique technology)
204	Technological Opportunities (mission or support, if P^3I is planned)
303.2.5	Manpower and Personnel Tradeoffs (if P^3I is planned)
303.2.6	Training Tradeoffs (if P^3I is planned)
303.2.8	Diagnostic Tradeoffs (if P^3I is planned)

4.8.3 Product improvements

This section discusses some key LSA considerations in product improvement projects (PIPs). The purpose of product improvement can be performance improvement, supportability improvement (including R&M improvement), cost reduction, or a combination of these factors. Historically, most PIPs have been performance-oriented. The increased use of the performance concept in DOD Directive 5000.1 provides a new avenue to address supportability concerns.

Four major issues dominate LSA considerations in product improvement efforts:

1. Has the use or application changed?

2. Are the new additions already developed or in development?

3. Are there supportability problems with the existing system?

4. Are any LSAR data bases available and usable?

If the application, mission, and/or use changes significantly as a result of the planned improvement, consideration should be given to updating or preparing the Use Study, task 201. If the change in use has an impact on R&M of new or remaining parts of the item, the R&M will require updating, which can affect goals, tradeoff results, and the existing LSAR data base. The new design may make a change in maintenance concept and levels desirable. It could have this as a design objective.

If the new items are still to be developed or in development, some degree of design influence is possible (questions 2 and 5).

The third issue, existing supportability problems, may be a part of the plan or it may have to be "sold" by the LM. Existing LSAR data bases should be evaluated for use as is or updated to reduce LSAR costs. Table 4.27 summarizes specific PIP task-tailoring considerations. Like many generalizations, if the underlying assumptions change, the conclusions can be affected.

4.8.4   NDI tailoring

General. The general advantages of procuring nondevelopmental items (NDIs) to satisfy military needs was discussed briefly in Chap. 2. This section discusses the possible use of LSA in the acquisition of NDIs. The term *nondevelopmental item* implies one of the following:

1. A CFE/GFE item that is already in the inventory or in development

2. A standard commercial off-the-shelf item not currently procured

3. A commercial item that requires some modification to meet service needs.

Key principles. There are several key points to remember when deciding what LSA, if any, can be usefully employed in the acquisition of NDIs. First, most of the U.S. military services have manuals/procedures pertaining to the acquisition of NDIs. It is therefore important to tailor and integrate application of LSA with these procedures to preclude duplication. Size any LSA effort to supplement the NDI procedures only as necessary. Second, the amount of LSA needed is dependent on the specific circumstances. The amount done also may depend on who does the selection, the government or a contractor, as part of a larger program. Third, mission-system design influence through LSA is obviously very limited, if not impossible, in NDI acquisitions. However, what may be possible and desirable is the use of LSA to influence item selection from a supportability viewpoint. Fifth,

TABLE 4.27   PIP Tailoring

Task number	Task title	Needed	Are new additions developed?		Problems with existing	
			No	Yes	No	Yes
101	LSA strategy	Selective				
102	LSA plan	Yes				
103	Reviews					
	a. Design	Selective	Yes			
	b. Programs	Yes				
	c. LSA/LSAR	Yes				
201	Use Study	If PIP is for a new mission				
202	Standardization		Selective			
203	Comparative analysis	Selective				
204	Technological opportunities		Yes		Yes	
205	Design factors	Yes				
301	a. FRI	Selective	Yes			
	b. TI	Selective	Yes			
302	Support alternatives					
	a. Concepts	Selective	Yes		Selective	
	b. Contractor support	Selective				
303	Tradeoffs	Selective	Readiness LORA	Readiness LORA Are data available?		
401	Task analysis	Update				
402	Early fielding analysis	Yes, if major item or large quantity				
403	Postproduction support analysis	If dynamic technology				
501	T&E	Yes	Yes	Yes		
	LCC		Yes			

because of the shorter acquisition cycle for NDIs, the requirer frequently needs interim contractor support upon initial fielding until the service organic support is in place. Sixth, the requirer may need to have the hardware contractor conduct a limited LSA effort to identify the support resources to be acquired for the selected NDIs.

Logistics in the NDI process.   While the procedures for NDI acquisition used by different organizations may differ in detail, the following steps are usually applicable from a logistics viewpoint:

Step 1. Input/review of the operational requirements

Step 2. Collection and analysis of germane logistics data

Step 3. Preparation of logistics input to the NDI selection/evaluation criteria

Step 4. Acquisition of the logistics products needed to support the NDI item selected

Step 5. Assessment of the supportability of the fielded item as necessary

TABLE 4.28   NDI Tailoring Checklist

1. Are the necessary Use Study (task 201) data in the Operational Requirement document (step 1)?
   a. Yes; go to next question.
   b. No; consider Use Study to develop additional data needed.
2. Does the prospective NDI acquisition warrant consideration/specification of use or standard ILS support items (test, support, training equipment, etc.) (step 2)?
   a. No; go to next question.
   b. Yes; consider subtask 202.2.1, Supportability Requirements/Characteristics.
3. Are data needed to select support-related performance/specification (R&M, diagnostics, etc.) characteristics (steps 1 and/or 2)?
   a. No; go to next question.
   b. Yes; consider subtasks 203.2.1 or 203.2.2 and 203.2.3.
4. Has the item the NDI will replace had support problems (step 2)?
   a. No; go to next question.
   b. Yes; consider subtasks 203.2.4 and 203.2.5.
5. Is a new ILS element technology search warranted (step 2)?
   a. No; go to next question.
   b. Yes; consider task 204, Technological Opportunities, but restrict study to ILS elements.
6. Is there freedom to select the support concept (step 2 or 4)?
   a. No; go to next question.
   b. Yes
      (1) Will a LORA suffice to define the support concept?
          (a) Yes; go to next question.
          (b) Consider tasks 301, Functional Requirements Identification, task 302, Support System Alternatives, and subtask 303.2.2.
7. Are there sufficient data available to support the item (step 2)?
   a. Yes; go to next question.
   b. No; consider tailoring tasks 301, Functional Requirements Identification, and task 401, Task Analysis.
8. Is an operational availability analysis needed (step 2)?
   a. No; go to next question.
   b. Yes; consider subtask 303.2.4.
9. Will the fielding of the item significantly affect the existing support structure (step 2 or 4)?
   a. No; go to next question.
   b. Yes; consider a highly tailored task 402, Early Fielding Analysis.
10. Are problems with sources of support after production ceases likely (step 2 or 4)?
    a. No; go to next question.
    b. Yes; consider task 403, Postproduction Support Analysis.
11. Will the item be tested prior to the final acquisition decision (step 5)?
    a. No; go to next question.
    b. Yes; consider subtasks 501.2.1, 501.2.2, 501.2.3, 501.2.4.
12. Will follow-on support assessment after fielding be needed (step 5)?
    a. No.
    b. Yes; consider subtasks 501.2.5 and 501.2.6.

SOURCE: Derived from NAVSEA TO300-ACQ-PR0-010, "Systems and Equipment LSA Procedures Manual."

Tailoring LSA for government-selected NDIs.   Table 4.28 contains a series of questions that can be used to assist in determining what, if any, LSA would be useful in NDI acquisitions. If there are any requirements for the hardware contractor to prepare data (questions 7, 8, 9, 10, and 12) or to perform a LORA, task 102, LSA Plan, and subtasks 103.2.3 and 103.2.4, Program and Design Reviews, also should be required.

Table 4.29 shows the organizations that would normally perform any LSA work that needs to be done as a result of the answers to the questions in Table 4.27. Ideally, the ILS manager should have input into the preparation of the operational requirements. This could be done based on any Use Study (task 201) and/or Comparative Analysis (task 203) work performed.

Logistics data needed can be obtained (1) by performing the LSA tasks identified and/or (2) from data collected from a market survey as part of the market investigation. The LSA tasks performed by the requirer should be documented in some minimal form. Chapter 5 has suggested approaches and departure point forms of possible use.

The market investigation is usually performed to determine the availability of NDIs that meet the criteria established. The market investigation is performed to obtain detailed information about the candidates to assess their feasibility not only from a technology viewpoint but also from the standpoint of reliability, supportability, and cost-effectiveness. It also determines any testing needed and what, if any, logistics support is available. The ILS manager should obviously participate in the market investigation.

Market investigations may vary from informal telephone inquiries to comprehensive industry-wide reviews. Maximum use should be made of existing data. If insufficient data exist, the ILS manager can send a Request for Information to potential contractors. This request should ask for specifics about existing logistics support available for the item. Table 4.30 contains a sample market survey questionnaire. The list of questions should be tailored to the target suppliers. The question of maintenance concept, for example, needs more explanation to suppliers not previously active in defense work.

TABLE 4.29   Suggested Performing Organization

Question	Performing organization
1, 2, 3, 4, 5, 6, 11	Requirer
7	HW contractor
8, 9, 10, 12	Either

SOURCE: Derived from NAVSEA TO300-ACQ-PR0-010, "Systems and Equipment LSA Procedures Manual."

TABLE 4.30 Market Investigation Survey: Logistics Considerations

Information needed
1. What is the recommended support package? 2. What is the maintenance concept? 3. Provide a list of repairable items. 4. Provide a list (and copies) of available documentation (manuals, training materials, drawings, parts lists, etc.). 5. Identify tasks required to operate, maintain, and repair the item if not detailed in existing documentation. This will include operational, preventive and corrective maintenance, and overhaul tasks. 6. Identify the type of skill and quantity of personnel required to operate, maintain, and repair the item. Include the expected annual maintenance burden for each skill level. 7. Identify any training aids/devices which may be required. 8. Provide list of recommended support and test equipment, special tools, and calibration requirements. 9. Identify proprietary items and data rights. 10. Recommend procurement processes to acquire spare parts, test equipment, special tools, etc. 11. Estimate average time to process an order for spares from receipt of order to shipment. 12. Describe proposed warranty procedures. 13. Describe the Configuration Management and Configuration Control procedures applicable to the article under consideration. 14. When was the item placed on the market and total unit sales.

SOURCE: Derived from NAVSEA TO300-ACQ-PR0-010, "Systems and Equipment LSA Procedures Manual."

Contractor NDI selection.   In many cases, a contractor selects NDIs (see contractor design decision process in Chap. 5). In some cases, the hardware contractor may do primarily system integration (i.e., develop a system composed primarily of NDIs). While, conceptually, the contractor should use a process similar to that described in this section for government use, in practice this may not occur. However, the government may wish to review, at least in a generic sense, the supportability factors that the contractor will use in selecting CFE NDIs. The generic decision process selected by the contractor could be the same one it uses in performing hardware and software tradeoffs.

The contractor has a concern similar to that of the government for supportability of NDIs. Contractor-furnished NDIs must be integrated into the delivered system, and this presupposes working NDIs. Production-line delays due to nonfunctioning NDIs may be covered by warranty, but time lost is not covered. It is therefore in enlightened self-interest for the contractor to ensure the reliability and supportability of the NDIs, since the alternative of buying extra units to ensure availability is expensive.

NDI summary.   The ILS manager should (1) have input to the NDI selection process and (2) acquire the support needed for the item selected. A tailored LSA process can support these objectives. The amount of LSA effort required varies with the specific circumstances.

Key concerns in NDI procurement include configuration management to ensure that future replenishment will be interchangeable and continued availability of identical end items or items that are downward compatible. A responsive repair facility is needed if organic support is not planned. The commercial marketplace responds to competition by making changes to upgrade products. Such change can adversely affect the DOD because it is not usually a major customer. This can make the issue of proprietary rights in manufacture a significant consideration for a subsystem integral to a DOD system. The remedy of placing the data package in escrow for availability to the government should the manufacturer stop production has been employed in many programs. Interim contractor support is frequently necessary and should be a part of the considerations when NDIs are chosen. Regardless of the amount of LSA effort needed, the growing emphasis on NDIs, and their obvious benefits, make the use of analysis highly desirable.

4.8.5 Technology demonstrations

Technology demonstration projects are projects designed to demonstrate technical feasibility. No decision to produce such items has been made. Consequently, there is significant reluctance to spend project funds on detailed logistics resource requirement efforts. This eliminates consideration of the LA 400 series tasks immediately.

The 200 series tasks can take a different course. Task 201, Use Study, could consider alternative possible uses. Task 202, Mission Hardware, Software, and Support System Standardization, would aim at the use of standards in areas not involving the new technology. Technology projects frequently use what is handy in lieu of what is best when selecting items to be associated with the new technology in building the model to be tested. Task 203, Comparative Analysis, could identify support drivers and problems if the project can be expanded to cover these areas. Task 205, Supportability and Supportability-Related Design Factors, would not necessarily be required unless a major project objective is to demonstrate specific supportability-related parameter values. The 200 series tasks may be oriented toward the concept of readiness or supportability-related design definition studies, as illustrated in Appendix 6.

Support-system design task 301, Functional Requirements Identification, and task 302, Support System Alternatives, could take the form of design studies to explore the feasibility of innovative support concepts such as two- versus three-level maintenance. The following subtasks from task 303, Evaluation of Alternatives and Tradeoff Analysis, should be considered for design studies:

303.2.3	Systems Tradeoffs
203.2.4	Readiness Sensitivities
303.2.5	M&P Tradeoffs
303.2.6	Training Tradeoffs
303.2.8	Diagnostics
303.2.12	Facility Tradeoffs

The application of task 501, Supportability, Test, Evaluation, and Verification, should be confined to the collection of key supportability factor data. Management tasks should be minimized and included as part of the engineering management wherever possible.

In summary, the keys to LSA in technology demonstration projects are

1. No 400 series.
2. Selective use of 200 and 300 series with a possible design study orientation.
3. Minimization of LSA management 100 series paperwork with integration of management considerations with project management plans.

4.8.6 Accelerated programs

Accelerated programs are occasionally employed because of the understandable desire on the part of management to reduce the length of the acquisition cycle. Since accelerated programs reduce development time, they usually attempt to use items that are already developed or will complete development shortly. Schedule compression reduces the time available to do LSA design influence. Use of existing components reduces the opportunity for LSA design influence and the chances that innovative support concepts can be used. Design influence efforts thus tend to shift to preparation of supportability criteria for use in the selection of existing components to be used. The identification of supportability P^3I efforts should be explored since new technology possibilities cannot be exercised in the development time frame. The specific impacts on LSA task selection are as follows:

1. Task 202, Standardization, assumes greater importance.
2. Task 204, Technology Opportunities, is of less interest unless P^3I is planned involving ILS elements in task 303, Evaluation of Alternatives and Tradeoff Analysis (subtasks 303.2.2, 303.2.5, 303.2.6, 303.2.8, 303.2.11, and 303.2.12). In this case, doing these studies may have less benefit than normal.

3. Task 301, Functional Requirements Identification, and task 302, Support Systems Alternatives, have the least possible support impact. However, study of the contractor support aspect of subtask 302.2.1, Alternative Support Concepts, may be more important.

4.8.7   Competition

It is not unusual in CE and DEMVAL to have two or more contractors making competitive bids. In such circumstances, some programs managers are reluctant to fund a complete LSA effort by all the competing contractors, although this is what should be done. There are two other alternatives to a full-scale effort on all the awards made.

One alternative would be to ask the bidders what LSA tasks and subtasks they propose to accomplish as part of their proposal. This approach delegates all or part of the LSA tailoring responsibility to the contractors. A modified version of this approach would involve the specification of certain minimum tasks while requesting the contractors to identify other LSA activity to be performed. This approach could be combined with an emphatic statement in the instructions to offerers and/or in the award criteria to send a strong message that supportability is a major consideration.

The second alternative is to try to capitalize on the fact that either the requirer or the executor can do many of the LSA tasks. Thus a program office could reduce the cost of LSA by having a single organization, such as a support contractor, execute those tasks most suitable to requirer execution. Table 4.31 presents an example of such an approach.

4.9   Summary

This chapter has discussed in detail the following:

1. The LSA tasks and subtasks in terms of tailoring requirements.
2. Tailoring variations for various types of acquisitions.

The effectiveness of LSA can be improved by augmenting most standard tailoring guidance by focusing and enhancement of the LSA tasks and subtask statements. Many programs do not follow the classic acquisition process and therefore require special tailoring approaches to achieve cost-effective LSA requirements. Special emphasis should be placed on focusing of the tasks and subtasks selected.

TABLE 4.31   LSA in Concept Exploration and DEMVAL with Competition

Task	What prime "ought" to do	What service organization could do	Prime minimum requirements	Cost impact
102	Prepare LSA Plan	Prepare SOW	Min. plan requirements in SOW	Cheaper
103	Conduct Reviews R/M/Logistics internal designers discuss at govt. reviews	Support PM in reviews	X	Cheaper
201	Conduct Use Study Distribute Use Study Use Study field visit	Conduct Use Study	X	One versus many studies
202	Develop standardization criteria Use government decision criteria	202.2.1: Develop decision criteria for prime	X	Cheap; all using same criteria
203	Conduct Comparative Analysis: use results	Conduct analysis review	X	One versus many studies
204	Develop technology opportunities		NA	Let primes do as part of 303, etc.
205.2.3	Develop spec. requirements	Develop requirements based on 201, 203, and 303.2.4	NA	One versus many
302.2.1	Develop Support Alternatives (less last sentence) Use results	Develop baseline and alternatives	X	One versus many
303	Conduct tradeoffs			
303.2.1	Develop criteria, use models/government criteria	Use models/criteria	Identify models and criteria	
303.2.2	Develop Support Alternatives	Review		
303.2.3	Develop criteria/LCC system trades	Develop or approve criteria		
303.2.4	Conduct Readiness Sensitivity			
	Perform design alternative sensitivities	Develop system input data; perform system-level sensitivities	X	
303.2.7	Conduct LORA part of 2.2	NA	NA	
303.2.8	Develop diagnostics	Review	NA; make part of tradeoff criteria	
501	Develop T&E strategy	Perform test and evaluation	NA	One versus many

SOURCE: From Draft for SPAWAR, "Write Improved SOWs Easily (WISE)," October 1986.

Chapter

5

Alternatives to Executing LSA Tasks

5.0 Introduction

This chapter has six major sections concerned with the performance of the LSA tasks. The approach taken is to assume a full or "classic" development program and discuss the performance of the appropriate tasks and subtasks during each program phase, beginning with preconcept exploration. The approaches to task performance discussed are not necessarily the only approaches possible. The intent is to give the reader some departure points for consideration in deciding how to execute the task requirements. Many of the approaches are intentionally simplified. More complex programs can still use the concepts portrayed. Section 5.2 discusses an issue common throughout the initial acquisition phases, the use of a functional versus hardware approach in performing many of the 1A standard subtasks. The remainder of the chapter discusses task and subtask approaches and execution within each phase as follows:

The changing level and nature of LSA (5.1)

Preconcept exploration (PC) (5.2)

Concept exploration (CE) (5.3)

Demonstration/validation (DEMVAL) (5.4)

Full-scale development (EMD) (5.5)

Production (PROD) and postproduction (POSTPROD) (5.6)

Summary (5.7)

5.1 The Changing Level and Nature of LSA

In preconcept, no hardware (HW) exists. The need is not yet fully defined. However, even at this time, the technologists have ideas about possible technology to meet the need as hypothesized at this point in time. Because of the variability in the need and possible future solutions, analysis begins on a functional rather than a HW basis. By the end of Engineering Manufacturing Development (EMD), the need and solution have been explicitly defined in terms of a specific hardware/software (HW/SW) mission system and its ac-

TABLE 5.1 **Functional versus Equipment**

Level	Functional versus equipment system
System	NDI/functional (new design)
Subsystem	NDI/functional (new design)
Set	NDI/functional (new design)
Group	NDI/functional (new design)
Unit	NDI/functional (new design)
Assembly	NDI/functional (new design)
Subassembly	NDI/functional (new design)
Part	NDI/functional (new design)

SOURCE: From MIL-STD-280.

companying support system. This section is about the transition that occurs as a need moves between these two points. The term *function* in this chapter is used within the engineering context of functional versus hardware breakdown. Within integrated logistics, the term *functions* implies activities. After the need is defined and CE begins, designers initiate functional analysis and allocate the necessary functions to hardware, software, and personnel. While doing so, initial HW and SW solutions may be hypothesized. As design proceeds to more and more detail, more and more functions are translated into HW/SW requirements. Table 5.1 illustrates the concept.

As design proceeds to each lower level, the requirement is satisfied, if possible, by existing or in-development HW/SW. If the function cannot be satisfied in this way, a new design is initiated. In practice, interfaces and tradeoffs occur across each level. Initial decisions made for both nondevelopmental items (NDIs) and new design as a result of such considerations may change. The implications of this process for logistics support analysis (LSA) are obvious. LSA must generally begin initially on a functional basis. This allows the conduct of the use study and comparative analysis even though no system exists. Such an approach is also useful in reducing work load, since a functional system changes much less than an equipment-based system. By the end of engineering manufacturing development (EMD), the system must be completely on a hardware basis. LSA should transition in parallel with the design as it makes its transition from a functional basis to a hardware basis. The LSA proceeds down the indenture levels as the design proceeds down the indenture levels. As NDIs are selected at the various levels, the LSA may stop when an NDI is requirer-directed, depending on whether or not LSA has or is being done on the NDI selected. Failure to observe these principles will generally result in either (1) a failure to use all the available information (i.e., doing the analysis functionally when hardware information is available) or (2) needless analysis (analyzing options for decisions that have already been made). The transition concept just discussed is used throughout the remainder of this chapter.

5.2   Preconcept Exploration LSA

5.2.1   Objectives

The basic objective of preconcept exploration (PC) activity is to define the requirement or need. The basic objective of LSA in PC is to influence the definition of the requirement in a way that is favorable to future support. There are cases on record where a need, upon further examination, has proven to be unsupportable. The definition of need can include quantitative [mean time between failure (MTBF), mean time to repair (MTTR)] and qualitative (skills levels required) aspects for both HW and SW. This influence can be exerted by analysis of the impact on support-related factors, support constraints (the vehicle must fit within a specified cargo aircraft), support issues (what is the impact of the use of cryogenics at the organizational level of maintenance), and opportunities (is embedded training feasible and desirable). Factors, constraints, and issues may differ in nature, but the analysis approach is similar. The names and types of documents to be influenced, such as a need statement or an Operation and Organization (O&O) plan, may vary by service and country, but some program documentation of this type is normally required. The key point here is to identify the documents needed and to determine their schedule, since this affects the LSA schedule and possible analysis methods.

5.2.2   General analysis approach

The overall analysis approach is dependent on time, resources, and data availability. In PC, all these may be severely constrained. In PC there is no system. However, some general definition of possible concepts or systems is needed if analysis is to be performed. Therefore, some probable general options must be identified. The technologists can be helpful in this regard. These "solutions," or options, must then be evaluated for their logistics impact within the time, resources, and data available. Possible principles for doing analysis under these conditions include

1. *Level and amount of analysis:*   Identify the big issues (system and subsystem) first. With limited detail, breadth versus depth should be emphasized. Focus on issues that must be addressed in the near term if influence is to be achieved.

2. *Data availability:*   Expert opinion and rough order of magnitude (ROM) estimates can be used initially; harder data should follow when possible.

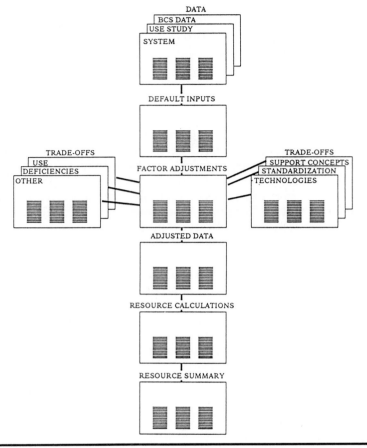

Figure 5.1   FECA Concept.

3. *Use of a simple, flexible model:*   This provides a framework or checklist for thinking, allows rapid manipulation of data as changes occur, and enhances credibility. The use of some analysis model that will address most of the factors, constraints, issues, and opportunities is suggested, if it is at all possible. Appendix 26 contains a full description of a simple departure-point model designed for Army use in PC and CE that uses a spread-sheet approach and can be run on a personal computer. The model could be modified for specific applications. The Defense System Management College (DSMC) Cost Analysis Strategy Assessment (CASA) model is another possibility. Larger models that can be run at higher levels without detailed data at lower indenture levels are also usable. The Army LOCAM model is an example of such a model. An "expert" opinion tradeoff technique is contained in Appendix 20. The sophistication of the model employed depends in

part on the complexity of the new system or concept. One U.S. service, for example, has employed complex simulation models in PC on aircraft programs. In PC, the model used need only provide comparative accuracy between alternatives, not absolute accuracy. As a general principle, the model used must be able to manipulate the types of data shown in Fig. 5.1. Appendix 26 contains a full description of such a departure-point model. Algorithms connecting inputs to outputs are obviously needed. In many cases, expert opinion must be employed as a data source, particularly in PC, since "hard" numbers may be difficult to find.

4. *Use of a team approach:*   This principle is true in PC as well as in other phases. A team approach should be used, if possible. In simple cases, this could be an LSA analyst consulting with members of other communities. As the project grows in complexity, teams need to be established. Formal identification of designated contributors is desirable in either case. The types of contributors needed in PC include personnel with operations, maintenance, logistics, cost estimating, and technology experience.

5. *Objectives for LSA should be set:*   Try to be more explicit than the general LSA objectives (influence design, etc.). Examples of greater specificity are (1) to identify 60 percent of the O&S costs, (2) to identify 80 percent of the contributors to unavailability on current similar systems, and (3) to identify the five largest contributors to O&S costs and/or system unreliability. These numbers are examples and can be varied. The objectives set may affect the resources needed for execution.

5.2.3   Possible tasks to be performed

MIL-STD-1388-1A identifies three tasks to be performed in PC. These are the LSA Strategy, task 101, the Use Study, task 201, and the Comparative Analysis, task 203. Table 5.2 summarizes the tasks/ subtasks that some experts believe to be useful in PC LSA.

Some have raised the issue that this is too much too soon. In some cases this may be true. However, one must ask the following questions: (1) Do I wish to influence the statement of need or requirement from a support viewpoint? and (2) What does it take to do this? If the answer to the first question is yes, the effort is not too soon, and some activity is in order. The second question must now be answered. The answer presented in this section represents an "ideal" answer. However, the answer is one of degree. Performance of part of the suggested activity should have some positive impact from a support perspective. The specific situation may dictate that the ideal is not possible for some reason(s). If this is so, our advice is to do what can be

TABLE 5.2   Possible PC Tasks/Subtasks

Task/subtask*	Major purpose
LSA Strategy Plan	Roadmap of what's to be done
Use Study (task 201)	Provides basis for estimating support burden of possible options; mission profile itself may provide design and support constraints
Comparative Analysis (task 203)	Provides initial set of support-related design targets; identifies drivers to focus analysis
Support Factors (task 205)	Refines new system support-related design targets; sensitivity analysis to assess risk
Model and Criteria Identification (subtask 303.2.1)	Provides analysis framework and rapid data manipulation in selected tasks
Standardization† (task 202)	Provides design constraints/requirements; evaluates impact of candidates
Technological Opportunities†	Identifies and evaluates benefits of possible new technology

*Can identify and evaluate design factors, constraints, issues, and risks.
†Optional depending on program considerations.

done, recognizing that the effort and probable impact will be reduced to some degree. The strategy/plan suggested is a combination of selected elements of an LSA Strategy (task 101) and LSA plan (task 102) in terms of the requirements in the 1A standard. It is the *plan* for doing the selected tasks in PC. The logic behind this is simply the fact that some formal plan needs to be developed to provide all the needed participants with a road map of what will be done if the work is to be done in a rational and timely manner. The use study is a key because it provides the basis for performing the other tasks. A diagram of the interrelationships of the possible tasks is shown in Figure 3 of the 1A standard. However, the tasks are, in fact, interdependent. While they could be performed in sequence in theory, there is considerable overlap in practice. An example of this will be shown in the task 101 discussion. Model and criteria definition in effect specifies subtask 303.2.1, Tradeoff Criteria and Model Definition. Some may question the need for a simple model for use in tasks 201, 203, and 205, but use of a model refines the analysis and adds credibility to the results. The logic for tasks 203 and 205 are stated in Table 5.2.

Some experts suggest performance of task 202, Standardization. Other experts suggest that task 204, Technological Opportunities, be substituted instead. Whether either is performed depends on the specific circumstances. The argument for performing task 202 in PC is that this is the best time to specify a requirement to use such things as standard test equipment and place constraints on manpower numbers and skill levels. The argument against is that premature standardization may cause serious design problems. Performance of task 202 in PC would be most appropriate on "accelerated" programs, i.e., a program with a shorter than normal development time that maximizes the use of existing items to achieve earlier than normal fielding. The argument for inclusion of task 204 in PC is that technology and needs interact. A "need" that is impossible to satisfy will not long survive. Thus a formal study of new possible technology could result in a better and more realistic definition of need. The argument for doing both tasks 202 and 204 reduces to a decision about formal versus informal performance in most cases, since both may be considered to some degree even if not formally required. Since doing LSA in PC is still not widely accepted, we do not introduce these tasks until CE. However, the methods discussed in CE could be employed in PC. Since the results of task 205 can provide useful inputs to requirements documents, its performance is discussed in this section.

5.2.4 Individual task execution discussion outline

The general outline used in the individual task discussions for all tasks except those in the 100 series is as follows:

1. Discussion
2. Analysis approach
3. Alternate approaches (if appropriate)
4. Documentation and use
5. Example (if appropriate)

To avoid duplication, frequent reminders are made to reader to review related task material in Chap. 4 if more information is desired on task cost, task tailoring and focusing, or other considerations, such as software.

5.2.5 LSA Strategy/Plan (Tasks 101/102)

Discussion. It is not uncommon to read LSA guidance that states that the ILS manager has developed a strategy once he or she has identified the tasks and subtasks to be required. This is a common practice, but obviously it covers only a portion of the requirements in this task in the 1A standard. This approach also loses the benefits of effective focusing. Additional requirements include identification of the performing organization, the availability of data, and the estimated cost to perform the tasks identified, as well as other data. The requirement in this task in the 1A standard to identify the costs/benefits must be interpreted very broadly. The requirement to identify the potential impact on design can be addressed generally. For example, a new start could be high, a DEMVAL "start" (a program whose first formal phase

TABLE 5.3   LSA Task 101/102 Strategy/Plan

Approach	LSA subtask
1. Identify the overall schedule	102.2.1.d, schedule
2. Assess resources:	
Manpower organization	101.2.1, performing organization
Data availability	
Timing (see 6 below)	
	101.2.1.b, Data availability
3. Develop LSA objectives	101.2.1, supportability objectives
4. Select tasks/subtasks	101.2.1, identify tasks and subtasks
5. Select task analysis method	102.2.1.c, task methods
6. Schedule LSA effort	102.2.1.d, task schedules

is DEMVAL) medium, a FSD "start" low, and an NDI none. A product improvement program (PIP) could be high or low in the area of new design depending on the phase most applicable to the amount of new design needed. The approach suggested and the relationship to standard 1A tasks 101 and 102 requirements is shown in Table 5.3. If you are interested in more information on task cost, task tailoring and focusing, or possible software considerations, please review Secs. 4.7.1 and 4.7.2.

General approach.   The following steps are suggested in developing and executing a "mini" strategy/plan for the PC LSA effort. The performing organization is usually the requirer, since the need is still being defined and a specific concept has not yet been selected. Thus HW contractors are not yet normally under contract.

Identify the overall schedule.   After identifying the type or types of overall program requirements documents to be developed, obtain their development schedule. Plan the LSA effort so that inputs are available to the top-level requirements, as shown in Fig. 5.2. A departure-point manpower-estimating table for the PC phase is presented in Table 5.4. The LSA strategy/plan discusses planned iterations to each draft. This can be done by doing a gross analysis initially using expert opinion input to the simple model used in the next phase.

Assess resources and develop LSA objectives.   Time, personnel, and data and model availability limitations affect the feasibility of LSA objectives. They need to be identified if the objectives selected are to be feasible and realistic.

Select analysis methods and set the LSA task schedule.   After model assessment and selection are completed, an analysis approach (gross, midlevel, complex) needs to be selected for each task and subtask. Hopefully, one or two models will serve all tasks.

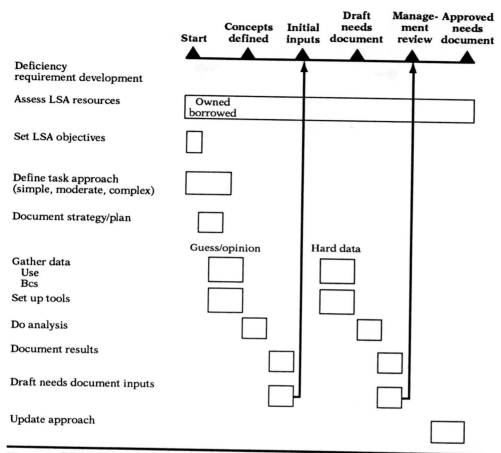

Figure 5.2 LSA Schedule, Preconcept.

Documentation and use of results. The results of the previous steps should be documented and distributed to interested organizations and participants. If new information or events necessitate changes, these should be made and the update distributed.

We suggest development of some standardized forms for each of the LSA tasks in PC for much of the report data. The reasons are twofold. First, use of such forms tends to focus on key points. Second, use of such forms reduces the tendency to prepare lengthy reports that contain lots of information available elsewhere and whose presence tends to obscure key points if, in fact, they were addressed. Departure-point forms for a combined strategy/plan are shown in Figs. 5.3 and 5.4.

Lawn mower example. The XYZ Company has been thinking of developing a new lawn mower design. Its current product is essentially a

TABLE 5.4   Manpower Estimating

	Time in weeks												
	Approach			Program size			Program complexity			Resource multiple (M/T)		Resource	
LSA tasks	Sim	Mod	Com	SMA	Med	LAR	Low	Med	High	Pre-CE	CE	M	T
1	2	3	4	5	6	7	8	9	10	11	12	13	14
101	1.0	1.0	1.0	0.7	1.0	3.0	0.8	1.0	2.0	1	2		
102	0.7	1.0	1.5	0.7	1.0	1.5	0.9	1.0	1.5	N/A	4		
103	1.5	1.0	2.0	0.3	1.0	3.0	0.5	1.0	1.5	N/A	2		
201	0.5	1.0	3.0	0.5	1.0	3.0	0.8	1.0	2.0	3	4		
202	0.5	1.0	3.0	0.5	1.0	3.0	0.8	1.0	2.0	3	4		
203	0.5	1.0	3.0	0.7	1.0	3.0	0.8	1.0	1.5	5	4		
204	0.6	1.0	3.0	0.5	1.0	2.0	0.5	1.0	2.0	N/A	4		
205	0.5	1.0	2.0	0.6	1.0	2.0	0.8	1.0	1.5	3	4		
301	0.6	1.0	2.0	0.5	1.0	2.0	0.6	1.0	2.0	N/A	4		
302	0.8	1.0	1.5	0.5	1.0	2.0	0.8	1.0	1.5	N/A	3		
303	0.5	1.0	3.0	0.5	1.0	2.0	0.8	1.0	1.5	N/A	3		
501	0.5	1.0	3.0	0.8	1.0	1.5	0.8	1.0	1.5	N/A	4		

Notes: Manpower (13) = approach × size × complexity × resource multiplier times (14) = manpower (13)/persons available: $t_R = MW/M$.

---

1. System or Project Name _____ Phase _____
   LSA point of contact _____ Phone _____
   System description:

2. Support-related opportunities requirements:
   Design:
   Support:
   Acquisition strategy:

3. Phase schedule: Start: _____ Draft: _____ Final: _____

4. Constraints:
   Time:
   Personnel:
   Data:
   Policy:

5. Critical issues for this phase:

6. LSA objectives:

   Overall   This phase

7. Selected approach overview:
   Sheet number: _____
   Narrative:

Figure 5.3   Task 101 Summary Sheet.

1. Approach number:
   Approach overview:

   | | Resource estimate (weeks) | | |
2. Task approaches:	ILS	Other Depts.	Time
Task 101			
Task 102			
Task 103			
Task 201			
Task 202			
Task 203			
Task 204			
Task 205			
Task 301			
Task 302			
Task 303			
Task 501			

   Total resources _____

3. Approach analysis:
   Objective achievement:
   Resource shortfalls:
   Data availability:
   Modeling support:
   Risk:
   Observations and recommendations:

Figure 5.4   Task 101 Data Sheet.

10-year-old design that is now losing market share to competitors. At the same time, XYZ has marketing information that the government is planning a project to develop a new mower because its current fleet is aging and is composed of mowers from many different manufacturers.

The XYZ Company decides to pursue both projects under one program management approach because it believes that high quality and durability are of increasing importance in its commercial development. Because of the emphasis on durability and quality, the XYZ Company program manager requests the company ILS manager to support the project by preparing inputs for the needs document. An initial draft requirements document is to be completed in 1 month. A final coordination draft is to be released in no later than 6 months.

The ILS manager decides that he needs a strategy/plan, a use study, a comparative analysis, and task 205. He eliminates field visits from the use study and uses expert opinion, hard data, etc. He reviews the task forms in this chapter for commercial application. He decides that he can use some directly, but that others require minor or major modifications. The forms in Figs. 5.3 and 5.4 will be used for task 101. Since these forms are keyed to the front-end cost analysis (FECA) model in Appendix 26, the ILS manager plans to modify this model to match the altered forms for use on the project. After looking at the schedule for needs documents, he decides to do a gross analysis based on expert opinion input to his model to develop input to the first draft and to refine this effort by obtaining further information and data for the input to the final draft. He schedules three half-day sessions with team members for the gross analysis. He requests team members from marketing, design, R&M, maintenance, supply, and pricing. He is careful to schedule the tasks so they have overlap as much as possible to maximize task interaction. After estimating the help he needs from other company groups, he puts the preceding information into a draft, coordinates it with the other organizations, and obtains the project manager's approval.

5.2.6 Use Study (Task 201)

Discussion.   The use study is of interest to other communities such as system engineering, design, and R&M, as well as the ILS elements. Parts of the use study may be done by other communities (R&M, HARDMAN, etc.). Table 5.5 matches the steps in the approach with the subtasks in 1A standard. If you desire more information on task cost, task tailoring, or software considerations, please review Sec. 4.3.1.

Analysis approach

Determine the range of options.   A use study is needed for each significantly different option. Discussions with the concept developers and technologists can help identify these options. Different options may have different uses. We did not include field visits in the approach in PC because we assume that the requirer and the executor are the same and have in-depth field knowledge. In later phases as HW contractors become involved, field visits may be desirable. However, most defense contractors have personnel with operating experience. They also may have field engineers on similar items who could be valuable sources of information on such factors as skill-level availability, and onsite Support and Test Equipment (S&TE) and handling equipment, and so on, available in operating organizations. The use study can be tailored by the level of detail and the data sources used (expert opinion, hard data, etc.).

TABLE 5.5   Task 201, Use Study

Approach	Subtasks
1. Determine the range of options	
2. Document system-level use	201.2.1, supportability factors
	201.2.2, quantitative factors
4. Define system functions and calculate annual use	201.2.1/201.2.2
5. Document	201.2.4, report

Document system-level and environmental data. *System-level use* is the use of the entire system. Examples of the data needed are listed in Sec. 4.3.1. In PC it may be necessary to create your own numbers. Environmental conditions provide a basis for adjusting failure rates and repair times and may be available as an input to data.

Define system functions and calculate annual usage. Functions should be defined as generically as possible to permit their use on more than one option. The use of each function should be defined for each mission. On large systems with different missions, some functions may not apply to all missions. Variations in basing modes also may need to be considered. You should be aware that one technology area's system is another's subsystem. A fire-control system on an aircraft or a tank is considered a subsystem. On a ship it might be called a system. The level-of-use information needed depends on the specific circumstances at hand. Sensitivity analysis, using the model selected, may be desirable when the source of the numbers is considered "soft." Use the formats of higher-level documents for similar data if possible.

Documentation and use of results. Documentation should follow the philosophy suggested in the strategy/plan discussion. While some narrative is necessary, standard worksheets are again desirable. A summary worksheet is shown in Fig. 5.5. Although it is defense-oriented, it can be modified for commercial use. Additional worksheets can be used to summarize peacetime and wartime use by mission and each function to determine annual use (such as hours, rounds of ammunition, etc.) or some other appropriate measure. Worksheets should be prepared for each significant option. A departure-point example is shown in Fig. 5.6.

The documented report provides data on supportability factors, constraints, and possible issues/opportunities. It should be distributed to interested requirer and executor communities in this phase and the next and used as

1. A basis for executing the other 1A standard tasks.
2. An input to higher-level requirements documents.
3. Input to the strategy for the next phase.

Lawn mower example. After talking with the design manager, the XYZ Company ILS manager decided that there were two possible options. The first was a mower using the normal blade. The second was use of the whirling string or nylon rope concept currently employed in many trimmers and weeders. At the work group's first

1. System or project: _____ Phase: _____
   LSA point of contact: _____ Phone: _____

   System description:

2. Use study approach:

3. System information:

System life (years)	_____ ☐		Surge rate	_____ ☐	
Quantity acquired	_____ ☐		Organization	_____ ☐	
Deployment (peace)	_____ ☐		Qty./org.	_____ ☐	
Deployment (war)	_____ ☐		Allow downtime	_____ ☐	
Threat level	_____ ☐		Operational avail.	_____ ☐	
			Remoteness of maint.	_____ ☐	

4. Climatic conditions:

Climatic conditions	% Fleet operations	% Fleet maintenance
Hot Basic Cold Severe		

5. System functions: ☐

Function	Narrative
1	
2	
3	

6. Narrative:

7. Data sources: 1. _____   4. _____
                     2. _____   5. _____
                     3. _____   6. _____

Figure 5.5   Task 201 Summary Sheet.

meeting, it quickly became apparent that the marketing representative could supply significant data. For example, he already had
data on industry-wide sales by total and by company. He did not
have (but could get) data by geographic location. The group decided
to examine such data when it became available to see if climatic
conditions by region provided any insights. The group also thought
that knowing the percentage of male, female, and teenage users
would help to determine how desirable a self-propelled feature
would be to function at the stage of "problem lawns," i.e., one system precludes the use of riding mowers. The group decided that
they could get some rough estimates by surveying other plant em-
ployees and neighbors. The marketing representative believed that

□ Peacetime   □ Wartime

Part 1   System option:

Part 2   System use:

Mission	(a) Operating time	(b) OT plus alert time	(c) Calendar time	(d) Number of missions	(a) × (d) = (e) Total OT	(b) × (d) = (f) Total OT + AT	(c) × (d) = (g) Total CT
Total	XXX	XXX	XXX				

Data source: _____

Part 3   System use by function for single mission

Mission	Function OP time	Function OP time	Function OP time	Function OP time	Rounds per mission		

Data source: _____

Part 4   Annual operating hours by function

Mission	Function	Function	Function	Function	Rounds per year	
Annual function usage						

Mobility

% Primary	
% Secondary	
% Cross country	

Data source: _____

Figure 5.6   Task 201 Use Record.

1. System or project: __*New Mower*__   Phase: __*Concept*__
   LSA point of contact: _____   Phone: _____

   System description: *Option 1*

2. Use study approach:

3. System information:

   System life (years)   _____ ☐
   Quantity production   _____ ☐
   Operational avail   _____ ☐

4. Climatic conditions (percent of population):   ☐

Hot	40
Moderate	50
Cold	10

5. System use: *Option 1*

Mission	Operating time	Prep. time	Number of missions	Total time
Cut grass	90	5	26	2470
Blow leaves	120	5	4	500
Total				2970

6. Operating time by function: ☐

Mission	Start	Cut	Unload	Stop
Cut grass	3	80	6	1
Blow leaves	3	108	8	1

7. Data sources: 1. _____   4. _____
                  2. _____   5. _____
                  3. _____   6. _____

Annual function usage				

Figure 5.7   Lawn Mower Use Record.

he could obtain better numbers from the trade association, but this would require several months. The group also needed data on how often per year a mower was used and how long each individual use took. Operation included start, run, unload, and stop. With this information and the data the marketing manager had on competitor prices, the group was able to draft a use study based on gross estimates. These data were recorded on the simplified use study form they developed (see Fig. 5.7).

TABLE 5.6   Task 203, Comparative Analysis

Approach	Subtasks
1. Allocate new system performance to functions	203.2.3, comparison system characteristics
2. Identify candidates and select baseline comparison (BCS) items	203.2.1, identify comparatives sytems; 203.2.2, BCS
3. Collect data and identify BCS characteristics	203.2.3, comparative system characteristics
4. Adjust BCS system (ABCS)	203.3
5. Identify ABCS drivers	203.5/6, Quantitative/qualitative drivers
6. Identify risks and document	203.8, Risks

5.2.7   Comparative Analysis (Task 203)

Discussion.   Other communities may do comparative analysis. Examples are R&M and HARDMAN/MANPRINT. They should be contacted to ensure that no duplicate effort is undertaken. In fact, the requirer may provide some HARDMAN/MANPRINT data. If so, use it. Placing values on the desired parameters and comparing them with the same parameters for existing systems provide a basis for various projections and adjustments that need to be made to support the risk analysis (see below). Table 5.6 summarizes the approach and the related subtasks. A summary example of a comparative analysis on a major system is presented in Appendix 18. If you desire more information on task cost, task tailoring and focusing, or software considerations, please review Sec. 4.3.3.

Allocate any known system performance to functions for the systems.   Performance parameters may vary with function. Move functions include speed, range, acceleration, etc. Shoot functions could include rate of fire, range, accuracy, etc. The concept developers may have such data. If so, use it. Placing values on the desired parameters and comparing them with the same parameters for existing systems provide a basis for various projections and adjustments that need to be made to set new system goals. Such an approach also supports risk analysis. Figure 5.8 is a departure-point form for allocation of requirements to subsystem/function. Possible diagnostic parameters are presented in table form in Appendix 22.

Baseline system.   Identify baseline-system candidates and collect data. Do not forget any standardization candidates identified. In turn, baseline comparison system (BCS) candidates may be standardization candidates. Maintenance failures data systems and Visibility And Management of O&S Costs (VAMOSC) data are a likely source. Acquisition O&S cost estimates and specification values of comparison systems may not be as realistic as field data but are better than nothing. As one author[3] suggests, round up all the data you can think of to per-

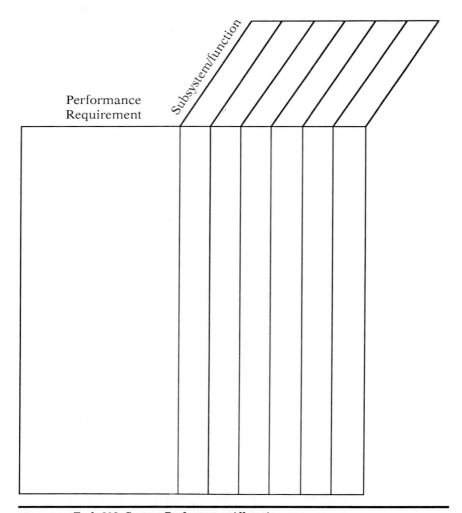

Figure 5.8 Task 203, System Performance Allocation.

form this task. However, ingenious steps may be needed to put the data into a useful form.

Adjust the BCS data. This is useful because it makes the BCS data more representative of the new system by making adjustments for differences in use, environment, technology, etc. An example of why such adjustments may be desirable is shown in Fig. 5.9. The figure indicates the variation in logistics data that exists for identical items installed in different systems. The figure contains Mean Time Between

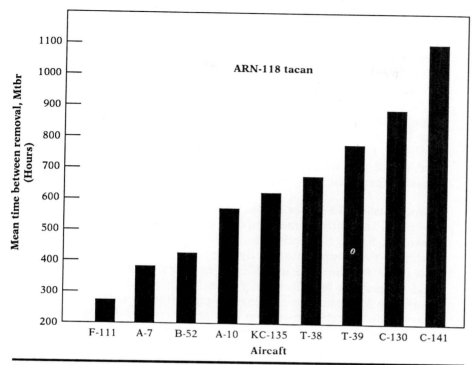

Figure 5.9   Removal Experience for Various Aircraft.

Repair (MTBR) data for the AN/ARN-188 TACAN installed in different aircraft. While the MTBR for the TACAN on the basis of all the data is 600 to 700 hours, the results are very different in some of the aircraft. The analyst must consider the factors that cause this variation in data and address them when estimating BCS parameters. Examples of such factors are (1) the environment (temperature, vibration) at the TACAN location, (2) installation within the crew compartment or in the avionics bay, (3) mission duration, (4) alternate means of terminal navigation, and (5) single or redundant installation. Some analysts feel that these adjustments can be made as part of task 205, which also discusses goal adjustment.

Identify the support drivers.   Identify the support drivers for the ABCS by inputting the ABCS system data into the model selected. The data can then be reviewed for each BCS/ABCS item and discussed with appropriate personnel. It is important to identify what attributes cause the driver. Use Pareto's principle—80 percent of the dependent variable x is caused by 20 percent of the independent variable y. The model also can be used to highlight the drivers that are most sensitive to variation and therefore require the most management attention.

Identify the risks in the BCS and ABCS data.   Different sources of data have different risks in terms of validity and accuracy. Most of the standard maintenance data systems in the United States are questioned in terms of accuracy because of the marginal motivation of the field preparer to fill out forms properly. Engineering estimates may have even less credibility. However, it is necessary to work with what is available. There is also risk in BCS selection. If the comparison is poor, use of the item decreases credibility, although this can be compensated for partially by adjustments from the BCS to the ABCS.

Documentation and use of results.   The documentation should contain the usual narrative and employ standard data sheets. A BCS worksheet for each function should be used. Another data sheet can document the Representative System (RS) data. However, since the data items for the BCS and RS are essentially the same, one sheet could be used for both to readily identify differences in the BCS and RS. Figure 5.10 contains a departure-point form of this type. The reasons for the adjustments to the RS data should be documented on a sheet addendum form in the narrative. Another worksheet useful to document driver/problem data is shown in Fig. 5.11. A summary worksheet also can be used for each function, driver, risk, etc. See Fig. 5.12 for a departure-point form.

The results are used in a similar fashion as the Use Study results. This includes

1. Input to task 205.

2. Input to higher-level requirements documents.

3. Distribution to interested communities (this phase and the next).

4. Focus and refinement of the LSA strategy for the next phase, including tailoring the SOW if HW contractors will be involved.

Lawn mower example.   At the second meeting, the XYZ Company work group discussed the comparative analysis to be made. They quickly decided they could use the forms in Figs. 5.10 and 5.12 with little modification. Item 5 on the form in Fig. 5.12 was modified by deleting "Manpower" and changing "Support costs" to "Support costs surrogate." The group could estimate many of the support costs by surrogates, parametrics, or expert opinion. Since the group had data on unique parts and assemblies delivered to their dealers, it could approximate spares costs. Labor costs at shop and factory were approximated by using labor-hour estimates multiplied by the known fre-

Figure 5.10 BCS/RS Function/Subsystem Data Sheet

1. System or Project: _____ Phase: _____
   LSA point of contact: _____ Phone: _____
   System description:

2. Function/subsystem title:
   Function/subsystem description:

3. Function/subsystem performance criteria:

4. Candidate BCS item name:

5.

	BCS factor	Data source	Adjusted BCS factor	Data source
5a. Application data				
Use rate:				
Use equipment:		☐		☐
Quantity produced:		☐		☐
Notes:		☐		☐
		☐		☐
5b. System/equipment data				
Crew size (operator)				
% OP crew appl. to func. (.1-1)		☐		☐
Unit cost ($M)		☐		☐
1000 kg unit wgt., metric tons		☐		☐
Number of subassemblies		☐		☐
MTBF (unscheduled) (hrs)		☐		☐
MTBM (SCH) (HR)		☐		☐
MTBM (scheduled) (hrs)		☐		☐
Organizational		☐		☐
Intermediate		☐		☐
Depot		☐		☐
MTTR (unscheduled) hrs on equip.		☐		☐
MTTR (unscheduled) hrs off equip.		☐		☐
Diagnostic effectiveness (%)		☐		☐
Equip. complexity (1–3.3 = complex)		☐		☐
% Scheduled maint. performed at:		☐		☐
Organizational		☐		☐
Intermediate		☐		☐
Depot		☐		☐
% Unscheduled maint. repaired at:		☐		☐
Organizational		☐		☐
Intermediate		☐		☐
Depot		☐		☐
Crew size (maint.)		☐		☐
Training locations, OPS		☐		☐
Training weeks, OPS		☐		☐
SE complexity (1–3.3 = complex)		☐		☐
Lines of code		☐		☐
Software complexity		☐		☐
Qual. of SW doc. (1–3.1 = poor)		☐		☐
SW stability (1–5.0 = no SW)		☐		☐
Pol. consumption (kg/hr)		☐		☐
Consumable value/unit		☐		☐
kg/unit consumable		☐		☐
Expenditure rate (unts/evnt)		☐		☐
Events/operating hour		☐		☐
% NDI		☐		☐

(Continued)

171

Figure 5.10   BCS/RS Function/Subsystem Data Sheet (*Continued*)

5. (cont.)	BCS factor	Data source	Adjusted BCS factor	Data source
5b. System/equipment data (cont.)				
Bat. dam. suscep. index (1–10)		☐		☐
Bat. dam. rep'r. diff. index (1–10)		☐		☐
% Embedded training		☐		☐
A_0 objective		☐		☐
5c. O&S data [operating unit/yr ($M)]				
Operator hours		☐		☐
Maintenance hours		☐		☐
Maint. man/hrs		☐		☐
Replen. spares/repair parts		☐		☐
Training operators		☐		☐
Training maintenance		☐		☐
Training equipment supp.		☐		☐
Support equipment		☐		☐
Support equipment supp.		☐		☐
POL		☐		☐
Consumables (Ch. VII, IX)		☐		☐
Transportation		☐		☐

6. Data sources: 1. _____ 4. _____
              2. _____ 5. _____
              3. _____ 6. _____

quency of assembly/part replacement. Fuel costs and the costs of owner manuals could be similarly estimated.

The engineering representative presented his draft of the form in Fig. 5.8, and this is shown in Fig. 5.13. With the new O&S cost factors and the performance requirements in hand, the group was able to make major substitutions and simplifications in blocks 5b and 5c of Fig. 5.10.

5.2.8 Supportability and Supportability-Related Design Factors (Task 205)

Discussion.   A key objective here is to set design objectives and constraints for the next phase. A possible useful distinction here is made in Ref. 32. The distinction is between (1) design-related parameters, (2) logistic constraints, and (3) logistics issues/opportunities. Design-related parameters are directly related to the new system envisioned (R&M, Built-In Test (BIT), etc.). Constraints are boundary conditions caused by external factors (such as cargo plane limitations, Intermediate (I)-level maintenance capabilities, etc.). Some constraints (hours per day, etc.) are impossible. Others (hours in a day a person works) may simply be hard to change.

Item name: _____

Subsystem/function: _____

1. Performance:

2. Technology involved: • Developing   • Mature

3. Support problems/constraints:

4. Logistics resource drivers, BCS item

5. Relationship to new system:

6. Risks in using data:

Figure 5.11   BCS/RS Driver/Problem Data Sheet.

Logistics issues can flow from both parameters and constraints. Much of the work from the other tasks can be consolidated directly into the task 205 report. This includes

1. Support constraints and/or risks from task 201
2. Constraints or issues of a standardization nature from task 202
3. Equipment parameters, issues, and/or risks from task 203

1. System or project: _____ Phase: _____
   LSA point of contact: _____ Phone: _____

2. System description:

3. Task approach:

4. Function/subsystem:
   a.
   b.
   c.

5. Quantitative logistics drivers
   a. Support costs
   b. Manpower
   c. Operational availability

6. Qualitative drivers/problems/constraints

7. Risks
   a. BCS data
   b. RS data

Figure 5.12   Task 203 Summary Data Sheet.

The same analysis model employed in the other tasks should be used if possible. Table 5.7 shows the relationship to the approach and the 1A standard subtasks involved. If you desire more information on task cost, task tailoring and focusing, or software considerations, please review Sec. 4.3.5.

Analysis approach.   The first action is to review the results of the other tasks. Some direct flow into the task 205 report can occur here, as shown in Table 5.7.

Establish support-factor goals.   This involves use of the model to examine each equipment factor for each function to determine its support impact. This is particularly important for factors identified as drivers in task 203. Sensitivity analysis and expert opinion can be used to review each factor and select goals for each. The system readiness parameters, such as A_0, of the system should be checked. System readiness analysis methods that contain factors not under the control of the contractor, such as mean logistics delay time (MLDT), should not be contractual obligations. Some adjustments between factors may be necessary. Once the final set of goals is determined, the model probably can be used to project an initial O&S cost and A_0 or other appropriate measures. A departure-point worksheet for this analysis is presented in Fig. 5.14.

Requirement	Subsystem/function	Pre-ops	Start	Cut/run	Unload	Stop	PM(SCH)	CM(UNSCH)	Other
1. Power				5HP					
2. Ave fuel consumption				1PT/hr					
3. Cutting power (RPMS)				750					
4. Rollability				9					
5. Steering				M					
6. Electrical		12VDC			12VDC				
7. Weight									40 lbs
8. Maintainability							1/2hr		
9. Reliability MTBF									200hr

10. Diagnostics
  Gas
  Oil
  Battery
	2 qts
	2pts
	12VDC								

11. Catching capacity					2 cu/ft				
12. Ao									.98

* = Ease of movement: 0-10; 10 Excellent
M = Manual

Figure 5.13 Task 201 Lawn Mower Sheet.

TABLE 5.7   Task 205, Design Factors

Approach	Subtasks
1. Review the result of other tasks 2. Establish support factor goals 3. Analyze constraints 4. Analyze issues 5. Analyze risks 6. Document	205.2.1, characteristics 205.2.2, sensitivities 205.2.5, constraints; 205.2.6, NATO

1. System or project name _____ Phase _____
   LSA point of contact _____ Phone _____

2. Representative system factors

Factor	Subsystem/ function 1	Subsystem/ function 2	Subsystem/ function 3	Subsystem/ function 4	Subsystem/ function 5
Name					
Unit cost					
Unit wgt					
No. LRUs					
MTBPM					
MTBF					
MTTR/on					
MTTR/off					
Diag. effect.					
SE complexity					
Lines of code					
SW complexity					
Qual. of SW doc.					
Bat. damage index					
Sustainability					
Repair difficulty					

3. Summary analysis

Factor evaluated	Range	Results	Recommended objective

Figure 5.14   Task 205 Supportability Factor Evaluation Sheet.

Analyze the constraints. Constraints may arise in the preceding step. If these are potentially movable, estimate the impact and develop opportunities to remove or alleviate the impact. Estimate the impact of the opportunities if the constraint is changed by exercising the model in a "what if" analysis. Make recommendations as appropriate.

Issue analysis. Issue analysis is similar to the analysis of constraints. The list of drivers from task 203 can be a major source of issues. Constraints and issues may involve a topic (I-level maintenance) rather than a specific system factor. In some cases, the model may not be able to handle the issue, and expert opinion must be used.

Risk analysis. The objective here is to identify and analyze the most significant support risks and determine if they must be examined in the next phase or held until later. For example, the impact of a new technology not achieving its reliability expectations can be estimated using the model. It is also desirable to assess the probability that the risk will occur. Focus on the high-impact, high-probability-of-occurrence items. The expected-value concept (impact, if it occurs, times the probability of occurrence) can be used to rank the individual risks. Actions that can be taken in the next phase to reduce risks should be developed for risks with high expected value.

Documentation and use of results. The final report should take the same approach as used in the other tasks. While separate worksheets for factors, constraints, issues, and risks can be developed, it is possible to use a combined worksheet. See Fig. 5.15 for a departure-point form. A departure-point summary sheet with the recommended factors, constraints, issues, and risks is presented in Fig. 5.16.

The results of this task provide the primary input to the higher-level requirements documents. In later phases this could include both government and contractor documents. The report also should be distributed to the interested communities in both this phase and the next and should be used to tailor the LSA strategy and focus the LSA tasks and subtasks for the next phase.

☐ Constraint   ☐ Issue   ☐ Risk

1. System or Project Name _____ Phase _____
   LSA point of contact _____
   Phone _____

2. Constraint/issue/risk _____

3. ILS element(s) impacted _____

4. Estimated impact
   A. O&S cost _____
   B. Readiness _____
   C. Supportability _____
   D. Other _____

5. Risk level (if appropriate) _____

6. Improvement opportunities (if appropriate) _____

7. Narrative (as appropriate) _____

8. Readiness action _____

Figure 5.15   Task 205 Data Sheet.

1. System or Program: _____ Phase: _____
   LSA point of contact: _____ Phone: _____

   System description:

2. Task 205 approach:

3. Supportability and Supportability-related design factor objectives:

4. Logistic constraints:

5. Supportability issues:

6. Supportability risks:

Figure 5.16   Task 205 Summary Data Sheet.

Lawn mower example.   The XYZ Company work group then addressed task 205. They decided they could use Figs. 5.14 through 5.16 with minor modifications. The performance factors and allocations developed for Fig. 5.8 were substituted for the factors in block 2 of Fig. 5.14. Factors to be evaluated in block 3 of this form were determined to be MTBF, MTTR, A_0, and ease of movement (rollability index). Ease of movement is a function of weight and the quality of the wheel subsystem. The engineering department had parametric assessments on this tradeoff for use. Since durability is a primary goal, various R&M and A_0 tradeoffs were explored. Another goal was to reduce mean logistics delay time, since this is usually a driver of A_0. Further tradeoffs and testing in this regard were planned.

Only minor changes were made to the forms in Figs. 5.15 and 5.16. "Production cost" was added to block 4 of Fig. 5.15. In addition, "Surrogate" was added to line 4a and "A_0" was substituted in line 4b in both Figs. 5.15 and 5.16. A possible constraint to the general approach planned was whether dealer shops doing local repair would agree to stock key unique assemblies and parts to the desired A_0 level. A second issue in this regard was the dealer receptiveness to increasing the range of items stocked if a design-to-discard approach was used. Marketing was requested to contact dealers to determine their actions to these new approaches. The impacts of the concepts were to be evaluated in the next cycle of the project.

5.2.9   Preconcept summary

This section has described a specific approach to the performance of LSA prior to concept exploration. The primary objective of this effort is to influence requirements from a support viewpoint. A secondary benefit is to provide data to help tailor and focus the LSA effort in the next phase. The keys to cost-effective accomplishment are

1. Limited detail.
2. Team approach.
3. Use of simple analysis models to quantify results and improve credibility.
4. Availability of analysis in a timely manner to increase the likelihood of influencing higher-level requirements documents.

5.3   Concept Exploration (CE)

5.3.1   Introduction

Various options to meet a stated need are explored during the concept exploration (CE) phase. Initially, design is conceptual. The end point is one or more system specifications that define the concept(s) to be evaluated during the next phase. The primary objective of LSA in this phase is to influence the selection of the final concepts/systems from a support and readiness viewpoint. CE probably provides the largest LSA opportunity for major influence of the design. Some LSA analysis begins to be performed at the ILS-element level and is integrated at the system level in this and later phases. As CE progresses, early ILS-element resource estimates can be refined. The general principles cited in Secs. 3.1.2 and 5.2.1 are again applicable to the performance of LSA in this phase.

5.3.2   Tasks in this phase

The following tasks, focused as appropriate, are frequently performed in this phase and constitute a mix of tasks previously performed and some tasks that are formally introduced in CE:

1. Updates to and extensions of the tasks performed in preconcept exploration [LSA Strategy, Use Study, Standardization (optional), Comparative Analysis, Technological Opportunities (optional), and Goals and Thresholds]
2. LSA Plan (task 102)
3. Standardization (task 202)

4. Technological Opportunities (task 204)

5. Functional Requirements Identification (task 301)

6. Support System Alternatives (task 302)

7. Evaluation of Alternatives and Tradeoff Analysis (task 303)

8. Test, Evaluation, and Verification (task 501)

Some of the aspects of the new tasks (202 and 204) were considered informally in Sec. 5.3.

The work may all be done internally by the requirer and/or support contractor. A second possibility is for the requirer to act as a system integrator but contract with various hardware suppliers for detailed studies of the various competing concepts. The third possibility is performance of all the work by one or more hardware (HW) contractors. If there is more than one hardware contractor, the issue of duplication may arise. One way to reduce such duplication is for the requirer to have one contractor perform selected tasks for all the other contractors. The issuing of performing organization in such circumstances was discussed in greater length in Chap. 3. Regardless of the exact organizational arrangements, the system-level LSA effort must be updated as appropriate and the LSA effort extended to subsystem level generally and below subsystem on a selective basis.

Timing is the key to successful LSA in concept formulation. If a number of contracts are to be let to HW contractors in this phase, development of LSA contract requirements is an early decision point. No matter which approach is taken, the LSA effort should again be planned to provide results to the various high-level requirements documents that are to be influenced.

5.3.3   LSA Plan (Task 102)

Discussion. The "mini" plan/strategy developed in the PC phase should be updated. Plan requirements in 1A standard should be tailored to fit the activity in this phase. For example, little, if any, LSAR data are needed in this phase. The plan should provide for both system updates and extensions of the analysis to lower levels (at least to subsystem level and selectively below). The LSA strategy developed near the end of PC and other PC results should be used to tailor and focus the LSA effort planned for this phase (see Table 5.8). If you want information on possible LSA plan requirements and costs, please review Sec. 4.7.2.

The full plan as described in 1A standard is normally invoked in DEMVAL and EMD. The plan for the production phase will normally focus largely on updates for engineering change proposals (ECPs) and the LSAR.

TABLE 5.8 Task 102, LSA Plan

Approach	Subtasks
1. Review the LSA strategy and the PC LSA activity	102.2.1c; what and how for each task
2. Plan and schedule the LSA	102.1d; schedule other selected requirements in
	102.2.1a through o
3. Document and distribute the results	

Plan approach

Review existing material. Review the LSA strategy and PC LSA activity, including issues and risks, and define the LSA work desired during this phase. Schedule the LSA activity (this could be both government and HW contractors), and examine the interfaces within the LSA and related external activity (e.g., the master schedule, program or design reviews, R&M activity) for compatibility. The plan should incorporate, not duplicate, other analyses to be cost-effective. LSA task execution should again overlap significantly.

Document, coordinate, and distribute the plan. This is self-explanatory. However, note that there may be two sets of plans, one for the government (or system integrator) at the system level of activity and one for each of the HW contractors.

Execute the plan. Update as necessary.

Lawn mower example. The XYZ Company still planned to pursue a commercial and a government version of their new mower. The ILS manager decided to do the tasks cited previously for this phase and laid out plans and schedules comparable with those in the preconcept "mini" strategy/plan. He looked at the LSA plan requirements in the LSA standard (task 102.2.1) and decided to concentrate on the requirements in subsections b, c, and d. He was careful to specify the analysis models to be employed and a summary of the approach to each task in the plan.

5.3.4 Program and Design Reviews (Task 103)

Discussion and approach. Requirer as well as internal and subcontractor reviews of the status of the program, design, etc. are a key management activity in any phase of a program. The approach to incorporating LSA into such reviews varies only in the nature of the specific review and the appropriate amount of LSA to incorporate. This is true internally in both the requirer and executing organizations. The key point is that LSA activity and issues must be included in such reviews for LSA to be a significant contributor. Both the requirer and executor ILS managers should have summary charts of the LSA effort ready for use at short notice. The other key point is that LSA reviews, to the degree possible, should be scheduled to coincide with other reviews. The candidate data to include in such reviews are the obvious ones, such as

1. Overall status of the effort by task
2. Status of individual task accomplishment, including objectives, planned activity, progress, and any problems
3. Any major issues that require management and resolution
4. System alternatives under consideration
5. Support alternatives under consideration
6. Supportability assessment of proposed designs
7. Evaluation and tradeoff results
8. Comparative analysis results
9. Review of supportability design requirements
10. Schedule or analysis problems affecting supportability

Lawn mower example.   The XYZ Company ILS manager initially only had to be concerned with internal reviews. The program manager decided that he would have program reviews once every 2 weeks. The ILS manager ascertained the general format for presentations at these reviews and drafted a sample summary chart and a chart format covering the status of each of the individual tasks (see Figs. 5.17 and 5.18). He obtained the program manager's approval of these formats and planned to present these and the types of information described in task 103, as appropriate, at each review. As the reviews began taking place, he found that the program manager liked to find that any problems identified had been resolved before the reviews took place.

5.3.5   Use Study Update (Task 201)

Discussion and approach.   The use study concepts discussed in Secs. 4.3.1 and 5.2.6 again apply. The analysis obviously should be extended at least one level below that in PC and further on a selective basis. The system-level analysis will probably also need to be updated. Changes that may occur at system level are as follows:

1. *System definition:* This may change. As further detail becomes available, the range of options may decrease, but the level of detail for each option may increase.

```
┌─────────────────────────────────────────────────────────────────────┐
│  Program name:  _____   Date:  _____               │
│  LSA Opr:  _____                                  │
│                                                                       │
│                                                                       │
│                    LSA Task                      Status               │
│                                                                       │
```

	LSA Task	Red	Amber	Green
101	LSA Strategy			✓
102	LSA Plan		✓	
103	Program and design reviews			✓
201	Use study		✓	
202	Standardization		✓	
203	Comparative analysis		✓	
204	Technological opportunities			
205	Supportability and supportability-related design factors		✓	
301	Functional requirements identification			
302	Support system alternatives			
303	Evaluation of alternatives and tradeoff analysis			
501	Supportability test, evaluation, and verification			

Critical Issues:

Figure 5.17   LSA Status Overview.

2. *Data:* New and more accurate data will become available. This can cause interaction in both previous and ongoing work in other tasks. Changes in use study data will affect the results of other tasks.

3. *New or different issues:*   These may arise from a variety of sources. As the analysis descends to another level or as system definition changes or expands and new data are obtained, the analysis changes and new issues probably surface.

Program: _____ Date: _____

LSA task No: _____ Manager: _____

1. Task objective:

2. Planned activity:

3. Progress:

4. Problems and planned resolution:

5. Help needed:

Figure 5.18   LSA Task Status Report.

However, the general approach used in the PC work is applicable to the analysis approach used in this phase. If the comparative analysis results are available, BCS items can be reviewed to identify any support problems as part of the use study update.

The worksheets and documentation used in the preconcept work are still applicable. The use of the results is also similar, although the names of the requirements documents probably will change.

5.3.6   Standardization (Task 202)

Discussion.   The benefits of intelligent standardization and use of standards are without question. Table 5.9 illustrates the suggested approach and related 1A standard subtasks. If you desire more information on task cost, task tailoring and focusing, or software considerations, please review Sec. 4.3.2.

Analysis approach.   The key to doing this task effectively in CE is to restrict the scope and level of the effort. Two ways to do so are to focus

TABLE 5.9   Task 202, Standardization

Approach	Subtasks
1. Identify and list candidates	202.2.1, support constraints, and 202.2.3, mission system candidates
2. Roughly rank the candidates	
3. Evaluate impact (including risk) of higher-ranked candidates	202.2.2, estimate impact
4. Develop recommended actions and document results	

(1) on ILS possibilities and (2) on higher-level candidates appropriate to the phase (e.g., fasteners are normally not really a burning issue in CE). As candidates are identified, list them and identify the source of the suggestion. The use study may provide candidates, particularly if field visits were made. The comparative analysis BCS developed also can be a source of possible candidates. Other possible sources of information on candidates are

The DOD Standardization Directory

The Army Adopted/Other Items Selected for Authorization/List of Reportable Items (Army)

Item Manager and Storage Site Assignment by Federal Supply Class (AF)

AFLC Maintenance Engineering Materiel Listing (AF) of Preferred Avionics Components

Army TMDE Register

Army Preferred Items List

Support Equipment (AF)

Avionics Preferred Support Equipment (NAVAIR)

Preferred List of Common Support Equipment (NAVAIR)

NAVAIR Design Selection List of Repairable Assemblies

Then categorize each as

1. Drop.
2. Do further analysis during this phase.
3. Implement.

Those coded 2 or 3 can be ranked judgmentally in order of magnitude of good or bad impact. The candidates can then be evaluated by using the data from the use study, the BCS or ABCS system developed in task 203, and an analysis model such as the one used in PC, combined with expert opinion or engineering estimates. The model also can be used to perform the risk analysis when needed.

The last issue is how far down the list to go. While a fancy economics formula could be set forth, in practice, resource limitations usually provide the answer to this question.

Documentation and use.   This task lends itself to the use of standard forms for documentation. Besides the usual narrative, an individual worksheet can be used for each candidate. A sheet to summarize all candidates is also useful. The same two sheets also can be used to document technological opportunities in task 204. Departure-point forms are presented in Figs. 5.19 and 5.20.

Task 202 results may interact with those of task 203 and serve as an input to task 205. The task 202 results can sometimes be direct inputs to higher-level documents. They also should be inputs to any task 202 execution in the next phase. The analysis methods employed in CE can be applied to later phases, although the level of detail will change. However, new approaches may be desirable to limit the work load. These are discussed in later phases.

Sheet no: _____   ☐ Task 202   ☐ Task 204

1. Opportunity:

2. Analysis aspects of systems influenced:

Function/ subsystem	Factor	Percent change	Acquisition impacts	O&S impacts	Design impact

3. Technology (task 204 only):

4. Problems imposed by opportunity:

5. Problems resolved by opportunity:

6. Potential schedule impact:

7. Comments:

8. Recommendations:

9. Actions required: Action item closed date: mm dd yy

Figure 5.19   Task 202/204 Opportunity Analysis Data Sheet.

☐ Task 202   ☐ Task 204

1. System or project name _____ Phase _____
   LSA point of contact _____ Phone _____

System description:

2. Task approach:

3. System hardware opportunities:

Issue	Sheet	Impacts	Action required

4. System software opportunities:

Issue	Sheet	Impacts	Action required

5. Support system opportunities:

Issue	Sheet	Impacts	Action required

Figure 5.20 Task 202/204 Summary Sheet.

Lawn mower example. The XYZ Company ILS manager reviewed the forms in Figs. 5.19 and 5.20 for possible use. He quickly noted that "System software opportunities" in Fig. 5.20 were not applicable, but he decided he could use the forms as is. At the group meeting, the engineering representative stated that a large number of standardization candidates were available from the company's own and other lawn mower designs. The group decided to do a selective NDI market survey on these candidates at assembly level. To alleviate dealer costs in low-sales areas and encourage owner repair, the group decided to recommend a design with no nonstandard test and support equipment or tools. The group also recommended that standardization be one of the factors in the tradeoff weighting criteria engineering was developing.

The XYZ Company had originally decided to use the Gemini tester as a design constraint. However, study showed that this tester was phasing out of production. However, a new digital tester, which was cheaper and more widely useful, would become available in the immediate future. When a check with the government program manager revealed that he planned to use the new tester, the XYZ Company decided to use it for depot repair of the commercial version being developed.

5.3.7 Comparative Analysis Update (task 203)

Discussion, approach, documentation, and use.   The same approach and documentation used in this task in PC can probably be used. However, if HARDMAN/MANPRINT analysis is required in this phase, it may be desirable to integrate these requirements into the approach. See Appendix 3 for a summary of the HARDMAN methodology. The system level may need to be updated, and the analysis is extended at least one level further than in PC and to lower levels as appropriate on a selective basis. The items to be used to construct the lower levels may change. As in the Use Study, as new information becomes available, the analysis results at all levels may change and new issues may surface. Use of the results is similar to that in PC, although the names and detail in the requirements documents will change.

Lawn mower example.   The XYZ Company decided to consider options for the commercial and the government versions of the new mower. The commercial options include both a gasoline-powered and an electric version. New data were collected on the two approaches. The system-level and subsystem-level analyses were updated to accommodate this decision, and the analysis of both options was extended to one level below the subsystem level in this area.

5.3.8 Technological Opportunities (Task 204)

Discussion.   While this task was not done formally in PC, such considerations were included informally through consultation with technology development personnel in identifying system options for analysis. The CE phase is the ideal time to make such an assessment. In most cases, various specialists have already considered the use of various new technologies when they were assessing the need and the feasibility of a new development. New technologies may be selected to improve performance (e.g., composites) and/or to reduce the support burden (e.g., use of BIT to improve diagnostic effectiveness). Work load can be controlled by the number of opportunities identified and the method of evaluating impact. The approach is very similar to that used in task 202 and is shown in Table 5.10. If you desire more infor-

TABLE 5.10   Task 204, Technological Opportunities

Approach	Subtasks
1. Identify candidate system and support-related technologies	204.2.1a/c, possible technology approaches to improve mission/support systems
2. Determine technology status	
3. Evaluate impact on candidates	204.2.1b, estimate impact
4. Identify risks	
5. Document	

mation on task costs, task tailoring and focusing, or software considerations, please review Sec. 4.3.4.

Analysis approach

Identify system- and support-related technologies.   This should include the technology and the aspects of the new system that it affects. Technologies that address the support drivers identified in task 203 are particularly good targets. Refer to Sec. 4.3.4 for candidate technology areas. There are many other sources of information, since the U.S. DOD maintains a sizable logistic R&D effort that can provide information on support possibilities. These include information on both government and industry technology programs. Review of the annual DOD Bibliography of Logistic Studies and Related Documents yielded 75 studies of interest on a major aircraft program.

Determine the technology status.   The status of the technologies identified in the preceding step needs to be evaluated. Will the technology be available when the new system is produced? If not, can it be phased in at a later date? What cost, performance, and support risks are associated with the technology? Does its use introduce new support problems (e.g., composites have unique repair needs)? These questions on status must be answered as part of this task.

Evaluate the impact of the candidates.   The impacts and risks of using the technology must be evaluated. This can be done through the use of expert opinion. However, if possible, the analysis model(s) selected should be used for this purpose. Risks can be evaluated in the same manner as the standardization candidates by using sensitivity analysis to do "what if" analyses that examine the change in the baseline support burdens and system parameters. A_0 sensitivity also should be checked.

Documentation and use of results.   The results of this task should be documented in a manner similar to that in other tasks. In addition to the usual narrative, the use of standard forms similar to the standardization forms is suggested (see Figs. 5.19 and 5.20).

The results are input into the task 203 update and the task 205 analysis. The other uses described in the other tasks also apply.

Lawn mower example. After deciding to use the data sheets in Figs. 5.19 and 5.20, XYZ Company work group identified a number of technology opportunities of possible use. The first of these was the use of composite material to lighten the weight of the mower. The second was the use of an electrical or electronic starter in lieu of a pull rope. This would remove a major source of irritation to children, women, and older people who frequently have difficulty in using pull ropes. The third was the use of specially hardened cutting blades to reduce the frequency of blade sharpening or replacement. The group noted that some high-quality kitchen knives are now sold with guarantees of free sharpening of the blade for the life. Use of this technology could greatly reduce the O&S costs of mower blades. The group decided to consult with manufacturers in these areas to determine the status of the technology in terms of their use and the projected costs of these versus other approaches.

The composite technology for the mower frame is currently available. It would increase acquisition costs but would provide greater safety protection on motor casing failures. The electronic starter technology also is at hand. However, it would increase acquisition costs slightly. There were no data on O&S costs impact. Further investigation of this was initiated. The hard-blade technology is available in other applications, but its feasibility in this application is unknown. Furthermore, the number of repairs/replacements caused by natural wear and tear versus those caused by "nicks" from hitting obstacles such as rocks is unknown. However, since long-life blades were deemed to be a very attractive feature by users, the group decided to recommend initiation of prototype testing to evaluate application of this technology to mower blades. It also decided to look for data on wear and tear versus damage-type incidents for consideration for possible warrantees and O&S costs impact. This type of information would be necessary for the life-cycle cost analysis the government would need anyway.

5.3.9 Supportability and Supportability-Related Design Factors (Task 205)

Discussion, approach, documentation, and use. This task was discussed in PC. The same general approach is again applicable in this phase. However, the analysis is expanded to include not only updates to tasks 201, 202, and 203 but also the results of tasks 204, 301, 302, and 303, which are first formally initiated in this phase. Interaction between the update and new information from tasks 301, 302, and 303 may require frequent update of the ABCS in task 203 and the work in task 205. As more detail is developed in task 205 in CE and later phases, design guidelines (desirable features) as well as design requirements may well emerge.

5.3.10 Functional Requirements Identification (Task 301)

Discussion. The objective of this task is to identify the operation and maintenance functions that must be performed for each concept option under consideration. This functional analysis provides the basis for developing the alternative support concepts in task 302. The functions identified will be used to define the operator and maintenance tasks to be performed. The work of other communities should be drawn on to the extent possible. System Engineering, the R&M FMECA, the RCM analysis, and the HARDMAN/MANPRINT communities all may have work that is directly applicable.

System engineering assigns mission functions to hardware, software, and personnel. System functions are tied directly to the mission(s) of the system. The interfaces between functions and mission HW/SW ultimately selected may be complex in some cases. Equipments selected to satisfy a function at one level may affect the functions needed below the equipment. The approach is summarized in Table 5.11. If you desire more information on task costs, task tailoring and focusing, or software considerations, please see Sec. 4.4.1.

Analysis approach

Review existing material. The level of detail is usually the same level as in the use study, the comparative analysis, and task 205. The functional requirements identification done by system engineering, HW/SW design, and R&M should be used to reduce the work load and ensure consistency.

TABLE 5.11 Task 301, Functional Requirements

Approach	Subtasks
1. Establish level of detail	202.2.1, support constraints, and 202.2.3, mission system candidates
2. Develop operations, maintenance, and replenishment network	
3. Define functions for step 2	301.2.1
4. Define methods of satisfying the functions	
5. Identify unique/driver functions and risks	301.2.2, unique/driver functions
6. Document	

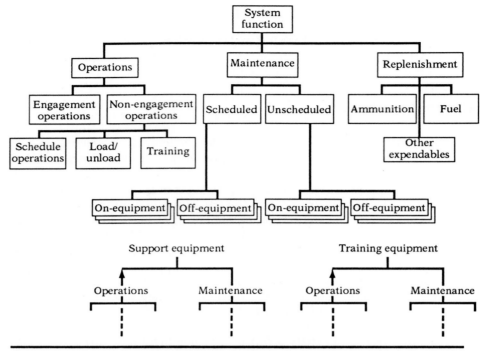

Figure 5.21   Network Flow. (From Army Logistics Management Center, LSA Course, 1986.)

Develop a network and define functions.   This helps to organize and systematize the analysis. This network should include operation, maintenance, and replenishment functions. Figure 5.21 illustrates such a flow diagram. A simple list of functions to be performed should be prepared for each block in the network. For each category of function, develop a list of factors that are significant to the function. For example, skill level, failure frequency, and manhours per failure are possible maintenance factors (Note: check to see if maintainability or MPT has done some of this). In CE, this work usually involves use of gross analysis or expert opinion, since detailed task analysis is much more costly and is usually performed in later phases. These factors are used to prepare estimates of manpower and possibly other support resources for later use. The data can be updated as LSA, HARDMAN, and other related studies are performed.

Identify functions that are unique or are cost drivers and the risks of satisfying these requirements.   Unique functions usually arise from the use of new technology or concepts. The driver analysis should be related to the drivers identified in task 203 to the degree possible. Review all of the functions and related factors. The risk analysis should focus on areas of perceived high risk. Planned use of new and sophisticated Built-in-Test (BIT) is an example of high risk. It again may be necessary to use expert opinion in the initial acquisition phases.

Alternate approaches. There are two commonly used approaches to functional requirements identification. The first is the block-diagram approach. This approach is illustrated in Figs. 5.22 through 5.25. Figure 5.22 depicts a mission accomplishment first-level block diagram for an airborne missile, whereas Figs. 5.23 through 5.25 show functional block diagrams associated with operations and readiness, maintenance of readiness, and repair. Mission essential functions are defined first. The support functions that are necessary to maintain the mission essential functions are then defined. These functions can be used in later phases to develop a requirements allocation chart (see Fig. 5.26).

A second approach is to identify maintenance functions. This approach can be used when the preliminary drawing structure is available. The functional requirements identification indenture level should be consistent with design definition progress. Figure 5.27 is an example of this approach.

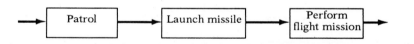

Figure 5.22 FRI Block Diagram: Maintain Readiness. (From Army Logistics Management Center, LSA Course.)

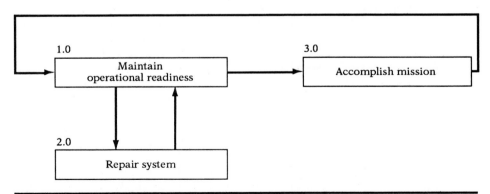

Figure 5.23 FRI Block Diagram: Repair and Readiness. (From Army Logistics Management Center, LSA Course.)

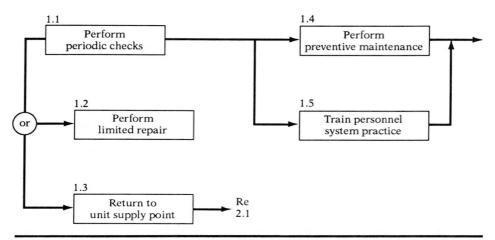

Figure 5.24 FRI Block Diagram: Maintain Readiness. (From Army Logistics Management Center, LSA Course.)

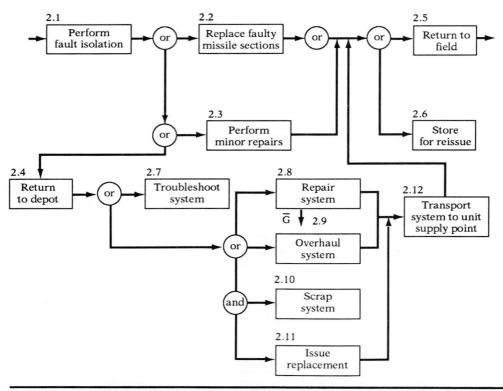

Figure 5.25 FRI Block Diagram: Repair. (From Army Logistics Management Center, LSA Course).

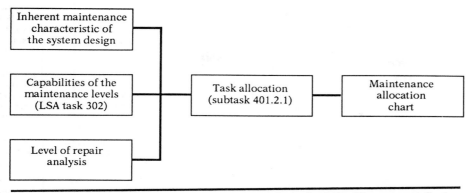

Figure 5.26 Maintenance Allocation Process. (From Army Logistics Management Center, LSA Course).

Subsystem	Component	Assembly	Maintenance actions					
			Inspect	Check	Calibrate	Align	Repair	Replace
Interface subsystem	Video performance		X	X	X	X	X	X
		Surface scan conv.		X	X	X	X	X
		Raw scan conv.		X	X	X	X	X
		Video switch		X	X	X	X	X
		Video mixer		X	X	X	X	X
	Data interface (software only, hardware pro-vided as GFE)			X			X	X
	Control interface (software only, hardware pro-vided as GFE)	Control status maint. subsystem		X			X	X
		Control display group AN/UV-21		X			X	X
		Combat system		X			X	X

Figure 5.27 Functional Requirements Identification.

Documentation and use.   Documentation is similar to that for the other tasks. Beyond the usual narrative and the new block diagrams, standard forms exist for documenting (1) operator functions, (2) maintenance functions, and (3) replenishment functions. See Figs. 5.28 through 5.30 for departure-point forms.

4. Operator functions	1:	2:	3:
A. Skill required			
B. Event			
C. MH/event			
D. Events/yr			
Peace:			
War:			
E. MH/yr			
Peace:			
War:			

1. Project:

2. Concept option:

3. System function/subsystem:

5. Critical/unique skill requirements:

6. Unique training requirements:

7. Risks:

Figure 5.28 Operation Function Data Sheet. (See reference 32.)

As stated previously, this task provides the basis for developing alternate support concepts in task 302.

Lawn mower example. The XYZ Company work group had considered functions of the mower and its subsystems at the top level when it initiated the use study. Now it looked at this subject in greater depth. The normal functions and subsystems of lawn mowers were obviously well known in the industry.

At system level, the functions usually are ship, assemble, prepare for use, operate, maintain, and store. Operation functions break down into start, cut grass, unload, and stop. With this knowledge, the work group examined the forms in Figs. 5.28 through 5.30. The group quickly decided that these forms were unnecessarily complex for their use. They devised a new form, shown in Fig. 5.31, that had the added advantage of addressing the first three subtasks in this task in one form. The R&M and design representatives were asked to collect ex-

1. System:

2. Concept option:

3. System function/subsystem:

4. ☐ Scheduled ☐ Unscheduled
   ☐ Off ☐ On ☐ Off ☐ On

5. Event (function/task)
   A. Frequency:
   B. Skill:
   C. Clock hours:
   D. Crew size:
   E. MH/event:
   F. TMDE requirement:
   G. Facility requirement:
   H. Supply support:
   I. Maintenance level:

6. Critical/unique skill requirements:

7. Unique training requirements:

8. Risks:

9. MTTR:

10. MTTR (max):

Figure 5.29   Maintenance Function Data Sheet. (See reference 32.)

1. Project:

2. Concept option:

3. System function:
   ☐ Ammo ☐ Fuel ☐ Other

4. Functions: Stock   Store   Transport   Load   Unload
   A. Resource determinant:
   B. Skill:
   C. MH:
   D. Facilities:
   E. Support equipment:
   F. Potential problems:

5. Critical/unique skills:

6. Unique training:

7. Risks:

Figure 5.30   Replenishment Function Data Sheet. (See reference 32.)

Functional requirement	Unique	Drivers			Risks
		Supply	Cost	Readiness	

Rev.: _____
Date: _____

Figure 5.31   Model XYZ functional requirements list.

isting data on this subject, prepare the analysis, and enter the data on the new form. These sheets were used as an input to task 302.

5.3.11 Support System Alternatives (Task 302)

Discussion.   Review the discussion of this task in Sec. 4.4.2. Since a large number of concepts are possible, particularly in CE, a matrix of support-concept options for each system option aids the analysis process. Examples of a method of recording such options is presented in Figs. 5.28 and 5.29. The various options are evaluated in subtask 303.2.2. The analysis model already discussed could be used for this purpose. The "plans" required in subtask 302.2.3 are system- and operations-oriented and do not duplicate the various ILS-element plans required in the acquisition process. Such support planning almost always starts with the maintenance concept. Table 5.12 summarizes the analysis approach. If

TABLE 5.12   Task 302, Support System Alternatives

Approach	Subtasks
1. Establish a baseline support concept for each support function in task 301	
2. Develop viable alternate concepts	302.2.1 through 302.2.4, alternate support concepts/ plans
3. Identify risks	302.2.5, risks
4. Document	

you desire more information on task costs, task tailoring and focusing, or software considerations, please review Sec. 4.4.2.

Analysis approach

Establish a baseline concept. The task approach described in this section is derived from Ref. 32. The work load can be controlled by the number of concepts developed and the amount of detail used in defining the alternatives. If a baseline support concept has not been developed, this can be done by defining how each function in each block of the network developed in task 301 will be performed. It may be possible to simplify this process by using the standard service-support concept as a departure point. For support functions, start with the maintenance function and add a concept for each ILS element to support the maintenance concept. A maintenance concept involves decisions on task allocation, task complexity, and task scheduling. Allocation affects the distribution of skills and equipment in the field. Task complexity addresses the ease of repair action and affects the repair policy. There are three basic repair policies:

1. Nonrepairable or discard at failure

2. Fully repairable or piece-part repair

3. Partially repairable

There is usually a high degree of commonality between the system concepts for similar functions. It may be necessary to fine-tune the baseline to ensure consistency between individual ILS-element concepts.

Develop alternatives. This is simple after a baseline has been established. We suggest starting with maintenance concepts and alternating them incrementally over the full range of options. Other ILS elements must be adjusted for each change in maintenance concept. A large number of alternatives can be identified in this fashion.

1. System: _____   LSA opr: _____   Date: _____
   Concept: _____   Phone: _____   Function: _____
   Operator concept: _____   LCN: _____   Sheet no: _____

2. Operator function	3. Manning concept	4. Skill	5. Training concept

6. Critical operator skills:

7. Risks:

Figure 5.32   Operator Manpower Concept Data Sheet. (See reference 32.)

1. System: _____   LSA opr: _____   Date: _____
   Concept: _____   Phone: _____   Function: _____
   Maint. concept: _____   LCN: _____   Sheet no: _____

2.	Scheduled (preventative)			Unscheduled		
	Organization	Intermediate	Depot	Organization	Intermediate	Depot
Maintenance Percent Skill Manning concept						
Maint. training						
Support equipment						
Supply support						
Technical data						
Facilities						
Transportation						
Software support						

3. Critical issues:

4. Risks:

Figure 5.33   Maintenance Concept ILS Impact Data Sheet. (See reference 32.)

1. System: _____ LSA opr: _____ Date: _____
   Concept: _____ Phone: _____ Function: _____
   Replen. concept: _____ LCN: _____ Sheet no: _____

2. Replenishment resource	Transportation				
	Conus	OS	Battlefield	Storage	Loading

3. Critical operator skills:

4. Risks:

Figure 5.34   Replenishment Concept Data Sheet. (See reference 32.)

Documentation and use. Documentation can take the same general approach used in the other tasks. Specific data can be developed for operator, maintenance, and replenishment functions, with an appropriate summary sheet. Figures 5.32 through 5.35 present departure-point forms. The results of this task are inputs to subtask 303.2.2, Support System Tradeoffs.

Lawn mower example. As the XYZ Company LSA work group considered alternate support concepts, it first reviewed the forms in Figs.

1. System: _____ LSA opr: _____ Date: _____
   Concept: _____ Phone: _____ LCN: _____

2. Support elements	Function/ subsystem	Function/ subsystem	Function/ subsystem	Function/ subsystem
Operator personnel				
Maintenance: Organizational Intermediate Depot				
Replenishment				

3. System support considerations:

4. Risks:

Figure 5.35   Support Concept Overview. (See reference 32.)

5.32 through 5.35. The group quickly decided that the form in Fig. 5.33 could be used to record the various support concepts developed. Although the form called for extraneous data, it provided a convenient way to assess the ILS-element impact of alternate maintenance concepts. The group had already decided that repair at the owner or local shop level was highly desirable from the user's viewpoint. The group therefore decided to consider such concepts as design for discard and to investigate approaches that would minimize the need for depot support. The group decided to use the support concept for the company's current mower as a baseline and consider new alternatives. Accordingly, the group filled out worksheets appropriate to this approach.

5.3.12 Evaluation of Alternatives and Tradeoff Analysis (Task 303)

On major systems, a large number (300 to 400) of tradeoffs are performed at various levels. This is therefore an excellent task for use of the principle of focusing mentioned in Chap. 4. Conversely, however, it is also a task where new subtasks could be added if a desired study cannot be specified by focusing one of the subtasks in the standard. If you desire more information on task costs, task tailoring and focusing, or software considerations, please refer to Sec. 4.4.3.

The analysis approach is taken from Ref. 32. Table 5.13 summarizes the approach and related subtasks.

Existing models should be used whenever possible, since such models have usually been accepted as valid. Clever application strategies can enable the use of many existing models. For example, many large models can be run using data only at higher levels. However, simple models can sometimes be developed at surprisingly low cost in time and labor, particularly if they are actually derivatives of existing models (see Sec. 7.4). Use as few models as possible. One major program performed all the subtasks in task 303 through the use of a life-cycle cost (LCC), system engineering tradeoffs, system readiness, and level-of-repair model. In the CE phase, several simple models may suffice in many cases. If all else fails (and not all issues lend themselves

TABLE 5.13   Task 303, Evaluation of Alternatives and Tradeoff Analysis

Approach	Subtasks
1. Define issues for analysis	Issues could fit any level of detail in the 303 subtasks
2. Define evaluation criteria	303.2.1, tradeoff criteria, 303.2.1, tradeoff criteria
3. Select/construct models	303.2.1
4. Do the analysis	
5. Document	

to modeling), use of expert opinion may be necessary. Different readiness modeling approaches are summarized in Table 7.13. A case example of a readiness analysis that illustrates its usefulness to the LM is contained in Appendix 21. Most of the U.S. services employ standard complex simulation models for readiness analysis. However, the simpler and less costly techniques can be used for many situations.

Analysis approach

Define the issues to be analyzed. Defining the issues to be analyzed is the key to focusing this task. Doing this leads to efficient definition of criteria and specification of the models to be used or developed. A good source of issues is the results of the other LSA tasks. An example of a subsystem or subsystem issue is determination of the optimal support concept.

Define the evaluation criteria and select/construct model(s). Evaluation criteria are usually at an end-objective level. For example, common high-level support objectives are lower LCC or O&S costs, readiness (A_0, sortie rates per day, etc.), and manpower quantities and skill levels. However, system engineering tradeoff criteria and weighting criteria for factors on major programs can become quite complex (see the example in Chap. 7). If there is more than one criterion, weighting may be necessary.

Do the analysis. It may be necessary to make several runs of the analysis. This is true of a Markoff chain or a simulation analysis of readiness or a sensitivity analysis. See the discussion of readiness modeling in Chap. 7. It is usually desirable to conduct sensitivity analysis to test input variables with considerable uncertainty. When input variables change as a result of other analyses, the model(s) should be run using the new variable values.

Alternative analysis approaches. The analysis approach advocated in the preceding subsection essentially uses the subtasks in task 303 as a shopping list. Specific issues to be analyzed are compared with the subtasks in task 303 to see if, with appropriate focusing, the subtask is useful to the analysis desired. However, in practice, the requirer seldom does such focusing in preparing requirements. A common practice of requirers is to specify the task 303 subtasks as specified in Table III of 1A standard for major systems or equipments. In some cases, the requirer will eliminate some subtasks not of obvious interest. The comparative evaluations (subtask 303.2.9) and the energy tradeoffs (subtask 303.2.10) are examples. A common approach by the executor is to use a simulation model for the readiness analysis (subtask

303.2.4), the LORA (subtask 303.2.7), possibly an LCC model, and tie the remaining subtasks to the system engineering model tradeoffs. This approach integrates most of the task 303 subtasks into the system engineering tradeoff process. Possible use of an LCC model on various subtasks is discussed in more detail in Chap. 7.

The analysis approach advocated in the preceding subsection tends to reduce the generality of requirer subtask call-outs by requesting only subtasks that can be related to the issues of interest (although new subtasks could be added if necessary). If the requirer does not focus the subtasks, the executor or proposed executor could use the results of any gross analysis it has performed and any issue focusing it perceives useful as a proposed departure point for task accomplishment in the proposal to convince the evaluators on the proposal review group that the proposer is knowledgeable and will do a good job of execution.

Documentation and use.   See the sample worksheets in Figs. 5.36 and 5.37 for suggested documentation. Additional narrative is needed to supplement these worksheets. These tradeoffs are used in making any adjustments to the goals set in task 205.

Lawn mower example.   The XYZ Company work group reviewed the forms in Figs. 5.36 and 5.37 and decided to use a slight modification of Fig. 5.36 but devised the simpler form in Fig. 5.38 as a substitute for the form in Fig. 5.37. The group concluded that the major issue for both the commercial and the government versions was a highly reliable, low-maintenance, low-life-cycle-cost mower. The group decided that it would conduct two iterations of task 303. The first would be a

| 1. System: _____   LSA opr: _____ | | | | | | |
| Concept: _____   Phone: _____ | | | | | | |
2. Tradeoff/evaluation issue	Sheet no.	Performance	Production cost	O&S cost	Risk	Recommendation

Figure 5.36   Tradeoff Analysis Summary Data Sheet. (See reference 32.)

1. System: _____ LSA opr: _____ Date: _____

   Concept: _____ Phone: _____ Sheet No: _____

2. TRADE-OFF ISSUE:

Functions Affected	Factors Affected	% Change	Cost difference over life of system		
			System costs	Baseline	Issue

4. Benefits:

5. Burdens:

6. Risks:

7. Recommendations:

Figure 5.37 Tradeoff Analysis Data Sheet. (See reference 32.)

1. System: _____
   Concept: _____
   LSA opr: _____
   Phone: _____
   Date: _____
   Sheet no: _____

2. Tradeoff issue:

3. Functions affected:

4. LCC factors affected, cost difference:

5. Benefits:

6. Burden:

7. Risks:

8. Recommendations:

Figure 5.38 Lawn Mower Tradeoff Analysis Data Sheet.

gross analysis. The second iteration would, if necessary, use a more sophisticated model such as the government CASA model (see Chap. 7). The standard A_0 equation would be used in the gross analysis. If necessary, a simple Markov chain model also would be used in the second iteration. If the government required a LORA, the group would use the model specified for use by the government. Any other analyses required would be accomplished using the system engineering tradeoff model. This required addition of some support factors to that model, but the system engineers agreed to accept the suggested additions. The next step was to collect the input data (this was refined for the second iteration) and exercise the models as appropriate.

The group identified the following specific tradeoffs to be conducted:

- Impact of design to discard on spares requirements
- High A_0 goal on R&M
- Cost impact of sparing to high A_0
- Weight versus ease of movement (alternate wheel assemblies)

5.3.13 Supportability Test, Evaluation, and Verification (Task 501)

Discussion.   The subtasks suggested for this phase are summarized in Table 5.14. If you desire more information on this task, please review Sec. 4.6.1.

Analysis approach

Identify and evaluate risks.   Review the results of the other tasks. Much of which follows may have been done already as part of those tasks. All the risks previously analyzed should be assembled in one list. Each factor, issue, and risk should be analyzed for probability of occurrence, the impact if the condition occurs, and timing. Probability of occurrence can be crudely analyzed by comparing the BCS, the RS, and the projected goal for the factor in the new system. Review the larger differences and determine if the increase is attainable. If the

TABLE 5.14   Task 501, Support Test, Evaluation, and Verification

Approach	Subtasks
1. Identify and evaluate risks	501.2.1, T&E strategy
2. Identify risk-reduction methods and select best approach	
3. Develop test plans and system	501.2.3, objectives and system support package and criteria; 501.2.2, support package
4. Document	

change in technology is at hand, large differences may be achievable. Large differences at the system level are more difficult to achieve than at lower levels. Such reviews can show which risks have the greatest probability of occurrence.

Impact if the event occurs can be assessed using the models selected for the other tasks. Sensitivity analysis performed in other LSA tasks should be reviewed to identify impacts. Additional analyses may be necessary. Combining probability of occurrence with impact is a simple statistical value, a useful approach to ranking risks.

Timing is also important. If the factor involved can vary widely before significantly affecting other factors, it may be appropriate to delay testing until later phases. If the factor is very sensitive, some risk-reduction study or testing in CE or DEMVAL may be possible and desirable.

Identify risk-reduction methods and select the best approach.   There are usually a number of actions that can alleviate or reduce each risk, ranging from do nothing (at least until later), to paper studies, to testing of various types. Each has a different consequence in terms of costs and the amount of risk reduction. A key action to reduce costs involves the integration of specific support-test issues into the overall test program. Only recommend separate testing as a last resort.

Develop test plans and a system-support package (SSP).   Once the desired tests and investigations have been identified, more detailed test plans (e.g., time, place, sample size, test criteria, test procedures) and the resources to support the testing must be developed. The resources are documented in the SSP. The SSP includes such things as any draft operating and maintenance procedures, spares and repair parts, any special or common test equipment or tools, calibration requirements, and so on. This information should go in the Test and Evaluation Master Plan (TEMP).

Documentation and use.   Besides the usual narrative, some standard forms are again useful. A form on each risk could list function, risk factor (e.g., MTBF), the testing, evaluation, and verification (TE&V) options for the function, the resources (normally time and money), and the residual risk for each option (see Fig. 5.39).

The SSP package form for each test should cite the logistics resources needed (including quantities required, the release and need dates, etc.) and the status/problems/actions (see Fig. 5.40).

The summary sheet should identify each of the issues, the test objectives for each issue, the priority of the test approach, and the date of input to the TEMP (see Fig. 5.41). Task 501 results provide the supportability inputs into the overall system T&E master plan. The evaluation of results provides a basis for any necessary changes in program support planning.

| 1. Program: _____ LSA opr: _____ Date: _____ | | | | |
| Concept: _____ Phone: _____ | | | | |

2.	Function/system/subsystem LCN	Risk	TE&V option	Resources	Residual risk

Figure 5.39   Task 501 Supportability Test, Evaluation, and Verification Strategy Data Sheet. (See reference 32.)

Lawn mower example.   The XYZ Company work group first reviewed the forms in Figs. 5.39 through 5.41 and decided that the form in Fig. 5.40 could be used if the company conducted prototype testing at various geographic sites. This was under consideration as a "plus" on the government RFP. The group decided to combine the other two forms into the simple form in Fig. 5.42. The group made a list of all the risks developed to date and went through the procedure described above. The largest risk area was obviously the long-wearing blade. Both feasibility and unit cost were a concern. The group decided to address the

| 1. System: _____ ILS point of contact: _____ Date: _____ | | | | | | |
| Test: _____ Phone: _____ | | | | | | |

2. Test issue

:3. Test objective:

4.	Logistics resource required	Quantity required	Need date	Release date	Source	Status	Problems/actions

Figure 5.40   Task 501 Support System Package Data Sheet. (See reference 32.)

1. Program: _____ LSA opr: _____ Date: _____
   Concept: _____ Phone: _____

2. Support issue	Test objective	Priority	Test approach	Input to TEMP

Figure 5.41   Task 501 Supportability Test, Evaluation, and Verification Summary Data Sheet. (See reference 32.)

1. System: *New mower* LSA OPR: _____ Date: _____
   Concept: *No. 1* Phone: _____

2. Support issue:

3. Test objective:

4. Priority:

5. Test approach options and costs:

6. Risk assessment:

7. Recommendations:

Figure 5.42   Lawn Mower T&E Data Sheet.

first concern by having different manufacturers build test samples. They also asked each of the manufacturers to perform paper studies that did tradeoffs on blade life versus cost. The reliability of electrical starters in various climates also would be addressed. Lastly, the effectiveness of Teflon additives was to be tested in hot and cold climates.

5.3.14   Summary

This section has described a specific approach to performance of LSA activity in the CE phase. The primary objective of this activity is to influence the concept(s) and specification(s) selected for the DEMVAL phase. A secondary objective is to identify possible tests or studies to reduce risk. A third objective is to provide data to more effectively tailor and focus the LSA effort required in DEMVAL. The major keys to success are the same as in preconcept LSA. The major difference in CE is greater (but still limited) detail and depth of analysis and formal (rather informal) execution of additional LSA tasks and subtasks.

5.4   Demonstration/Validation

5.4.1   Introduction

The objective of Demonstration/Validation (DEMVAL) is proof of principles and concepts by test. If competition exists, selection of the best concept will occur. In the Demonstration/Validation phase, performance characteristics are at least tentatively established, but much of the design is still flexible. Debugging takes place. Major changes in direction are not uncommon as various options are explored. Partial prototypes may be built and possibly tested in an operating environment.

At first glance, LSA in DEMVAL is simply a further downward extension of LSA in CE. However, this is not entirely true. The activities at the system and program levels are more complex. The objective of LSA begins to shift from pure requirements and concepts influence to detailed design influence, support-system refinement, and the initiation of detailed logistics resource identification. More sophisticated models may be required. Some LSAR data will probably be generated. These and other general changes are discussed in Sec. 5.5.2.

5.4.2   Major issues

More detailed and sophisticated models. As more detailed information becomes available, it may be possible and desirable to use more complex models. Instead of a simple A_0 equation, a Markov chain model that considers the impact of probabilities may be desirable. In more complex programs, a sophisticated complex simulation model may be useful. Chapter 7 discusses modeling in more detail.

Cost of the LSAR. The LSAR generated by task 401, Task Analysis, is a special issue since, depending on how costs are incurred, it may be the most costly task in the LSA. Whatever the requirer has specified must be done, but how can the necessary detail be developed most cost-effectively? Ways to reduce LSAR costs are discussed in Sec. 5.4.8.

5.4.3   Tasks to be performed

The tasks suggested in MIL-STD-1388-1A to be performed in DEMVAL are those performed in CE plus the selective application of task 401, Task Analysis. The same interactions that occurred between tasks and subtasks in CE will occur in DEMVAL. The level of detail obviously increases. Contributing specialists for tasks described in detail in CE remain the same.

5.4.4 LSA Plan (Task 102)

The approach to this task is generally the same as in CE, with more detail as appropriate. We have reviewed few plans that adequately address task execution in terms of specifically how the tasks will be performed (models, etc.) and which communities contribute to task execution. Most plans resort to generality about subcontractor pass-through, yet major subcontractors frequently design major elements of the system. This is discussed in more detail in Chap. 3. Another common failing is the lack of specific discussion of the use of the task results within the contractor's organization. The executor may wish to address these subjects in the proposal or plan to help differentiate its approach from those of competitors.

5.4.5 Program Management, Design, and LSA Reviews (Task 103)

The approach in DEMVAL, for whatever reviews are conducted, is identical to the approach used in CE, except that the level of detail and the specifics obviously change.

5.4.6 Use Study (Task 201)

The purpose and general approach in Secs. 5.3 and 5.4 are unchanged except for the depth-of-level issue discussed in the next paragraph. The existing use study should be updated if any new information that affects these levels becomes available. Examples are changes in planned mission(s), use, and environment. A bomber aircraft originally planned for a high-altitude mission profile may change to one with a low-altitude profile. The threat may change from high intensity, high technology to lower-intensity, third-world scenarios.

The chief issue is the level to which to perform the study. While past LSA practice generally has been to document one set of A table data at the system level, lower levels may be needed. On a ship project, for example, A table data are needed for each subsystem/equipment because these are used differently on different missions. Selected use and LSAR A table data may be employed selectively below this level. A possible approach is to review the R&M analysis and selectively develop A table data where individual item usage levels are significantly different. Accurate usage drives logistics resource requirements. It is therefore important to devote considerable attention to this issue.

While the requirer or system integrator may maintain a functionally based study, the individual contractors (if there is competition) should transition to a hardware orientation as this becomes possible.

5.4.7   Standardization (Task 202)

Standardization is somewhat independent of the level of detail available. It proceeds as generally discussed in Sec. 5.3. DEMVAL is a very appropriate time to consider imposition of requirements for use of standard test equipment (if not already done), restrictions of fasteners (as done on one major aircraft program), and similar possibilities. Standardization occurs "naturally" in the performing design organization as selection of HW/SW is made to fulfill needed functions. Subtask 202.2.2, support impact analysis of mission system HW and SW standardization recommendations, can be achieved effectively by integrating support considerations into the tradeoff/evaluation process used to make such decisions. The problem here is to focus the effort on the most important areas, since the work load can become limitless with less probable benefit as lower levels are reached. Generic standardization is a feasible alternative at lower levels. The third subtask, recommendation of mission HW and SW approaches, if required, should be assigned to design personnel. One approach to the third subtask is to make standardization an issue at all design reviews. Another is to include standardization as a factor in the system engineering tradeoff analysis. LSA, at most, should be a record keeper and report preparer for this subtask, except in terms of evaluation of support impact (the second subtask).

5.4.8   Comparative Analysis (Task 203)

Update.   Consistent with use study updates due to change in threat, etc., the previous comparative analysis should be updated and extended as necessary. The BCS and ABCS and driver analysis are revisited.

Function analysis versus hardware.   As in the use study, as hardware is identified at each functional level, significant changes in actual or projected HW and SW data are made. If NDIs (GFE or CFE) are directed, the analysis should go no further unless it is necessary to consider modification to improve selected logistics drivers. The remaining analysis should focus on new design areas.

Level of detail.   The needed level of detail in the analysis varies with the purpose of the task objectives. It is not necessary to go beyond several levels to satisfy the need for data to help select goals for the system level. However, for driver analysis it is necessary to go at least one level further generally and then use Pareto's principle that a few of the items cause most of the problem. In summary, descend further very selectively and only as far as necessary.

A "different" approach.   Another approach to driver analysis that was employed successfully on one program can be derived from any R&M failure or R&M predictions made on the new system if acquisition costs of individual components, LRUs, etc. are available. The approach is as follows:

1. Multiply unit cost by the failure rate. Rank the results. High-cost, high-failure-rate items are good O&S cost driver candidates. Low-cost, high-failure-rate items are good readiness improvement candidates.

2. Multiply failure rates by repair times. Rank the results. High-failure, high-repair-time items are good readiness/manpower driver candidates.

5.4.9 Technological Opportunities (Task 204)

This task is much the same in approach as that executed in the preceding phase. The design community should handle the mission HW/SW aspects. LSA should focus on the support impact of mission-system candidates and on the identification of support technologies and concepts. The candidates identified in the preceding phase should be reevaluated in this phase. New candidates should be evaluated in the same manner.

5.4.10 Supportability and Supportability-Related Design factors (Task 205)

There are five major considerations in this task in DEMVAL. The first is simply an update of work performed based on new or additional information. The second is the possible use of more sophisticated models commensurate with the added detail now available. Updated information, rather than further levels of detail, is needed to refine goals. However, further levels of detail are still needed in the model to do the driver, sensitivity, and risk analyses. The third consideration is possible use of proprietary HW/SW as these are fitted to functions. This could be made part of the system engineering tradeoff process. A fourth consideration is the functional versus hardware issue. The rule to substitute HW and SW for functions again applies, except for system-level analysis done by the requirer or system integration contractor because of competitive executing organizations. The fifth consideration is the level of detail, data availability, and analysis model requirements for the other tasks that employ a model.

5.4.11 Functional Requirements Identification (Task 301)

This analysis will naturally proceed as functions are turned into HW/SW/personnel candidates in accordance with the path discussed earlier. The HW and SW paths can have major differences. The HW and SW designers will probably use different functional analysis techniques, but the underlying concept is similar. The system task identification (task 301.2.4) can begin if the FMECA and any RCM analysis or HARDMAN/MANPRINT analysis have been done. Some task identification and analysis may be needed to support DEMVAL. Major parts of the HW/SW selected that are not critical to the decision to proceed to EMD may not resemble the final design. On the other hand, a "working" model may be built and taken into the field or shipboard for actual testing and use. If so, there may be a requirement for limited ILS support for this activity. A review of the overall plan and the TEMP will help determine what, if any, logistics support is needed. The work performed in this task can be tailored to help meet this requirement.

5.4.12 Support System Alternatives (Task 302)

The work done in this task in DEMVAL involves (1) update of work done in the preceding phase, if any, and (2) expansion to lower levels of detail as design progresses. The update considerations are the same as in the other task updates. As function requirements identification proceeds down the indenture level, NDIs and new design should be treated differently. GFE NDIs currently in production should have a support concept and plans. The major issue here is compatibility of these concepts with the support concepts and plans for the new system. Some iteration or change may be necessary for consistency in the overall concept. The support concepts and plans of GFE or CFE that are currently in development also should be identified and examined for compatibility with the new system. Since there is usually infrastructure pressure to use standard concepts, any departure from these must be justified. Since support concept and plan alternatives can be perpetuated almost forever, focus should be on cost, readiness, and manpower drivers. The work in task 203 in this regard and the identification of new or unique functions in task 301 are a good starting point.

5.4.13 Evaluation of Alternatives and Tradeoff Analysis (Task 303)

The big issues here are what models, what analysis, and at what levels. As you proceed down the indenture levels, the possible tradeoffs change and the numbers of tradeoffs will probably increase. The need for models, at least in theory, can increase in direct proportion. See the model discussion in Sec. 5.2.1. A key step is to make sure that support is properly represented in the system engineering model, since this integrates support considerations into the heart of the system engineering process. There will probably be two system engineering models, one for HW and one for SW. The system engineering model may be tailored to the specific project, since different programs have different factors that can have different weights. Some companies do not have formal system engineering models. If the requirer does not ask to see the model, the performing organization may wish to advise the requirer of the model and weights used to preclude the possibility of a later surprise reaction by the requirer. The executor must respond to the requirer, who may or may not tailor and focus the task 303 subtasks required (most do not). Blanket invocation of all the subtasks is the executor's nightmare. In such situations, the executor must focus the subtasks or perform needless work and justify its lack of utility. The level of execution is dependent on the model, the issue, and the data available. The results of this task at the system level are fed back into the goal setting in task 205.

5.4.14 Task Analysis (Task 401)

Overview. The subtasks in this task are listed in Table 5.15. A full discussion of the execution of this task is contained in this section, although much of the work is done in EMD and possibly production. A discussion on how much of this task should be done in DEMVAL occurs in Chap. 4.

This task generates the bulk of the LSAR data. The analysis is performed on each LSAR candidate identified. The logic for selection of

TABLE 5.15 Task 401, Task Analysis

401.2.1	Task analysis
401.2.2	Analysis documentation
401.2.3	New/critical support resources
401.2.4	Training, requirements and recommendations
401.2.5	Design improvements
401.2.6	Management plans
401.2.7	Transportability analysis
401.2.8	Provisioning requirements
401.2.9	Validation
401.2.10	ILS output products
401.2.11	LSAR updates

LSAR candidates is usually included in the statement of work. A common definition for such candidates follows:

LSAR candidates shall include: systems, subsystems, end items, components, assemblies, support and test equipment, and training equipment that require documentation of operational logistic support parameters and requirements and all items for which the government does not have existing maintenance capability. Maintenance capability, as used in this context, includes but is not limited to: trained personnel, transportation and handling, logistics technical data, support and test equipment, supply support, and facilities. The selection of LSAR candidates shall be governed by the following procedures:

The contractor shall prepare an initial list of LSAR candidates in consonance with the criteria below. The list shall include LSA control number, national stock number/manufacturer's part number, and item name as available. The initial list of candidates shall be included in the initial LSAP submission and shall be augmented by the contractor as design engineering progresses. The following material shall be candidate for LSAR:

a. Contractor-furnished installed equipment items that can or will be inspected, repaired, maintained, or overhauled separately as part (on-equipment maintenance) of the system/equipment

b. Contractor-furnished installed equipment items that can or will be tested, repaired, maintained, or overhauled as part (off-equipment maintenance) of the system/equipment

c. Contractor-furnished, non-installed GFE items to include support and test equipment and training equipment

d. Installed and non-installed GFE items when such analyses are required to interface GFE with contractor-furnished equipment or when usage/environment will be different and/ or to determine total support requirements of the contract end item(s)

e. Installed and non-installed GFE items for which government-furnished data are inadequate or incompatible, and are not economically repairable, and where such data are necessary to fulfill contract requirements

f. Connecting and installation hardware, bracketry, standard hardware items, bulk material, and simple parts, which are not economically repairable, shall be included in the LSAR documentation of the next higher assembly.

Specific subtasks and related data tables are shown in Table 4.15. A large number of requirer inputs are needed to perform this task, such as identification of levels of maintenance to be documented, description of the capabilities of the personnel (target audience) who will maintain the system, personnel skills and quantity limits, and annual operating hours. See the listed inputs to this task in MIL-STD-1388-1A.

Use of an automated LSAR data system facilitates the generation of input data and output reports. These reports, their potential uses, and users by LSA top-level objective are listed in Table 6.1. While automated LSAR systems have made significant improvements in ease of recording and the quality of LSAR data, further improvements are still being made in the LSAR. A common complaint with the LSAR is that it does not satisfy all ILS-element requirements. While it may never satisfy all such requirements, significant progress is being made in this regard. A part of this problem is a "people" problem; e.g., certain members of some ILS teams still like to use the old way, that is, get their data from their own reporting requirements. This too is gradually being overcome. LSAR data systems are discussed in depth in Chap. 6.

The task analysis analyzes the operator and maintenance tasks listed in task 301 to identify the exact procedures and the resources needed to perform them. Task analysis develops

1. Source data for the preparation of logistics documentation (technical manuals, training programs, provisioning, etc.)
2. Logistics resource requirements for each task
3. New or critical resource requirements
4. Transportability requirements
5. Support requirements that exceed goals or constraints
6. Data to develop design alternatives to reduce support resources needed

The analysis approach that follows is derived primarily from Ref. 33. The approach may be simplified in practice when the task is performed by experienced personnel.

Analysis approach

Task analysis (subtask 401.2.1).   This subtask requires a detailed analysis of each operator and maintenance task to identify

1. Operator and maintenance procedures
2. Logistics resources required to perform the task
3. Task frequency, elapsed time, and manhours
4. Level of maintenance to perform each task

The four-step process, with resultant data tables and products, is summarized in Table 5.16. Each task identified by task 301 is analyzed. Task analysis is a four-step process:

TABLE 5.16 Task Analysis Steps, Data Record/Table, and Products

Step in task analysis	Data table	Data product
Task allocation allowance	CA, task requirements	Maintenance chart
Frequency determination	CA, task requirements	
Task design/description	CB, subtask requirement	Source data for technical publications and training
Resource enumeration	CC, sequential task description	
	CA, task requirements	Provisioning, training, manpower, PHST&T requirements
	CG, task support equipment	
	CI, task-provisioned item	
	CK, task inventory	

1. Task allocation
2. Task frequency determination
3. Task design/description
4. Resource enumeration

Task allocation is the process of assigning operator and maintenance tasks to the various levels of support. Allocation decisions must consider three factors, as was shown in Fig. 5.26. The design complexity influences the level of support assigned to a maintenance task. Complex tasks with lengthy procedures, extensive skills, longer times, and special tools generally go to higher levels. Tasks of less complexity go to I or O level. The maintenance concept defines the present maintenance capabilities at each level of support. The allocation planning laid out in the maintenance concept should be followed to ensure consistency with current support capabilities. The LORA determines two things: (1) whether to discard or repair upon failure, and (2) the cost to repair at the different levels in order to minimize cost of repair. The result of allocating maintenance tasks to the various maintenance levels is a maintenance allocation chart. This document can be generated from the LSA-004 report.

The second step in the process determines how many times per year a task is performed. *Task frequency determination* addresses both operator and maintenance tasks. They are recorded in the LSAR data tables. Operator, replenishment, and functional alteration task frequencies depend on the amount of system use and the types of missions. Predictions of the frequencies of such tasks are based on information in the use study on anticipated operating hours and are recorded in the A data tables. Preventive maintenance tasks are scheduled based on the results of the RCM analysis. The RCM determines whether anticipated failures will be fixed by (1) hard time replacement, (2) condition monitoring, or (3) on condition. The disposition of preventive maintenance, provided by the logic, is the basis for scheduling a preventive maintenance task. Corrective maintenance task frequency is predicted using a mathematical formula. The formula considers the failure modes requiring a maintenance task, the part failure rate, and the number of hours of operation per year.

The third step, *task design/description,* is the development of procedures explaining how to perform the tasks. Task description is a two-step process:

1. Task design
2. Task description

Task design requires the creation of maintenance logic diagrams. Such diagrams lay out the steps in a maintenance task. They show the maintenance functions that must be performed and the order in which they occur. A maintenance diagram is presented in Fig. 5.43, which illustrates the task activity at one level.

Task description is development of the procedures that are the narrative for technical manuals. Each step on the maintenance logic diagram can be broken down into elements or elemental activities. These can be the smallest units into which a task can be divided. The elements are arranged in proper sequence. A narrative is then written stating what must be done to perform each step and/or appropriate element. Table 5.17 provides a breakout of possible elements. The amount of detail developed in the process just described should be adjusted appropriately to the level of detail required for the technical manual and training material needed and for needed accuracy in time estimates. Task descriptions are documented in the CC data table.

The elements can be the basis for maintainability time predictions. By summing up the elements, a predication of the elapsed time and the manhours required can be made. This information is recorded on the LSAR CD data tables and can be used in preparation of technical manuals and training materials.

The final step in the task analysis process is *resource enumeration.* As each procedure is developed, this step lists the skills, manpower, tools, parts, support and test equipment, and facilities needed to execute the procedure. The resource requirements for each maintenance task are recorded in the LSAR CD, CG, and CI data tables. The E through H data tables are used to identify the resource requirements in the C data tables in more detail. By totaling the resource requirements for each task, the total logistics support resource requirements can be identified. Resource determination provides the basis for provisioning, training, support and test equipment, and facility requirements.

Subtask 401.2.2 documents the results of the task analysis in the

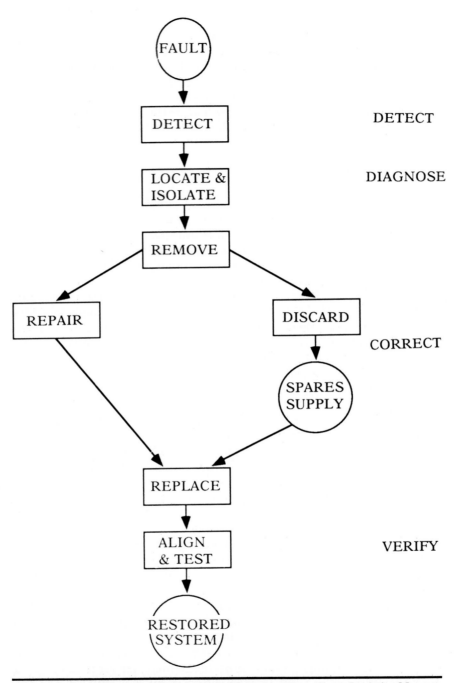

Figure 5.43 Corrective Maintenance Task Diagram. (From Army Logistics Management Center, LSA Course, 1986.)

TABLE 5.17 List of Categories and Elemental Activities Active Repair Time

Category	Elemental activity	Activity no.
Preparation	System turn-on, warm-up, setting dials and counters as necessary. Activity #1 plus time awaiting particular component stabilization. Opening and closing radome. Gaining access and reinstalling covers (other than radome). Obtaining test equipment and/or tech orders. Checking maintenance records. Procuring components in anticipation of seed. Setting up test equipment.	
Malfunction verification	Observing indications only. Using test equipment to verify malfunctions inherently not reproducible on ground. Performing standard test problems or checks. Testing for pressure leaks. Attempting to observe elusive or nonexistent symptom(s). Using special test equipment designed specifically for this equipment. Making a visual integrity check.	
Fault location	Fault self-evident from symptom observation. Interpreting symptoms by mental analysis only from knowledge/experience. Interpreting displays at different settings of controls. Interpreting meter readings. Removing unit(s), subunit(s) and checking in shop. Switching and/or substituting unit(s)/subunit(s). Switching and/or substituting parts. Making a visual integrity check. Checking voltages, continuity, waveframe, and/or signal tracing. Consulting tech orders. Conferring with tech reps or other maintenance personnel. Performing standard test problem(s). Isolating pressure leak. Using special test equipment designed specifically for this equipment.	
Repair	Replacing unit(s)/subunit(s) Replacing parts. Correcting improper installation or defective plug-in connection(s). Making adjustments in aircraft. Making adjustments in shop. Baking magnetron. Precautionary repair activity (include so-called fault location, part procurement, and repair times spent when symptom not verified). Repairing wiring or connections.	
Final malfunction test	Functions checkout following completion of repair.	

SOURCE: From Army Logistics Management Center, LSA Course.

LSAR. Subtask 401.2.3 identifies those logistic support resources required to perform each task that is new or critical. New resources are those which require development in order to operate or maintain the system. Examples are new or restructured personnel skills (CD and G data tables), new or restructured training devices (C data tables), new or special transportation systems (H and possibly J data tables), new computer resources (C, E, and H data tables), and new maintenance techniques or procedures (CC data tables). Critical resources are not necessarily new but require special attention because of such factors as schedule constraints, cost implications, or known scarcities. They may be documented in the C and H and in some cases the E, F, G, and J data tables.

Subtask 401.2.4 identifies training requirements and provides recommendations concerning the best mode of training and the rationale for the recommendations. The LSA-14 report, "Training Task List," identifies tasks requiring training and the rationale for the requirement. Subtask 402.2.5 analyzes the total logistics support resource requirements, determines which tasks fail to meet support-related design goals or constraints, and identifies tasks that can be optimized or simplified to improve support. Improved alternative approaches are developed. Some of the LSAR output reports can be used to support this subtask. One approach is to initiate action on the basis of individual maintenance task analysis (the C data tables). The other method is use of appropriate LSAR output reports. An example is the LSA-006 report that monitors REM requirements. The LSA-001 report gives annual manhours by maintenance level by skill specialty code. The SE utilization report is another. See Table 6.1, which describes the use of the various output reports.

Subtask 401.2.6 identifies management actions that can be taken to minimize risks associated with each new or critical resource identified in subtask 401.2.3. This should focus on the results of subtask 401.2.3 initially. Subtask 401.2.7 conducts a transportability analysis on the system/equipment and any sections thereof when sectionalization is required for transport. When the general requirements of MIL-STD-1366 are exceeded, document the transportability engineering characteristics in the LSAR. Sensitive or fragile items may require special handling or packaging due to the type of material used or the design. The LSAR J data table is used to document transportability requirements. It is normally done primarily at the system level for special requirements.

Subtask 401.2.8 develops the provisioning technical documentation for those support resources requiring initial provisioning. The H data tables are used to record this information. The LSA-36 report, "Provisioning Requirements," provides a format for this information. The provisioning parts list (PPL) lists all the piece parts in the equipment.

The tools and test equipment list (TTEL) contains information about unique tools and test equipment. The repairable items list (RIL) and the long lead time items list (LLTIL) are also subsets of the PPL. Bulk items such as wire, adhesive, solder, etc. are listed in the common bulk items list (CBIL).

Subtask 401.9 validates the key information in the LSAR through performance of operations and maintenance tasks on prototype equipment. Validation requirements are coordinated with other system engineering demonstrations and tests, such as maintainability demonstrations, reliability and durability testing, etc. Validation is documented by a Supportability Assessment Report (DI-S-7121).

Subtask 401.2.10 involves the preparation of various output summaries. The government identifies the summaries it wishes to receive in the SOW. However, the executor should generate additional output reports for use internally (Table 6.1 identifies possible contractor uses of the various LSAR output reports).

Subtask 401.11 involves the update and expansion of the LSAR as new and more detailed information becomes available.

Documentation and use.   Most of the subtasks are documented in the LSAR automatic data processing (ADP) system. However, subtasks 401.2.5, 401.2.6, and 401.2.9 are documented in the data item descriptions (DIDs) listed in Table III in MIL-STD-1388-1A. The uses of the data are very extensive. Many have already been mentioned in this discussion. They will vary with the nature of the individual program.

Cost-effective LSAR.   If the program is large enough to warrant development/purchase/lease and use of an automated LSAR ADP system, this pays handsome dividends in reducing cost and improving data quality. Using such systems, a data element is entered once and used many times with reduction in errors in data transfer. For example, the logistics control number (LCN) is entered once for each item in most such systems, compared with about 200 times if the government 2A ADP system is used. Many companies and contractors who sell/lease LSAR ADP systems also have developed linkages to key LSAR inputs such as the FMECA. As the CALS program matures, additional linkages will occur.

The team approach to the development of the LSAR input was discussed in Chap. 3. This approach is essential for both cost-effectiveness and data accuracy.

Automated LSAR systems, including the DOD system, allow significant use of cross-referencing to identical tasks employed on different LCN candidates. Use of this practice can save time, reduce errors, and increase consistency in task write-ups.

Another way to reduce the cost of this task is the development and use of an LSAR procedures manual to describe in detail the procedures and methods to be used in executing this task. Table 3.1 contains a departure-point outline for such a guide. This is particularly useful because (1) many different specialists contribute input data and (2) many HW contractors use contract personnel to perform much of this work, since generation of the data creates large peaks and valleys in work load. Having specific procedures reduces variations in data generation.

Another possible way to reduce work load is to draw on the work of other communities. The Human Engineering (HE), maintainability, training, and HARDMAN/MANPRINT communities all may do task analysis. The Maintenance Requirements Card (MRC) and Maintenance Index Page (MIP) required by the U.S. Navy duplicate much of the LSAR, as shown in Table 6.7. The work of the maintainability community also may be useful. For example, J. P. Lowery, in an article in the 1985 winter issue of *Logistics Spectrum,* the magazine of the Society of Logistics Engineers, discussed the fact that maintainability engineers frequently prepare maintenance logic diagrams as part of their prediction work. These could be obtained and used as the basis for preparing detailed task descriptions. One should note in this regard that in many cases the task description is sufficiently simple that preparation of a maintenance logic diagram may not be necessary. An open issue in many government organizations is which community uses the other community's data. Failure to resolve this issue frequently results in preparation of duplicate data requirements. Most contractors face up to this issue, since they cannot afford to generate duplicate data in competitive situations. They therefore select one community to develop the data and have other communities use it to satisfy duplicate government requirements.

Another way to reduce work load is the use of LSARs that may be available on like or similar items. While the data may need modification (R&M may be different for the new application) and update, we have seen several cases where as much as 50 percent of the cost of the LSAR was saved.

5.4.15 Supportability Test, Evaluation, and Verification

Discussion, approach, documentation, and use.   This task in this phase has two major aspects. The first is the earliest possible initiation and execution of any risk-reduction studies or tests recommended and approved at the end of the last phase. An example of the types of studies that can be undertaken in the reliability area is shown in Fig. 5.44, which shows the degree of measurement various kinds of reliability testing provides in terms of certain characteristics.

	RDT & E	RQT	System DT & E	IOT & E	PRET	FOT & E	Operation
Design	O	O	O	O	O	O	O
Parts	◇ O	◇ O	◇ O	◇ O	O	O	O
Workmanship	◇ O	◇ O	◇ O	◇ O	O	O	O
Operational concept and environment			◇	◇		O	O
Maintenance concept and environment				◇		O	O

Figure 5.44 Reliability in Total Test Program. O good measurement; ◇ some measurement; RDT&E: Research and Development Test and Evaluation; RQT: Reliability Qualification Testing; System DT & E: System Development Test and Evaluation; IOT & E: Initial Operational Test & Evaluation; PRET: Periodic Reliability Evaluation Test; FOT & E: Follow on Test and Evaluation; Operation: Stable Fleet Operations.

The second aspect is an update and extension of the work in the preceding phase to new requirements as a result of the work done in the other LSA tasks in DEMVAL, with a major focus on performance testing planned near the end of DEMVAL. The analysis approach, documentation, and the use of results are similar to those in CE.

5.4.16 DEMVAL summary

This section has described a specific approach to the performance of LSA in DEMVAL. The objective of LSAR in this phase begins to shift in emphasis from design influence to possible detailed support resource identification. Data to help tailor and focus the LSA effort in the next phase are again generated.

The keys to a cost-effective effort are

1. Timeliness of results

2. Team approach

3. Focus of the effort on the drivers and issues

4. A timely and efficient transition from a functional to a hardware approach with focus on the new design effort

5. A cost-effective approach to any LSAR effort.

5.5 Engineering Manufacturing Development (Formerly FSD)

5.5.1 Introduction

Objectives. The primary objective of the program in engineering manufacturing development (EMD) is to develop and test the detailed design. Prototypes are usually produced for testing purposes. The primary objective of the LSA program is to fully define the support system and to identify detailed support requirements for all the ILS elements. A secondary LSA objective is to exert design influence to the degree possible at this point. Most design influence opportunities are at the lower indenture levels, which are developed in EMD. The LSAR is the major source of data for both these objectives.

Major issues.   There are several LSA issues that arise in EMD. The first is the growing amount of detail available. In some cases, this level of detail must be addressed (e.g., task 401, Task Analysis). Descending to too much detail in others (tasks 201, 202, 203, etc.) may create mountains of paper of decreasing cost-effectiveness. Tradeoffs may in some cases go down to the piece-part level, but few would advocate use of all the models at this level. What is needed are a few models that are widely applicable across and down descending levels of analysis. Approaches that address the detail cost-effectively are needed.

A second issue is task 402, Early Fielding Analysis. Depending on approach, this task could involve great detail and therefore large costs. Task 402 is therefore a great candidate for tailoring and focusing.

A third issue is whether to do task 403, Postproduction Planning. Although Table III in the standard says to do it in production, the former DOD instruction on this subject suggests EMD. Subtask 303.2.1 actually suggests its consideration in all tradeoffs.

5.5.2 Tasks in this phase

The guidance in Table III of the standard is somewhat confusing on this subject. A task 102 update, task 103, task 401, and task 402 are all listed as absolutes (although Table III codes task 102 as selective for equipment). The other tasks and subtasks vary in coding. If updates are necessary to consider major changes to the overall program direction, many of the 200 and 300 series tasks and subtasks would require update.

While the executor does not have to consider such changes imposed by the requirer until a change is made in the contract, it must consider internal change. Such internal change could (but not necessarily) affect some or most of the 200 and 300 series tasks.

The level of detail can and will also expand in some tasks, depending on the level of detail developed in DEMVAL. In many cases, tasks 301, 302, and 303 will or could go to lower levels. One forgets that LSA is an interactive process at one's peril. We will mention possible approaches to several subtasks that could get paper-heavy and focus the main discussion on task 402. In the interest of brevity, we will not repeat contributing specialists when the task has been discussed in detail previously.

5.5.3 Standardization (Task 202)

As detailed design develops, the possibility of more support-element standard equipment will probably expand. The first subtask in task 202 is therefore worth performing. The fourth subtask, risk, could be added to the first in major areas. To eliminate the need for the second and third subtasks (and reduce a potential mountain of paper), standardization could be added to the system engineering tradeoff model used at the various indenture levels.

5.5.4 Functional Requirements Identification (Task 301)

As the detailed design develops, more operation and maintenance functions and associated tasks are identified and need to be documented in the LSAR. This is simply a continuation of the task 301 work begun in DEMVAL. Focus on the new, unique, driver, and high-risk functions is one approach. Since maintenance tasks will be identified from the FMECA and RCM, FRI may only be needed to identify operator and servicing type tasks.

5.5.5 Support System Alternatives (Task 302)

As task 301 changes and expands, so can the task 302 subtasks. Since this can again generate lots of paper at lower levels, some way of addressing this task cost-effectively is needed. As we descend down the indenture level, more and more items will have the same support concept. If the item is an LCN candidate, the concept is recorded in the BB data table. Thus attention should be on the exceptions rather than the common. An LOR model could be used to evaluate the alternatives developed.

5.5.6 Evaluation of Alternatives and Tradeoff Analysis (Task 303)

While some task 303 subtasks are normally system level (such as subtask 303.2.4, Readiness Sensitivities, and subtask 303.2.9, Comparative evaluations), most of the analysis can be performed below system and subsystem level if the proper models are chosen. A potential issue in doing so is the question of whether new models are necessary to handle the increased detail. The key again is to use general models fitting analysis of many issues and as few special models as possible. The other issue is how far down the indenture level to go. The discussion on functions versus hardware in Sec. 5.1 gives at least a partial answer to this question. Inclusion of support factors in the system engineering model used at the different levels is another cost-effective approach.

5.5.7 Task Analysis (Task 401)

The execution of this task was discussed in detail under DEMVAL. Since most of the LSAR is normally developed in EMD, use of cost-effective execution methods is even more important than in DEMVAL. During EMD, the LSAR is used to develop the detailed ILS-element resources and the output reports and/or tapes the requirer has specified. The LSAR is also a unique source of design feedback at the new and lower indenture levels addressed. The LSAR has executor internal uses as well as a means of satisfying customer requirements. An astute customer will require the executor to identify the performer plans for internal use. See Table 6.1 for possible internal uses of the various output reports. Various ad hoc reports are also possible if the executor uses one of the contractor-developed LSAR ADP systems for 2A standard. Since these ad hoc capabilities vary from system to system, a detailed discussion of them is not feasible in this book. With the advent of use of 2B standard, ad hoc capabilities will increase significantly.

5.5.8 Early Fielding Analysis (Task 402)

Discussion. Performance of task 402 (see Table 5.18) may require considerable collaboration between requirer and executor. The intent and objective are good, but this task is an obvious example of a task that would benefit from focusing.

The third subtask can be related to subtask 303.2.4, Readiness Sensitivity. Execution of this could be viewed as an update of this task using the latest data available (possibly from the LSAR). It also could use the data from task 501 as a basis for update. Some service program offices in the United States use support contractors to do such analysis.

The fourth subtask can involve specialists in survivability, vulnerability, and battle damage. If the studies in subtask 303.2.11 were re-

TABLE 5.18   Task 402, Early Fielding Analysis

Subtask		Possible analysis approach(s)
403.2.1	New system impact	Detailed review of impact at O and I levels
402.2.2	M&P impact	Include in the first subtask evaluations
4.2.2.3	Readiness impact of shortages	Use readiness simulation model with updated input data—"what ifs"
402.2.4	Combat resource requirements	Use same model with wartime input values
4.2.2.5	Problem-resolution plans	Use wartime use study data for alternate LSAR "what if" reports

quired and performed, determining changes in combat usage is possible. An example of why differences can occur in combat is the repair of composites. The needs for such repair in wartime will obviously vary significantly from those in peacetime.

The exact nature of the work to be done and the analysis approach taken can affect when the subtasks are done. If the execution is based to a large degree on the availability of task 401 data, the task obviously cannot be performed until late in EMD.

Please review the discussion of this task in Sec. 4.5.2 if you desire more information on task costs, task tailoring and focusing, or software considerations.

Analysis approaches.   Inputs to this task or subtasks include the Use Study, the Evaluation of Alternatives and Tradeoff Analysis, and the Task Analysis, as well as other requirer-furnished information. A possible approach to the first subtask is to evaluate the impact at various organizational levels using the LSAR data. This could include

1. Interactions and commonality of support requirements for the new and existing systems and support asserts

2. Maintenance manpower and skill levels and support work centers

3. Availability of support equipment, facilities, and utilities required for new system maintenance functions

4. Mobility and deployment, including procedures, equipment, and resources to transport the new system

At organizational level, this could include examination of existing resources, equipment lists, and manpower and personnel resources. This approach can involve a lot of detail. Depending on the detail developed in the use study and/or the comparative analysis, this subtask could amount to an update or extension of these studies. The subtask is more complex at I and D levels because of burden and resource sharing at these levels. This is an excellent opportunity to look for useful data from another program's documentation, such as the LSA reports, maintenance plan, manpower plan, depot maintenance plan, etc. The trick is to focus on the essential as opposed to the mountain of detail. An examination of consumption rates of the new versus the old system is a good place to start. A comparison of system usage, performance characteristics, and operating impact on the infrastructure of the new system versus the old is important.

The second subtask at the O, I, and D levels can be performed within the first subtask. Higher-level projections cannot be performed unless the requirer provides a lot of data about other systems, long-range manpower and skill level projections, etc.

The third subtask involves the impact of shortfalls on readiness. If a complex simulation model was employed in subtask 303.2.4, new data and/or "what if" projections can be used to evaluate impact. If not, the models that were used should be examined for possible use. If all else fails, you may be able to play "what if" with the LSAR by varying key factors. The last resort is to be ingenious and devise a new approach.

There are alternatives to the fourth subtask. Table 4.19 summarizes the possible impact of battle damage on the ILS elements. If subtask 303.2.11, survivability/battle damage, was performed, the results can be used as a departure point for this subtask. The use study should provide differences in peacetime and wartime uses and may have useful information on wartime scenarios. Another approach is to apply wartime usage to the LSAR and then adjust the results based on any survivability scenario.

5.5.9 Test, Evaluation, and Validation (Task 501)

The purpose and approach to the first four subtasks performed here is similar to those described earlier. The item tested more closely resembles the production item. Testing more closely resembles use in later fielding. More factors are examined, more data are collected, and more analysis is performed. The readiness analysis in task 303 and other LSA task results, including the LSAR, are updated. The major LSA emphasis in EMD shifts to issues that surfaced in any of the first four subtasks performed in task 402.

Two methods are commonly used to project mature supportability parameter values from T&E results. The first method relies on comparison with predecessor systems. It assumes that the reliability growth rates of the new system are similar to those of its predecessor, as shown in Fig. 5.45. The second method combines the development and operational test reliability results with the improvement possible by fixing the cause of failures found during testing. This method, which is not dependent on the availability of data from a comparable

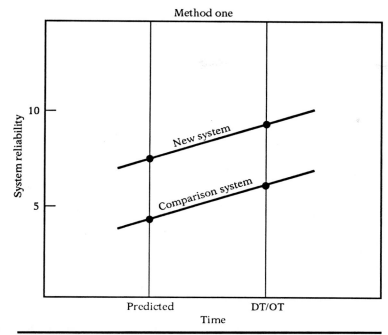

Figure 5.45 Projected Reliability Growth: Method 1.

system is shown in Fig. 5.46. Methods such as reliability growth prediction methods (Duane curve, MIL-HDBK-189, etc.) also may be used when appropriate.

5.5.10 EMD summary

The major LSA emphasis in EMD shifts to detailed logistics resource determination. As appropriate, LSA tasks performed may be updated or extended. A key objective in EMD should be cost-effective performance of task 401. Performance of task 402 may require considerable collaboration between requirer and executor. Significant focusing is needed to make its performance cost-effective.

5.6 Production/Postproduction

5.6.1 Introduction

As production begins, a production design is at hand. However, many design changes may occur as production problems or operational difficulties are experienced as the system is fielded. The LSA tasks previously performed must be updated as appropriate. After production of the end item ends, logistics problems can occur as the system remains

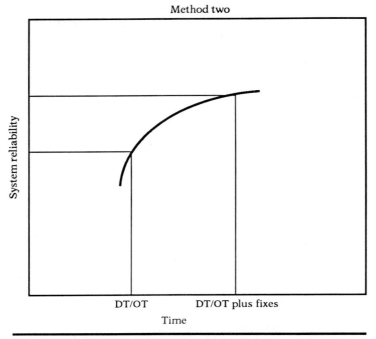

Figure 5.46   Projected Reliability Growth: Method 2.

in use for as much as 20 to 30 years (e.g., major weapon platforms such as aircraft, ships, and tanks).

5.6.2   Some major issues

Some major issues in this section are

1. What LSA updating should be performed?
2. Should the LSAR continue to be updated after end-item production ceases?
3. What should be done about planning for postproduction support problems?

5.6.3   Tasks in production/postproduction

Table III of MIL-STD-1388-1A cites a number of tasks and subtasks that may need to be updated or redone when design changes occur. Obviously, this work should only be done in the areas affected by changes. The object is to assess the impact on support concepts, plans, and resources. Such support impacts are supposed to be evaluated before an Engineering Change Proposal (ECP) is submitted. Thus an LSA effort focused on such changes can provide a basis for the support impact of the proposed change. If an ECP is approved, changes must be made in the detailed logistics support resource data to ensure that the correct data are available for acquiring the correct logistics resources. It is also possible to have high-level program changes, such as a new mission or change in mission profile, and/or changes in support concepts. The failure to achieve various goals for R&M and BIT, for example, may significantly affect support plans. In this event, appropriate portions of the LSA subtasks may need to be updated or reanalyzed. Remember, all projections prior to actual use are just that—projections. The LSAR is a collection of a large number of such estimates. Field experience for some items may vary wildly from such projections. Many programs experience some "bad apples." The LSAR data are one way to generate the new data needed for individual ILS-element data needed, such as technical manual changes.

A new task, task 403, Postproduction Support Analysis, is frequently invoked if it has not already been started. The fifth subtask in task 501, Postdeployment Support Assessment, also may be required to assess the performance of the logistics system against actual operation and use of the new system.

5.6.4 Postproduction Support Analysis (Task 403)

Discussion. The end result of this task is a Postproduction Support Plan to address the issues discussed in that paragraph. When should the plan begin? The answer is sooner than you would think for the reasons that follow. Such planning should begin as soon as plans begin for a major ECP or change in system type or model (e.g., from the F-14 to the F-14A). Many major items go through a number of such changes. The B-52 is now on the H model. Manufacturing sources may begin to disappear on the portions of the A model no longer in use on the B model. It may take several years for the budgeting for postproduction support (PPS) to be obtained. Major ECPS may have the same impact. Loss of manufacturing sources generally increases costs and results in longer lead times. Supply shortages and decreased readiness may result. The management options to PPS problems are to be reactive, i.e., fight fires as they occur, or to be proactive, i.e., plan ahead for such contingencies. The U.S. DOD has obviously taken the second approach, since this subject is addressed in the new DOD acquisition policy and this LSA task exists.

Task approach

General. Experience in this task is still in its infancy. The U.S. services have initiated PPS efforts on some systems. We participated in a review of most of these efforts when tasked to design a PPS data system for one system. While the "state of the art" is obviously evolving, there are a number of common threads to such efforts. Task 403 currently requires the following actions:

1. Assess the useful life of the new system.
2. Identify potential resource problems.
3. Develop alternatives to resolve or address these problems.
4. Develop an overall plan for PPS.

TABLE 5.19   System Postproduction Support Plan Outline

1.0 Introduction
   1.1 Background
   1.2 Purpose
   1.3 System X description
2.0 Postproduction support management
   2.1 Introduction
   2.2 PPS organization
   2.3 PPS support team
   2.4 Working group meetings
   2.5 Activity responsibilities
3.0 PPS planning activities and strategies
   3.1 General
   3.2 Supply support
   3.3 Maintenance
   3.4 Technical data
   3.5 Depot planning
   3.7 Configuration management
   3.8 RMA
   3.9 Continued system engineering
   3.10 PPS strategies
4.0 Conduct of the PPS program
   4.1 General approach
      4.1.1 Phase 1: Problem Identification
         4.1.1.1 Vendor Survey
         4.1.1.2 Criteria/priority
      4.1.2 Phase 2: Problem solution
         4.1.2.1 PPS data base
         4.1.2.2 Reprocurement identified problem sheets
            and status reporting
4.1.3 Phase 3: Problem correction
   4.1.3.2 PPS data requirements
   4.1.3.3 Data flow
   4.1.3.4 PPS data elements
   4.1.3.5 Program reviews
5.0 Level of repair program
6.0 PPS major milestones

A typical outline of current PPS plans is contained in Table 5.19. Major principles underlying an effective PPS program include

1. Up-to-date and in-effect plans (e.g., ILS, maintenance, CM, depot, etc.)

2. An operating system to collect and analyze RMA and performance data

3. Preparation and maintenance of a list of all system parts and vendors

4. Complete and current technical data reprocurement packages to support competition

5. Strict hardware configuration control kept up to date

6. Use of vendor surveys to determine part availability for procurement or repair

7. Establishment and use of a PPS data base

8. Use of alternatives strategies to solve specific PPS problems

9. Identification of PPS resource funding requirements

10. Ongoing analysis of RMA and quality data trends on out-of-production equipment

More specifics on the areas identified can be obtained by reviewing the PPS checklist in Appendix 25.

The PPS plan should identify the organizations involved in postproduction management. These include the requirer program office, the in-service engineering agent, and the organizations responsible for R&M analysis, supply support, maintenance planning, configuration management, drawing repository, and depot overhaul point. In the United States it is not unusual to have the prime contractor and possibly one or more support contractors involved. Postproduction supply support must be addressed at the individual item level. The availability and adequacy of technical data are key elements in PPS. Acquisition Method Coding Conferences (AMCCs) can be held to identify shortfalls in the existing technical data and to recommend a plan of corrective action.

The system maintenance plan must be updated to reflect ECPs and PPS analyses. A configuration-indentured index should be entered in the PPS data base to identify parts lists for all configurations and commonality across configurations. The RMA data base files from updated RMA analyses should be included in the PPS data base as necessary.

A number of PPS strategies are available for ensuring continued

TABLE 5.20   Spares and Repair Parts Actions: PPS Strategies

Develop a reprocurement technical package and alternate production	Restrict the issue of data-critical applications in support of combat-essential items
Withdraw from disposal	Phase out less essential systems employing the same parts
Procure life-of-type buy	
Seek substitute (interchangeable parts)	Restrict issue to system where no substitute is available
Redesign component to acceptable standard components if not interchangeable	Accelerate replacement of the system
Purchase plant/equipment and establish an organic depot capability	
Subsidize continuing manufacture	
Draw (cannibalize) from marginal, low-priority system	

support on out-of-production items. The specific strategy will vary with the individual case. Typical strategies are shown in Table 5.20.

A typical PPS program is conducted in three phases (see Fig. 5.47). As a part of the program, specific repairables and parts are identified as potential problems. Which of these are actual problems can be identified through use of various data bases and information sources discussed later. No one method is perfect.

The PPS data base can be used to support problem prioritization and analysis. To minimize the cost of the data base, criteria are needed to restrict the items entered. Table 5.21 contains suggested departure-point criteria for this purpose. The data base can include LSA data, LORA data, RMA data, procurement data such as unit costs and procurement lead times, and current supply availability data. Most systems employ a reprocurement-identified problem (RIP) sheet (see example in Fig. 5.48), and a RIP summary form. Many of the PPS data systems designed to date appear to be overdesigned in the sense they have many features that are not used in practice. A relatively simple system that can input data automatically from related systems to reduce data input work load can provide much of the data for the analysis required. For example, we drafted one PPS data base using an LSAR H record and a word processor. The system contained 34 input data elements and produced 6 different output reports. The H record accommodated about 70 percent of the data elements. The data system can be used to project yearly schedules for PPS investigations and to assist in PPS budgeting.

A general PPS analysis approach is shown in Fig. 5.49. More detailed analysis approaches should be designed for each specific type of PPS problem.

The PPS data system should document parts data and monitor di-

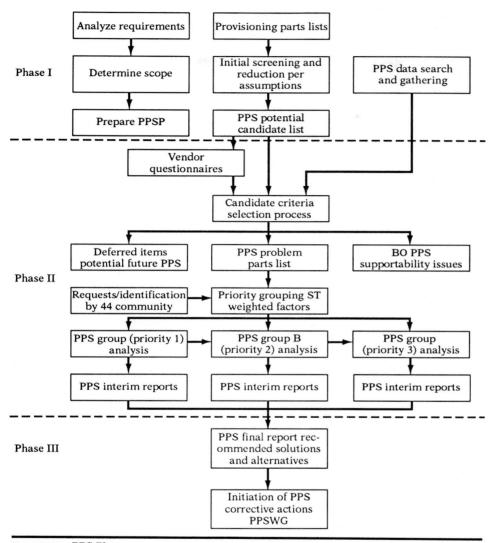

Figure 5.47 PPS Phases.

minishing manufacturing sources. The initial effort should focus on out-of-production and suspected problem items. ECPs and problem reports should be reviewed for PPS impacts. The criticality of PPS-listed items should be identified. Action should be taken to ensure adequate stock levels on critical items. A comparison of repair costs with replacement costs should be made where appropriate. LORA updates should be conducted on items approaching uneconomical repair. Sen-

TABLE 5-21   Selection Criteria for Inclusion in PPS Data Base

Sole source	Production termination
No longer installed in system	Unique to system
Identified as problem	Shortage of spares
Increasing production lead time	High usage
Mission essential	

Part number:	Cage:	Item name:	RIP control number:	Priority:	Date:
Depot:		National stock number:			
Manufacturer's name:		Initiating activity:			
Problem description		Recommended problem solutions		Problem disposition	

Supply posture as of: / /

Qty. on hand:
Back order: Firm Contract: Est. MDD: / / Qty:

Demand/qtr.:
Repair: PA: Est. MDD: / / Qty:
NHA:
Cnblz. act:

Remarks:

Figure 5.48   Sample RIP Sheet.

sitive points in the previous LORAs should be checked periodically to see if field results indicate possible changes in the repair-level decisions in those areas.

Phase 1: Problem Identification.   There is no one single source for problem identification. Problem identification is accomplished in a number of ways, including the following data sources:

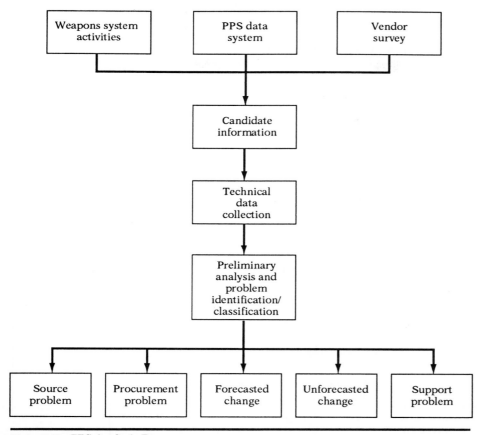

Figure 5.49 PPS Analysis Process.

1. Vendor surveys
2. Government-Industry Data Exchange Program (GIDEP) data
3. Other data sources

A vendor survey is a widely used approach. Figure 5.50 presents a sample form. Some vendors have a policy of notifying the customer when they plan to cease production. Another way is the review of GIDEP reports that cite parts going out of production. A third is for the system support organizations involved to identify parts with potential or current problems. Various data sources should be reviewed to ascertain

1. Existing and planned sources of supply
2. Expected lifetime of maintenance significant items

Drawing no.   Vendor part number

Description

Yes   No

1. Do you plan to produce the part through 2003? (system life)
   If not, will production capability be maintained through 2003?

2. Termination of part production:
   a. Has already occurred.
   b. Is planned to occur _____ (date)

3. When production terminates:
   a. Are customers notified through the GIDEP alert system?
   b. Do you provide a minimum of 6 months advance notice?
   c. Does advance notice include an opportunity for a lifetime buy of parts?
   d. Will you make a production warranty available?
   e. Future requests for production would:
      1) Include setup costs in unit price.
      2) Require separate setup costs (est. $ _____)
      3) Require a minimum quantity order (# _____)
      4) Require production lead time of _____months.
   f. Would you transfer manufacturing rights?
   g. Would you make manufacturing process, procedures, and test specifications available?
   h. Are special drawings, tooling, and/or equipment necessary to produce the part retained?
      If yes, for how long? _____

4. Comments:_____
   _____
   _____

5. If we have additional questions, whom may we contact?

   _____   _____   _____
        Name   Title   Phone #

6. Contact point if vendor has questions:

Figure 5.50 Vendor Survey Form.

3. R&M data

4. Costs associated with in-house and contractor manufacturing and repair alternatives

5. Supply and consumption data

6. Planned production improvements

The vendor survey should be performed to ascertain the status of vendor production and/or inventory of system parts, support equipment, and factory test equipment. The effectiveness of the vendor survey is largely dependent on aggressive follow-up. As vendor responses are received, the data base is updated to reflect date of receipt and is reviewed for completeness. Special attention should be given to critical items. To further focus the effort, problems can be prioritized to

TABLE 5.22 Prioritization Factors and Values

Primary factors	Value
No longer produced, insufficient assets	250
Currently produced, insufficient assets	200
Activity-identified problem	100
No longer produced, sufficient assets	075
Planned production stop in 2 to 5 years	050
Planned production stop prior to end of system service life	025
Added factors:	
Mission essential	075
System unique	050
Sole source/source controlled	025

determine which need the most immediate attention by using weighted values related to the severity of the problem. Departure-point weights are contained in Table 5.22.

Phase 2: Problem solution. The PPS data base developed in phase 1 can be used as the basis for detailed analysis of specific PPS problems and in-depth tradeoff studies of the various PPS strategies. In some cases, such as use of substitute parts identified in the data base, the necessary analysis may be minimal. However, more rigorous studies frequently will be needed to determine the cost and other consequences of the available strategies to provide solution options. As the analysis is completed, the RIP sheet (Fig. 5.48) can be finalized. The RIP sheet can be used as the primary means of documenting and tracking an identified support problem. Data sources for analysis of specific problems can include

Aperture cards/drawings	IPBs
Technical manuals	Training materials
	Usage data
Calibration data	Component reliability data
SM&R codes	Operating manuals
Technical bulletins	RMA data
APLs	Specifications
Test program sets	Configuration-indentured index file
Technical repair standards	

At a minimum, the PPS data base should include the following information:

Part number	Commercial and government entity (CAGE)
National stock number	Item nomenclature

Best replacement factor/MTBF	COG code
SM&R code	Unit price
Repair cost for each repairable	Next higher assembly
DOP/designation	Acquisition method
Predicted parts usage	

Additional data elements that are frequently needed for analysis and may or may not be in the data base include

Work unit code	Lead time
Manufacturer's name	Turnaround time of repair cycle
Survival rate (O or I level)	Depot survival rate
Not repairable this station%	Life-cycle recommended support quantity
Type version	Quantity of part in the end item
Quantity on hand	Quantity on back order
Number of NHAs available	Number of items available for possible cannibalization
Current contract number, if any	Purchase requisition number, if any
Quantity to be delivered under the contract/PR	Estimated delivery date for the contract/PR

Phase 3: Problem correction. After review of the problem analysis and the RIP sheets, the proper authority determines and initiates actions to implement the selected decision.

5.6.5 Test, Evaluation, and Validation (task 501)

As production items are fielded, one remaining subtask, the actual assessment of field data in accordance with the supportability assessment plan for deployment (the fifth subtask in 501), remains to be done. The exact approach to this subtask depends on the specific plan developed.

5.6.6 Postproduction summary

The normal support activities are carried out as production ceases. This may go on for as long as 10 to 20 years. General support and LSA activities in this time period were covered in the task 403 discussion in Sec. 5.6. One issue is who does the work, the government or a prime or support contractor. Various combinations of these organizations are possible and occur in practice. As the CALS concept of an integrated weapons data base is implemented in the coming years, interchange of data between such organizations will grow increasingly practical.

A last issue is how long the LSAR data base is to be maintained and updated. While the DOD theoretically requires this to be done, few programs appear to do so in practice. The use of the LSAR as input to the PPS data base is one possible method for such updates. A number of defense contractors would like to obtain updated service LSAR data bases of existing equipments to be used on new systems to eliminate all or part of the cost of regeneration for use on the new program. As new uses of the LSAR in postproduction are developed, greater use of life-cycle LSARs may develop. However, the cost of updating competes with end-use data (technical manuals, parts reprocurement, etc.) for scarce funds. Again, CALS will affect this in the future, since it may make both more economically feasible.

Postproduction support problems are being exacerbated by a number of different trends. Use of a postproduction support program and data base takes a proactive versus reactive approach to such problems. This allows better planning and budgeting and can result in lower cost, fewer time delays, and hence improved readiness.

5.7 Chapter Summary

This chapter has summarized a number of possible approaches to task execution. The executor in practice can use these approaches as a departure point for deciding on its particular approach. Requirers can use these example approaches as an indication of what they might receive as a result of requiring a specific task or subtask. As a final caution, readers must realize that accepted approaches to execution of many of the LSA tasks are still to be developed. Hopefully, this chapter will assist such development.

6

The Logistics Support Analysis Record

6.0 Introduction

The Logistics Support Analysis Record (LSAR) is a repository of data resulting from the application of LSA. It identifies the type, quantity, and distribution of the logistic resources and contains much of the information used in the development of those resources. The LSAR, historically, has been the automated part of this data base. With the rapid emergence of the Computer-Aided Acquisition and Logistics (CALS) initiative, much more of the LSA-related information (reports, drawings, etc.) will be automated; however, the LSAR will be a key element in the CALS network.

The objectives of LSAR data base are to

Identify the detailed ILS-element required resources

Eliminate duplicate data generation

Allow all ILS elements to have use of the same data

Facilitate summarization and analysis of the ILS data

Provide a source of design feedback at the detail level

This chapter discusses methods to develop and manage the LSAR effectively and efficiently. It specifically discusses

Background   (6.1)

Overview of MIL-STD-1388-2B   (6.2)

Contracting for the LSAR   (6.3)

Planning the LSAR   (6.4)

Managing the LSAR   (6.5)

Summary   (6.6)

6.1 Background

MIL-STD-1388-2 (series), DOD Requirements for a Logistics Support Analysis Record, establishes uniform requirements for the development and delivery of the LSAR. In the 1970s and early 1980s, each service had its own unique requirements for logistics data submission, particularly in the area of provisioning data. In 1984, with the release of MIL-STD-1388-2A, all the services agreed to use the LSAR as the automated logistics data base and for provisioning. This standardization produced many benefits to the LSAR. First, the Material Readiness Support Agency (MRSA) produced a program, written in COBOL, that was available to all services and contractors. This program, while functional, was cumbersome. Second, standardization of the LSAR allowed independent companies to develop LSAR programs that had front-end data entry and edit screens and were easier to use than the government program. MRSA still retains the prerogative to validate contractor-developed programs. A list of the currently validated vendor software is available from MRSA.

With the release of MIL-STD-1388-2B in 1991, the LSAR entered a new era. While the former version used an 80-column flat-file format, the new version uses a relational data base format. With the former file format, the logistician had to use the standard reports or develop (or pay to have developed) complicated programs in COBOL or some other programming language. This also made it difficult to interface with other data bases and modeling tools. Now the logistician can use a much simpler and more user-friendly fourth-generation data base management system (DBMS) language to accomplish the same purpose and develop ad hoc reports. The new LSAR structure encourages the independent development of LSAR systems and linkages with other data bases and supports the CALS objectives, some of which are reduction of redundant data entry, integration with other data bases, and digital delivery of data. MRSA still retains control over the file (table) and data-element formats and validates LSAR software, although there is no longer any government-provided software. The flexibility that this new structure provides will make the LSAR a more readily useful tool for the ILS manager and design community. The future of the LSAR will depend in part on the LSA manager's ability to convert its data into information that becomes the basis for sound logistics decisions.

Although MIL-STD-1388-2B does not require submission of the LSAR in an automated format, for a program of any size this is the only practical way to track and use the data. With MIL-STD-1388-2B, the logistician can develop the required tables using one of the popular data base programs (i.e., dBase, Paradox, Oracle, or even Lotus 1-2-3). Naturally, such an approach will not deliver all the features of a dedicated LSAR program, but it may be adequate for a very small program. It is becoming more and more evident that the LSAR is evolving into a more flexible tool as part of the CALS initiative, and the reasons for not automating in the future are negligible.

The 2B standard version of the LSAR is primarily a software change. If you are familiar with 2A through a software vendor's program, there may seem to be very little change from your present perspective. It is conceivable that the data-entry screens will not change significantly. The method of obtaining reports will not change, and if the LSAR program that you were using before allowed you to develop ad hoc reports, you will only see an increased ease and capability to obtain such reports.

The upgrade to the 2B standard was driven in large part by advances in data base technology that enhance its ease of use. The structure is easier for data base programmers to manage, more logistics documentation requirements can be satisfied directly from the LSAR (see Table 6.1), and some data elements have been added or enhanced.

Throughout this book, and particularly in this chapter, data have been talked about in terms of a particular data table. Most LSA and ILS-element managers will never see nor have to be concerned with these items. What they will need to understand is what data are in the LSAR and what uses can be made of it.

Throughout this chapter the LSAR will be discussed in reference to the MIL-STD-1388-2B version. Where this differs from the 2A standard release, it will be noted. It is important to realize in any discussion about the two releases that they both have the same goal: to be a more cost-effective integrated logistics data base. The 2B standard version is an improvement on this process achieved primarily by using more efficient data base management technology.

6.2 Overview of MIL-STD-1388-2B

MIL-STD-1388-2B, DOD Requirements for a Logistics Support Analysis Record, consists of the standard itself and six appendices (see Table 6.2). It is not the intention here to belabor points adequately discussed in the standard but rather to identify the subjects that are covered in the standard and to expand on them. This book will not be sufficient to obtain a complete, detailed understanding of the LSAR without the standard itself.

The main body of the standard provides an overview of the LSAR. One of the key points is the discussion of tailoring. The LSAR must be tailored to a level and depth consistent with the LSA requirements. For example, what hardware indenture levels and maintenance levels are to be documented? In addition, the relational format of the 2B standard makes this a convenient data base in which to store logistics information and facilitates the integration with engineering design, manufacturing, product support, and other program functions that heretofore may have been separate. The standard also includes general requirements for data submissions. This involves two key elements: data selection and data/report submission media. Data selection will be discussed in detail later in this chapter. The *only* practical way to accomplish data submission with the 2B standard is

TABLE 6.1 LSA Output Reports

No.	LSA Report	ILS Element	Product	Document	Contractor Use	Government Use
001	Manhours by Skill Specialty Code and Level of Maintenance	MX MP TG	DI-ILSS-81113A		Design Influence	Review
002	*Personal and Skill Summary*			This data is in LSA-001 in 2B		
003	Maintenance Summary	MX	DI-ILSS-81119A		Design Influence	Review
004	Maintenance Allocation Chart	MX TD SE	MAC DI-ILSS-81140A	MIL-M-63038B(TM)	Review	Review
005	Support Items Utilization Summary	SE MX SS	DI-ILSS-81141A		Justify Requirements	Review
006	Critical Maintenance Task Summary	MX TT	DI-ILSS-81142A		Design Influence	Review
007	Support Equipment Requirements	SE TT MX	DI-ILSS-81143A		ILS Resource Determination	Review
008	Support Items Validation Summary	SS SE MX	DI-ILSS-81144A		QA	Review
009	Support Items List	SS MX	DI-ILSS-81145A		QA	Review
010	Parts Standardization Summary	SS	DI-ILSS-81146A		Design Influence	Support Breakout
011	Special Training Equipment/Device Summary	TT SE	DI-ILSS-81147A		ILS Resource Determination	ILS Resource Determination
012	Facility Requirement	F	Facility Requirements* DI-ILSS-81148A		ILS Resource Determination	ILS Resource Determination
013	Support Equipment Grouping Number Utilization Summary	SE	DI-ILSS-81149A		ILS Resource Determination	ILS Resource Determination
014	Training Task List	TT MP	Training Task List DI-ILSS-81150A		Training Materials	Review
015	*Sequential Task Description*			This data is in LSA-019 in 2B		
016	Preliminary Maintenance Allocation Chart	MX TD SE	PMAC DI-ILSS-81151A	DI-L-7190	Maintenance Publications	Review
017	*Preliminary Maintenance Allocation Summary Tool Page*			This data is in LSA-016 in 2B		
018	*Visibility and Management of Operating and Support Cost (VAMOSC) Summary*					
018	**Task Inventory Summary**	MP TT MX	Task Inventory List DI-ILSS-81152	MIL-STD-1478	Technical and Training Pubs	Review

TABLE 6.1 LSA Output Reports (Continued)

No.	LSA Report	ILS Element	Product	Document	Contractor Use	Government Use
019	*Maintenance Task Analysis Validation Summary*			This data is in LSA-019 in 2B		
019	**Task Analysis Summary**	TD MX S	O&M Manuals* Task Skills Analysis DI-ILSS-81153A		Quality Assurance	Review
020	*Tool and Test Equipment Requirements*			This data is in LSA—004 in 2B		
021	*Task Referencing List*			Not applicable in 2B		
022	*Referenced Task List*			Not applicable in 2B		
023	Maintenance Plan Summary	MX R SS	AF Maintenance Plan DI-ILSS-81183A	AF 66-14	Quality Assurance	Review
024	Maintenance Plan	MX R SS	Navy Maintenance Plan DI-ILSS-80119C	OPNAVINST 5000.49A	ILS planning	Same
025	Packaging Requirements Data	PHST	Packaging Requirements Data DI-PACK-80120	MIL-STD-2073-1A	Pack and ship Information	Pack and Ship Information
026	Packaging Developmental Data	PHST	Packaging Developmental Data* DI-ILSS-81154A	MIL-STD-2073-1A	Pack and Ship Information	Pack and Ship Information
027	Failure/Maintenance Rate Analysis	R MX SS	DI-ILSS-81155A		Design Influence	Review
028	*Reference Number/Additional Reference Number Cross Reference List*					
029	*Repair Parts List*			This data is in LSA-030 in 2B		
030	*Special Tools List*			This data is in LSA-030 in 2B		
030	**Illustrated Parts List**	TD SS SE	RPSTL* DI-ILSS-81156A IPB* Stockage List Type Four*	MIL-STD-335(TM) MIL-STD-49502(TM) MIL-M-38807(AF) Marine Corps	Technical Manuals	Review
031	*Part Number/NSN/Reference Designator Index*			This data is in LSA-030 in 2B		
032	DLSC Submittals	SS	DLSC Screening DI-ILSS-XXXXX	DI-V-7016	DLSC Screening	Review
033	**Preventive Maintenance Checks and Services**	MX TD	PMCS DI-ILSS-81157A	MIL-M-63038(TM) MIL-M-63036 (TM)	Maintenance Manuals	Review

TABLE 6.1 LSA Output Reports (Continued)

No.	LSA Report	ILS Element	Product	Document	Contractor Use	Government Use
034	*Stockage List Type Four*			This data is in LSA-040 in 2B		
036	Provisioning Requirements	SS	PPL DI-V-7002A DI-V-7003A LLTIL DI-V-7004A RIL DI-V-7005A ISIL DI-V-7006A TYEL DI-V-7007A CBIL DI-V-7008A DCN DI-V-7009A SCPL DI-V-7192A	MIL-STD-1561	Provisioning Documentation	Provisioning
037	**Spares and Support Equipment Identification List**	SS SE MX	DI-ILSS-81158A		Supply Support	Same
039	**Critical and Strategic Item Summary**	SS MX	DI-ILSS-81159A		ILS Resource Determination	Same
040	**Authorization List Items Summary**	SS TD MX	COEIL BIIL AAL EDMSL DI-ILSS-81160A	MIL-M-63036(TM)	Emergency Repairs	Review
040	*Components of End Item (COEI) List*			This data is in LSA-040 in 2B		
041	*Basic Issue Items (BII) List*			This data is in LSA-040 in 2B		
042	*Additional Authorization List (AAL)*			This data is in LSA-040 in 2B		
043	*Expendable/Durable Supplies and Materials List (ESML)*			This data is in LSA-040 in 2B		
045	*Stockage List Type Three*			This data is in LSA-040 in 2B		
046	**Nuclear Hardness Critical Item Summary**	S MX	DI-ILSS-81161A			Review
050	Reliability-Centered Maintenance	R MX TT	RCM Report* DI-ILSS-81162A	MIL-STD-2173(AS) MIL-STD-1843 AMCP 750-2	Design Influence	Review

TABLE 6.1 LSA Output Reports (Continued)

No.	LSA Report	ILS Element	Product	Document	Contractor Use	Government Use
051	*Reliability Summary-Redesign*		This data is in LSA-058 in 2B			
052	*Criticality Analysis Summary*		This data is in LSA-056 in 2B			
053	*Maintainability Analysis Summary*		This data is in LSA-058 in 2B			
054	*Failure Mode Analysis Summary*		This data is in LSA-056 in 2B			
055	*Failure Mode Detection Summary*		This data is in LSA-056 in 2B			
056	**Failure Modes, Effects, and Criticality Analysis Report**	R MX SS	FMECA Report DI-ILSS-81163A	MIL-STD-1629	Design Influence	Review
058	**Reliability, Availability, and Maintainability Summary**	R S MX	RAM Report* DI-ILSS-81164A	MIL-STD-785 MIL-STD-470	Design Influence	Review
060	*LSA Control Number Master File*					
061	*Parts Master File*					
065	**Manpower Requirements Criteria**	MP TT	MARC Report DI-ILSS-81165A		Design Influence	Resource Planning
070	Support Equipment Recommendation Data	SE TD MX	SERD DI-ILSS-80118C	MIL-STD-2097	ILS Resource Determination	Use
071	**Support Equipment Candidate List**	SE TD MX	Support Equipment Candidate List DI-ILSS-81166A	MIL-STD-2097	Design Influence	Review
072	Test Measurement and Diagnostic Equipment Requirements Summary	SE TD MX	TMDE Registration DI-ILSS-80288A	DA Form 4062-R DA Form 4062-1R	ILS Resource Determination	Use
074	Support Equipment Tool List	SE TD MX	Support Equipment Tool List DI-ILSS-80289A	MIL-STD-2097	Identify New SE	Use
075	*LSAR Manpower Personnel Integration (MANPRINT)*					
075	**Consolidated Manpower, Personnel and Training Report**	MP TT	MANPRINT Report DI-ILSS-80290A		Design Influence	Resource Impact
076	**Calibration and Measurement Requirements Summary**	SE TD MX	CMRS DI-ILSS-81167A	MIL-STD-1839	Identify CMRs	Use
077	Depot Maintenance Interservice Data Summary	F SE MX	Joint Depot Maintenance Interservice Data DI-ILSS-80291B	JLC 28/29/30 Forms	Depot Planning	Use
078	**Hazardous Materials Summary**	S MX PHST	DI-ILSS-81168A		ILS Resource Determination	Use

251

TABLE 6.1 LSA Output Reports (Continued)

No.	LSA Report	ILS Element	Product	Document	Contractor Use	Government Use
080	Bill of Materials	SS	Bill of Materials DI-ILSS-81169A		Quality Assurance	Review
085	**Transportability Summary**	PHST	Transportability Report* DI-ILSS-81170A	MTMC PAM 70-1	ILS Resource Determination	Review
100	*Chronolog Information*		Data processing reports not applicable in 2B			
101	*Transaction Edit Results-Selection Cards*					
102	*Transaction Edit Results-LCN Master*					
103	*Transaction Edit Results-Parts Master*					
104	*Transaction Edit Results-Narrative Master*					
105	*Key Field Change Transactions*					
106	*Reference Number Discrepancy List*					
107	*LCN-Task Identification Code Cross Reference List*					
108	*Critical Data Change*					
109	*Unidentified Transactions*					
126	**Hardware Generation Breakdown Tree**	R TD SS	DI-ILSS-81171A		Design Influence	Review
150	*Provisioning Error List*		Data processing reports not applicable in 2B			
151	Provisioning Parts List Index	SS	PPLI DI-ILSS-XXXXX	MIL-STD-1561	Reference	Use
152	PLISN Assignment/Reassignment	SS	DI-ILSS-81172		Quality Assurance	Review
154	Provisioning Parts Breakout Summary	SS	DI-ILSS-8029B		ILS Resource Determination	Use
155	Recommended Spare Parts List for Spares Acquisition Integrated with Production	SS MX	SAIP DI-ILSS-80293A	MIL-STD-1561	ILS Resource Determination	Use

*The LSA report serves as source data for this requirement but does not satisfy it entirely.
NOTE: LSAR DID is DI-ISSS-81173; **boldface** indicates new reports in MIL-STD-1388-2B; *italics* indicates not contained in MIL-STD-1388-2B; MP: Manpower and Personnel; SE: Support Equipment; TT: Training and Training Support; R: Reliability and Maintainability; MX: Maintenance Planning; SS: Supply Support; F: Facilities; S: Safety; TD: Technical Documentation; PHST: Packaging, Handling, Storage and Transportation.

TABLE 6.2 MIL-STD-1388-2B Contents

Section	Title
1	Scope
2	Referenced Documents
3	Definitions
4	General Requirements
5	Detailed Instructions for Manual/Automated Preparation of the LSAR Relational Tables
6	Notes
Appendix A.	LSAR Relational Tables
Appendix B.	LSAR Reports
Appendix C.	Guidance for Assignment of Logistics Support Analysis Control Number (LCN), Alternate LCN Code (ALC), LCN Type, and Usable on Code (UOC)
Appendix D.	Application and Tailoring Guidance for the Logistics Support Analysis Record (LSAR)
Appendix E.	Data Element Dictionary
Appendix F.	List of Logistics Support Analysis Record Acronyms

through some automated means. Even though the standard discusses manual generation and submission of data tables and reports, this is not practical on even the smallest programs. This is roughly akin to saying that you can perform a spreadsheet analysis manually. It is possible, but why do it? For very small programs, the minimum automation should be submission of the tables in an ASCII format using a commercially available data base product.

Paragraph 4.2.2.2 of MIL-STD-1388-2B describes the minimum requirements for an LSAR automated data processing (ADP) system to be validated. These requirements are

1. Ability to accept the LSAR data tables

2. Data elements conform to the LSAR data element definition (DED)

3. Ability to output the LSAR data tables

4. Ability to produce *all* the LSAR reports (It was only necessary for the 2A standard systems to be validated for the reports that they wished to generate.)

5. Ability to output change-only data

6. Automated user comment capability

These requirements are the only requirements that the government places on the LSAR ADP system developer. However, these are *not* the only requirements that most end users will have. Other user considerations are ease of data entry, adequacy of on-line edits, integration with existing systems/data bases, ad hoc report capability, mainframe/PC/network versions, LSA *and* ADP expertise of the developer, support provided, and training provided. It is the responsibility of users to determine their requirements and select or develop a system that meets them.

Paragraph 5 of the standard, "Detail Instructions for Automated or Manual Preparation of LSAR Relational Tables," provides overviews of the data tables. These are discussed fully in the next subsection. DD form 1949-3 now includes space for the requiring activity to include their information for the A, E, and X tables. (*Note:* The DD form 1949-3 included in the original release of MIL-STD-1388-2B is only an example and is *not* the approved form.)

The standard itself and appendices A, B, and E address contractual material. Appendices C, D, and F are informational in nature.

6.2.1 Appendix A: "Logistics Support Analysis Record Relational Tables"

This appendix discusses the detailed requirements of the data table structure, relationships, and contents and identifies key data elements. For those familiar with the 2A standard, Table 6.3 provides a crosswalk of the 2A standard terms to the 2B standard terms. This appendix is divided into sections called *functional areas* (Table 6.4). The functional areas identify the table and data requirements for each area. The first letter of a table code corresponds to the functional area letter; i.e., the cross-functional requirements are in the X tables. As an example, Table 6.5 lists the tables in the cross-functional (X) area.

Within the tables the data-element code identifies each data element. This is an eight-position code in which the last characters identify the table. EIACOD*XA* is the End-Item Acronym Code from the XA table. If the data element migrates to another table as a foreign key, it retains its code. (*Note:* The table letter designation follows the record letter designation of MIL-STD-1388-2A as much as possible; i.e., the A tables in the 2B standard and the A records in the 2A standard contain roughly the same type of data.) The functional areas are as shown in the following subsections.

X Tables: Cross-Functional Requirements. These tables contain data that crosses or links multiple functional areas. They are required to establish a system in the LSAR. They capture the basic setup information

TABLE 6.3   MIL-STD-1388-2A to MIL-STD-1388-2B Structure Comparison

2A Standard	2B Standard
547 data elements	518 data elements
127 data cards (i.e., B07 card, Reliability)	104 tables (i.e., BD table, RAM Indicator Characteristics)
15 data records (i.e., C record, Operation and Maintenance Task Summary)	9 table groups (i.e., C tables, Task Inventory, Task Analysis, Personnel, and Support Requirements)
3 master files	No master files—each table is effectively a master file

TABLE 6.4   LSAR Functional Areas

Table	Functional area
X	Cross-functional Requirements
A	Operation and Maintenance Requirements
B	Reliability, Availability, and Maintainability; Failure Modes, Effects, and Criticality Analysis; and Maintainability Analysis
C	Task Inventory, Task Analysis, Personnel and Support Requirements
E	Support Equipment and Training Materiel Requirements
U	Unit Under Test Requirements and Description
F	Facilities Considerations
G	Personnel Skill Considerations
H	Packaging and Provisioning Requirements
J	Transportability Engineering Analysis

TABLE 6.5   LSAR Cross-Functional Requirements Tables

Table code	Table title
XA	End Item Acronym Code
XB	LCN Indentured Item
XC	System/End Item
XD	System/End Item Serial Number
XE	LCN to Serial Number Usable on Code
XF	LCN to System/End Item Usable on Code
XG	Functional/Physical LCN Mapping
XH	Commercial and Government Entity
XI	Technical Manual Code and Number Index

for the system, such as End-Item Acronym Code, level of repair analysis (LORA) data, some contract data, and Logistics Control Number (LCN) structure. There are also tables that allow the system to be managed by serial number, a functional to physical LCN mapping, commercial and government entity (CAGE) and associated data, and usable on code (UOC). These tables establish the basic keys for the LSAR.

A Tables: Operations and Maintenance Requirements. These tables contain information relating to the anticipated operation, the operation/maintenance environment, and maintenance requirements. These tables document the user's requirements, and the data should be provided by the requirer in the contract specification. These tables will only be completed for those items (primarily the end item and possibly major assemblies) which have operational and/or maintenance requirements imposed on them. *Note:* The R&M elements of MAMDT, MTTR, MTBMA, and MTBF are broken out in two categories: logistical and mission-related. The 2B LSAR standard uses the terms *technical* to describe logistical data and *operational* to describe operation-related data. The technical factors document all failures and resultant actions (i.e., failure of a redundant component). The operational factors reflect the operational reliability and maintenance characteristics that the system must demonstrate (i.e., computer failure that results in an aborted mission). Two key points to remember about technical and operational failures are (1) an operational failure will always be a technical failure, and (2) the technical factors are the factors that should be used in most instances for logistics planning. In other words, even though failure of a redundant component may not cause an operational failure, it will require use of a replacement part(s), maintenance manhours, and other logistics expenditures.

B Tables: Item Reliability, Availability, and Maintainability Characteristics; Failure Modes Effects and Criticality Analysis; and Maintainability Analysis. These tables contain most of the reliability, availability, and maintainability (RAM) characteristics of the LSA candidates. In particular, they contain the logistics design considerations, item function, maintenance concept, and the results of the failure modes and criticality analysis (FMECA), maintainability analysis, and reliability-centered maintenance (RCM) analysis. They also include a narrative data element in order to document potential redesign of an item due to the logistics analysis. One table cross-references failure modes to their repair tasks. Data are entered in these tables for each repairable item in the system.

C Tables: Task Inventory, Task Analysis, Personnel, and Support Requirements. These tables contain task narrative for operation and maintenance tasks. Even though LSA task 301 specifies the documentation of both operation and maintenance tasks, historically the LSAR has not often been used to document operator tasks, although it does have this capability. With the increased emphasis on concurrent engineering, documenting both these task types in the LSAR could benefit the development of total support of the end item. Other data found in these tables include the resources (personnel, support equipment, and provisioned items) required for each task. These tables provide the link between the tasks and the provisioned support items. (*Note:* These tables contain the information that was documented in the C, D, and D1 records of MIL-STD-1388-2A. There are no D tables in the LSAR.)

E Tables: Support Equipment and Training Materiel Requirements. These tables contain the data relating to existing or new support/test/training equipment. Much of this information is required for developing the Support Equipment Recommendation Data (SERD) report. These tables should be completed to document the design/ILS information for each item of support equipment used to support the system under analysis. They also can be used to document alternate support equipment.

U Tables: Unit Under Test Requirements and Description. The unit under test characteristics are captured in these tables. Here is the direct link between maintenance task requirements and fault-isolated replaceable unit(s). Tables will be completed for each reparable assembly that will be tested. They store the data used in the Calibration and Measurement Requirements Summary (CMRS) and identify Automatic Test Equipment (ATE) workstations. (*Note:* In MIL-STD-1388-2A this information was store in the E2 records.)

F Tables: Facilities Considerations. The data in these tables describe and justify any proposed special and/or additional facilities. Here the logistician can describe a baseline facility, modifications to existing facilities, type of facility, construction costs, environmental considerations, and new facility requirements. Tables can be completed for each new facility or modification required and can link such facilities or modifications with a task requiring such construction.

G Tables: Personnel Skill Considerations. These tables provide the ability to document basic skill information and new or modified skill requirements for support of the system. These tables reference the skills to the tasks that require them. G tables will be completed for each new skill required to operate/maintain the system.

H Tables: Packaging and Provisioning Requirements. These tables are primarily comprised of the data elements that were called the *Parts Master File* in MIL-STD-1388-2A. The provisioning, as well as the application, information is stored here for each component of the end item. These data are used and updated in the Defense Logistics Support Center (DLSC) screening process. These tables also contain packaging data, parts effectivity, and design-change information. The H tables are filled out for each part-numbered item within the system and allow for serial number and configuration tracking and control.

J Tables: Transportability Engineering Analysis. These tables document the transportability and shipping-mode information. These tables are normally only applicable to the end item, although in some larger systems tables may be prepared for sectionalized portions of the end item.

6.2.2   Appendix B: "LSAR Reports"

The 2B LSAR has 48 standard reports (see Table 6.1) that are defined in the military standard. These reports each have a Data Item Description (DID) associated with them and would be contracted for by including the appropriate DID on the contract data requirements list (CDRL). In MIL-STD-1388-2B, reports have a greater significance than they had in 2A standard. In the 2B standard ADP systems it will be unwieldy, if not impossible, to review the entire data base by looking only at the tables (i.e., master files in the 2A standard). This places a burden on the requiring authority to determine what information is required and what reports will be necessary to review/use that data. Appendix B of the 2B standard contains a crosswalk matrix of LSAR Data Tables to Reports. Once the requiring logistician has selected the reports desired, he or she can use this matrix to identify the required data elements and properly fill out DD Form 1949-3.

A contractor or subcontractor may wish to use more reports and include more data elements in the LSAR than specified by the requirer. Besides the standard reports, the relational data base structure greatly enhances the development and use of ad hoc reports. In the 2A standard ADP system the logistician was saddled with a cumbersome 80-column card format. While the same basic data are generally available from the 2A standard LSAR, it was a chore to use. It usually required the work of a programmer to develop any unique reports. With the 2B standard, not only is the development of unique reports simplified, but contractors can use the key fields to "hook" into other data bases and truly integrate the logistics and design functions. Managers who do not limit themselves to developing the LSAR as a deliverable to their customers, but as a tool to develop all their logistics products, will be the managers of the future.

Appendix B of the 2B standard also describes the report selection criteria. This is an important but often overlooked phase of the LSAR. Each report can be selected by LCN, alternate LCN code (ALC), UOC, and Service Designator Code. In addition, many of the reports can be selected by maintenance level and other criteria. At different times and for different reasons you may want to change the selection criteria. For example, you may want to look at the support items required at depot level, or you may have a problem with a particular component and wish to see the data on that one particular LCN. The more familiar you are with the LSAR, its reports, selection options, and data elements, the better you will be able to utilize it cost-effectively.

The DIDs for these reports, as well as a more complete description of them, may be obtained from the MRSA LSA electronic bulletin board. See Appendix 1 for an example of the FMECA DID and report description.

6.2.3 Appendix C: "Guidance for Assignment of LSA Control Number (LCN), Alternate LCN Code (ALC), LCN Type, and Usable on Code (UOC)"

Appendix C of the 2B standard provides guidance for establishing the LCN structure. The LSAR can accommodate both physical and functional LCN assignments. The *physical* LCN assignment (also called the *traditional*) corresponds to the physical breakdown of the equipment and usually is developed using the drawing tree. This structure is required to develop provisioning documentation. The functional structure may be used to document reliability and maintainability data against the appropriate system regardless of where an item is physically located. When both physical and functional LCNs are used, the physical LCNs should take precedence for data storage, and there must be a map of physical to functional LCNs (Table XG).

The standard discusses the three most effective LCN assignment methods: classical, modified classical, and sequential. The LCN-ALC-UOC combination contains the key fields that are necessary for report selection and data base management. If the LCN is properly constructed, the 2B data standard system will allow hardware to be broken out by hardware-generation breakdown, functional breakdown, work-unit code, and work-breakdown structure. The LCN identifies an item to its next-higher assembly and the function of this structure to allow for data "roll-up" and report generation. The 2B standard provides an excellent description of the LCN assignment requirements. Selection of method should follow the order that they are listed in the standard. That is, the classical method is the most convenient and easiest method to use and document (the 2B standard allows for more characters in the LCN so that this method may be used more often than it has been in the past on large programs). If your system is more complicated, use the modified classical method. And finally, as a last resort, use the sequential method.

The ALC is used to document alternatives, such as design or maintenance concepts, one of which will be selected for the final system. This is important to remember, particularly in provisioning, since the ALC is not a selection criterion in the provisioning summary report (LSA-036). As a rule of thumb, do not use ALCs if they can be avoided. The basic ALC is 00. (In the 2A standard, the basic ALC was a blank.)

Different configurations can be handled by using the UOC. Each configuration should be assigned a UOC at the system level. When an item is applicable to more than one configuration, multiple UOCs can be assigned to an LCN-ALC combination. The UOC is mandatory for record establishment and is the first key field criterion for report selection.

6.2.4 Appendix D: "Application and Tailoring Guidance for the LSAR"

Appendix D of the 2B standard provides some general guidance for the application and tailoring of the LSAR. It discusses the relationship of LSA tasks that have an impact on the LSAR and the selection of

LSAR data requirements by identifying data requirements for other program deliverables and by integration with other program elements. It also touches on the timing of deliverables, as well as alternate methods of LSAR delivery. Each of these topics will be discussed in greater detail later in this chapter.

6.2.5 Appendix E: "Data Element Dictionary (DED)"

This is the section of the standard that identifies and defines the data elements that may be contained in the LSAR. In addition to the data elements, it also contains tables that cross-reference the data to the associated tables. Each DED entry has a DED number, a data-element title, the field format, and a definition. In addition, there also may be data items, data codes, and role names. A data item is used when a limited number of descriptive items of information apply to a field. For example, LCN-TYPE may be either *physical* or *functional*. The data code is one or more alphanumeric characters that represent a data item. Following the previous example, a data code would be *P* for the data item *physical* and *F* for *functional*. A *role name* is a unique modifier of a data title when it is used in a particular data table. For example, *REFERENCED TASK CODE* has the same definition and format as *TASK CODE,* but its title is different owing to its use and position in the data tables.

6.2.6 Summary

The foregoing was intended as a brief summary or introduction to the LSAR and the 2B standard. It was not intended to be a substitute for the MIL-STD-1388-2B. No one could consider himself or herself knowledgeable in LSAR without reading and understanding the military standard. It is a fairly clearly written document, and little would be gained by reiterating all the information contained therein. Rather, this book will try to expand on the guidance found in the standard.

6.3 Contracting for the LSAR

Up to this point, this chapter has dealt with what the LSAR is and its uses. This section will discuss considerations in contracting for and delivery of the LSAR. It specifically discusses

Preparing LSAR requirements

LSAR delivery options

Conversion of the 2A standard data to the 2B standard

Reviewing LSAR data

Timing

Life-cycle maintenance

6.3.1 Preparing LSAR requirements

The LSAR can store a tremendous amount of data. The data elements within the LSAR were included to satisfy the data demands of the Army, Navy, Air Force, and Marines across a broad spectrum of program sizes and complexity. This means that there is no standard requirement for LSAR. Each LSAR should be unique.

The first consideration in contracting for an LSAR should be the desired use of the data. Table 6.1 identifies some uses of the LSAR reports to satisfy data requirements of the various logistics elements. After determining the reports required, you can use Fig. 14, "LSAR Data Tables to Report Matrix," in MIL-STD-1388-2B to identify the data elements you need to include on DD form 1949-3. When tailoring the LSAR requirements, it is important to note that it is not always necessary to get all the data elements that are included on a desired report. For example, supply support personnel may not be interested in seeing the maintenance task distribution or repair recycle times on the LSA-036 (Provisioning Requirements). These are not mandatory data elements to process the report and do not need to be included on DD form 1949-3. However, if these data elements are requested by another discipline, they will be on the LSA-036. Conversely, some data elements may not fall out on any of the standard reports, or the standard reports may not satisfy a particular user's requirements. Data elements may be called out on DD form 1949-3 solely for use in ad hoc reports. It is important to know what data elements can be included in the LSAR and how it can be used. With this in mind, the contractor should review DD form 1949-3 to (1) ensure that the customer has asked for sufficient data to generate the reports required, (2) determine how those data will be developed, and (3) determine other data elements that should be added for internal use.

Another consideration in the preparation of LSAR requirements is the amount of detail to be included. Some of the factors here include the level of maintenance and the hardware indenture. Questions pertinent to the level of maintenance data desired are: Will the data include all levels from crew through depot support? Will the data include maintenance and operator tasks? How much detail will be included in the task? On the hardware side, these questions should be answered: Will a hardware FMECA be required? Should the analysis only go to the lowest repairable level, or must it go to the piece-part level?

In the event that there have been previous LSAR efforts, the LSA manager must determine the applicability of those efforts to the present contract. Will such data be GFI provided to the contractor? Will the LSAR need to be updated or completely redone? In the past, previous LSAR efforts have generally not transferred into new contracts. Finally, if the data were developed under the MIL-STD-1388-2A requirements, should they be converted to the 2B standard format?

6.3.2 LSAR delivery options

MIL-STD-1388-2B allows the customer great flexibility in how the LSAR data may be obtained. The LSAR reports may be delivered either in hard copy or digital format, the LSAR data may be delivered in their data-table format either by some physical magnetic medium or by telecommunications, or the customer may have direct interactive access to the LSAR data. Because of this greater flexibility, there is a greater onus on customers to ensure that they do not cause unnecessary expense to the program by requiring a delivery option that does not meet their needs. The following paragraphs identify some of the considerations that should be taken into account when specifying the means of LSAR deliverables.

The LSAR hard-copy reports are the least flexible of the delivery methods. However, there may be times when this is the best method of delivery. Examples are in early phases of a program where there are insufficient data to perform additional analysis or when a customer has no resident data-processing capabilities. Digital-format LSAR reports are only slightly more flexible, but they may be adequate when the preceding conditions exist and there are requirements for multiple hard-copy reports.

Submission of the LSAR data tables in digital format delivered either in a magnetic medium (i.e., tape or disc) offers the customer much more capability to analyze the data if some internal means of processing the data (i.e., a validated LSAR system) is available. Personnel with the capability to produce reports (both the LSAR standard and ad hoc reports) and analyze the various required data elements for content and completeness are also obviously needed. Other reasons to select this method include additional internal analysis with the data, to combine with or include in other LSAR data files (such as subcontractor to contractor), or to allow for maintenance over the life cycle of a system. In general, if the customer has the capability of processing the LSAR data, receiving the data files is the most cost-effective means of delivery. The DID for delivery of the MIL-STD-1388-2B data tables is contained in Appendix 1.

The third method of delivery is interactive access to the contractor data base. This method will probably become more widely used with the new 2B standard and the CALS initiative. However, there are a host of potential problems to consider when your customer has access to your data base or when you as a customer want that access. Some of these considerations are:

Will the access be to standard and other predefined reports or include ad hoc reports?

How many data bases must the contractor maintain (working files, files ready to be approved by the customer, validated files)?

How many access lines must be available?

How much access time does the customer want/need?

If reviews are accomplished through the interactive access, how is this controlled (e.g., when two of the customer's ILS-element managers send conflicting comments directly to the contractor)?

Are the telecommunications standards, hardware, and software compatible?

What are the levels of access?

CALS-HDBK 59[19] discusses these topics in more detail, and Appendix 9 contains LSAR SOW language for use in a CALS environment.

Regardless of which of the preceding methods is used for initial and interim submissions, the final submission usually should be in some digital format. This will allow the data to be enhanced in later development stages, maintained by the customer for the life cycle of the item, and used in other programs for various analysis purposes.

6.3.3 Conversion of 2A standard data to 2B standard format

While there are numerous advantages to using the 2B standard LSAR data tables, if a program's files were originally developed using the 2A standard, it may not be a simple matter to convert those files to the 2B standard format. This will depend on the volume of data, the type of data, the phase of the program, and the intended use of the data.

If it is early in the program and the volume of data is insignificant, then the conversion should be seriously considered. In later phases when more data are involved, greater care must be taken in this decision. Although at the present time there are no validated software conversion routines, DOD and some commercial vendors are developing such programs. While these routines may convert a great deal of the data, they will *not* eliminate the need for manual manipulation of some of the data. The A, B, C, D, E, and H records will convert *if* the 2B standard table keys are contained in the 2A standard file. In any case, the X tables will probably have to be set up manually. The F, G, and J records will primarily require manual conversion. Another factor is that the two LSAR systems treat reference tasks differently. How the referencing was done could have a major impact on the amount of manual manipulation that will be necessary. Finally, there are data in the 2B standard version that are not recorded in the 2A standard version. If these data are called out in DD form 1949-3, additional analysis may be required to obtain these data.

All of this is not intended to discourage ILS managers from selectively converting their 2A standard programs to 2B standard format. If the program or data will be in use for any long period of time, the ease and flexibility obtained through 2B standard format certainly argues for the conversion. Each case must be considered on the basis of its cost-effectiveness. The ILS manager needs to carefully scope how much time and manual effort will be involved, since there will be no "black box" conversion in which 2A standard data are input and 2B standard data are magically output.

6.3.4 Reviewing LSAR data

Obviously, the LSAR cuts across many ILS boundaries. For this reason, it is not practical to have only one person review the LSAR. A team approach is needed. Each team member need not review the LSAR in toto, but only those areas which apply to his or her specific element. The LSAR should be reviewed not only as a data deliverable but also with emphasis on how the data are to be used and their adequacy for the task. For example, if task narrative will be used as the basis for the technical manuals and the same people who review the LSAR tasks also review the manuals, they will have a more incremental insight into development of the manuals. In such a case, when the technical manual review is finally held, it will be more a matter of checking the format, since the contents will already have been reviewed. In one large program, the technical manual review took only 3 days because the reviewers had also been a part of the LSAR review team. A good review of the LSAR not only ensures that the data are properly documented but also, and more important, identifies logistics problems in the system.

The LSAR may be reviewed in total or by using some valid sampling technique. It is important to review the LSAR as completely as possible in its early stages, even though not much data may have been entered, since this is the foundation on which the LSAR will build. If the LCN structure is not adequate or data are not being entered properly, these procedure and structure-type problems need to be addressed and corrected early. There have been cases where the LSAR has been delivered only at the end of the contract. When the data are wrong at this point, not only are there many years of data to correct, but also the issue of why to correct it surfaces because the program is at its end. The LSAR must be reviewed incrementally as it is developed, particularly if it is to be used to develop other data deliverables. For the initial reviews it may be beneficial to use a checklist type of approach until the reviewers become familiar with the data and format of the LSAR. A sample checklist is presented in Appendix 16. As the LSAR becomes larger, a 100 percent review could become prohibitively time consuming; however, this review should not be ruled out without careful consideration.

As on-line access to the LSAR becomes more common, more on-line review is anticipated. Just as in the hard-copy review, the customer's ILS manager must coordinate individual team member comments. The manager cannot permit the LSAR developer to respond to every whim of each member of the ILS team, particularly if they are in conflict among themselves. On-line review has been touted as a great time saver to the government and the contractor because the review will not have to take place at the contractor's plant. While this is beginning to happen, it is important to note that the LSAR should never receive its initial review at the contractor's site during the review meeting. The LSAR should be available to the ILS team members at least 30 days prior to the review meeting. The members of the team should provide their comments to the ILS chairman for consolidation before meeting with the contractor. Another option to consider is accomplishing the review via a teleconference. Finally, having the ILS team, or members, meet on a regular basis allows everybody to be aware of the program's progress and exchange ideas in an open forum.

6.3.5 Timing

The LSAR is an incrementally developed product. This means that not every report delivered will be meaningful early in the program. Not every data item will be included in the initial LSAR, nor will the level of detail be as developed as required for the final deliverable. However, it is necessary that the data be available in time for use by the requiring logistics element. For this reason, the LSA manager must be aware of data schedule requirements. One method to accomplish this is to identify which data deliverables will be developed either directly from the LSAR or to which the LSAR data will be an input. In Sec. 6.3.1, data deliverables were identified as a basis for establishing the required data elements in the LSAR. Using this same methodology, one can identify a schedule for having data available in the LSAR. Table 3.10 presents an example of this. The ILS manager must keep in mind that not all the required data nor the entire depth of a particular table needs to be completed at the same time. For example, the Long-Lead-Time List (DI-V-7004A) may be required early in the contract. This list would not usually include either all the parts on the final Provisioning Parts List nor all the data elements. Another example is the technical manuals. In this case, the LSAR task descriptions need to be developed to the level of detail necessary to support manual development.

6.3.6   Life-cycle maintenance

As the LSAR becomes a more acceptable and useful tool, more consideration must be given to maintaining this data base throughout the entire life cycle. The LSAR can be used in postproduction in a manner similar to its use in acquisition. At least one DOD service is now planning to do this. It also would be useful to support future analytical and design efforts. The effort and expense taken to develop these data should dictate that the LSAR remain valid throughout a system's life cycle. However, before this can happen routinely, there must be a plan to integrate the LSAR with other existing data systems.

6.4   Planning LSAR Development

In any major program, the LSAR will develop into a large data repository. For this repository to be an effective and usable data base, considerable upfront planning is necessary. One of the things that should be considered seriously is the development of an LSAR guide for the program, either as a stand-alone or as part of the LSA procedures guide (see Table 3.1 for an outline of this guide). This guide would contain information that is unique to a particular program in the development, entry, review, and use of the LSAR. Some sample items to include might be the program RAM objectives (the A table data), the LSA candidate list and LCN structure, and a list of the applicable LSAR data items (it is inconceivable that all data tables or elements would be used on a program). In addition, the guide could include the internal function group responsible for entry and review; an abbreviated data element dictionary that has been tailored by data element, definition, and code; and standard reports that may be selected and their applicable selection criteria. Appendix 24 provides an example of guidance for one area, task analysis documentation, that could become a standardization problem if further guidance is not provided.

6.4.1   LCN structure

The LCN structure should be developed as early as possible. Care should be taken to make it as flexible as possible. The 2B standard gives a good description of the various types of structures and how to develop them. Figure 6.1 taken directly from the standard, is the in-

LCN ASSIGNMENT. The LCN may represent either a functional or hardware generation breakdown/disassembly sequence of system/equipment hardware including support equipment, training equipment, and installation (connecting) hardware. As such, the LCN is a key field utilized to input data into the LSAR data system and to extract reports from the data system. Normally, development of the LCN structure and assignment of LCNs through the subsystem level should be accomplished prior to completion of the first LSAR data tables. Extreme care should be exercised in developing the structure, so that the least number of characters is used for each indenture level. This can be accomplished by identifying the maximum number of parts/assemblies which will be assigned a unique LCN at each indenture level. If the maximum number of items at a given indenture level is less than or equal to 36, then 1 alphanumeric character would suffice. If the maximum number of items is greater than 36 but less than or equal to 1296, then 2 alphanumeric characters would suffice and so on. No more than one position of the LCN should be used to identify the system. It is useful to develop an LCN structure for the entire system/equipment hardware. Care should be exercised in assigning the LCNs, since the order in which they are assigned will affect the order of FMECA data and task analysis information, and may affect the order in which it will be used on a repair parts lists and assignment of PLISNs. For example, if it is a requirement for attaching hardware to appear on a repair parts list prior to the assembly, these items would have to be assigned LCNs which are less in value than the one assigned to the assembly. In assigning the LCN early in the design of an end item, it is also advantageous to skip one or two LCNs, so that an additional item can be inserted later on due to design changes. This advance planning avoids the possibility of having to resequence at a later point in the program. The above guidance should be considered prior to assigning the LCNs.

Figure 6.1   Excerpt from MIL-STD-1388-2B, Appendix C.

troduction to the LCN assignment. Here we will only make a couple of points for emphasis:

Use as few characters as necessary for each indenture level.

Use only one character at the system level.

The LSAR will use an ASCII sort, which means that numerical characters will sort ahead of alphabetical characters.

Try to skip one or more characters within LCN indenture levels to allow for expansion later.

Remember that the LCN is used to identify an item to its next-higher assembly and provides the means for filing, sorting, and manipulating the LSAR data.

The development of this structure is one of the first and most important things that the logistician must do in establishment of the LSAR. Time spent early ensuring that the LCN-ALC-UOC structure is workable, correctly illustrates the system breakdown, and is as expandable and as flexible as possible will save many headaches later in the program. The classical structure should be the structure of choice until it is shown to be unworkable. The 2B standard data system has expanded the number of characters in the LCN from 11 to 18. This will enable the classical structure to be used with many more systems than it could have been in the past.

The role of an LCN is to uniquely identify a part/subassembly to its next-higher assembly. The LCN is sensitive to the position of the part in the hierarchy of the hardware breakdown tree. As a general rule, a part number may have one or more LCNs depending on its use in the system, but an LCN will identify only one part number (until the lowest indenture). Even though the LSAR allows the use of functional LCNs, at some point in the program a physical LCN structure must be developed so that the provisioning lists can be created. The LCN structure will determine the sequence of the various provisioning lists.

Software supportability and configuration historically have not been documented in the LSAR. However, the LSAR can perform these tasks. If the program manager decides to include software in the LSAR, it will be a major consideration in the development of the LCN structure. One way to accomplish this is to tie software program modules to the hardware by use of a separate subordinate LCN for the software where possible.

It is recommended that in establishing the LSAR the logistician avoid ALCs other than 00 as much as possible because this will result in a simpler structure with only two key fields (LCN and UOC) to remember and track.

6.4.2   Data-element selection

The next task that must be performed is to decide what data elements need to be included. This must be done even if you have a contract and the data items are required on DD form 1949-3. The LSAR is a data base to record logistics data, and the more that it can be consolidated and integrated with other existing data bases, the more effective the LSAR will be. Remember that the data required on DD form 1949-3 is what your customer wants as an end product. There may be more information that you can store there for your own use. The LSAR data cuts across many logistics and design elements. Therefore, in establishing your data requirements, consideration should be given to information needs to be shared with and by each of these elements. Another point to consider is how the data will be developed, stored, and accessed. For example, if the reliability group uses a software program in development of the FMECA, can the FMECA be transferred electronically into the LSAR? Or could reliability engineers enter the FMECA directly into the LSAR? Traditionally, much of this information has been stored in the LSAR, but conveniently entering it and retrieving it have been problems. With the relational LSAR there should be a quantum leap in the development of integrated software among the various design and logistics functions.

Another consideration is how much detail needs to be included with some of the narrative data elements. Task documentation is a major area of consideration in this regard. See Appendix 24.

If the LSAR is a contract deliverable, a clear understanding between you and your customer as to the detail that will be contained in it is critical. DD form 1949-3 identifies the LSAR data elements that the contractor will be responsible for delivering, but both the government and the contractor must clearly understand the level of detail to be included. Each data element called out on DD form 1949-3 should be discussed to eliminate any confusion about the data requirement and how it is to be documented. In addition, the LSAR reports required should be checked against the required data elements to ensure that they are called out on DD form 1949-3. It is not uncommon for reports to be required by the customer for which there are insufficient data in the LSAR to generate. Conversely, there may be cases in which not all the possible information on a report is required; in this case, the nonrequired data elements should be deleted from DD form 1949-3. Figure 6.2 identifies some of these concerns. This should be a topic for detailed discussion at the LSA guidance conference.

6.4.3 LSAR reports

The LSAR contains 48 standard reports using the 2B standard data system. In the development of the 2B standard, the authors tried to have the reports satisfy more of the established design and logistics data deliverable requirements than the 2A standard. An example of this is that the LSA-056 output report (FMECA report) satisfies and supersedes the reliability requirement that was formerly established by DI-R-7196. Appendix 1 has the new DID for the FMECA data submission and the report description for LSA-056. Many of the reports in the 2A standard that were administrative or not data callouts by a logistics element were not included in the 2B standard, e.g., LSA-101, Transaction Edit Results. This means that fewer reports are available in standard 2B, but the reports that are available are more applicable. As the 2B standard develops, it will be fairly easy to develop standard reports to satisfy more logistics and design-element data requirements. In addition, the ad hoc capability provides the means to develop reports for unique program requirements and to monitor the development of the data base. Table 6.1 lists the standard 2B reports and their potential uses.

Once the LSA manager has decided which reports are required, there are two issues of concern that need to be considered when talking about LSA reports. The first is that the LSAR is a growing data base which in its infancy may not contain enough data to produce meaningful reports. This fact should be recognized, and reports should

The LSAR is an automated portion of the logistic data base. It is important that the data be documented in a manner that can be used to identify logistics requirements for all the logistic and design elements. Some of the more common requirements are discussed below.

Maintenance Plan: The RAM Characteristics Narrative in the Table BB is used for the Item Function (RAMCNABB(A)) and Maintenance Concept (RAMCNABB(B)) in the Navy Maintenance Plan (LSA-024) report. These tables must contain enough detail to satisfy this requirement.

FMECA: The B Tables in the LSAR provide space to document the results of the FMECA. It is important to the logistic program that the FMECA be performed to the end item's hardware topdown breakdown structure, even though a functional FMECA may also be documented. The reason for this is that logistic support is provided to hardware items not functions. Performing and documenting the FMECA in this manner provides traceability from the identified failures to the logistic support required to correct them. The depth to which the FMECA and LSA will be done should also be defined, e.g., to the lowest repairable or piece part. This may be dependent upon whether or not depot level repairs are included in the SOW.

RCM: The BF table allows for the documentation of the RCM analysis.

LORA: The maintenance concept given in the specification may be changed by the results of the LORA. In addition the LORA will effect the assignment of SMR codes and maintenance levels for the maintenance tasks.

Task Frequency: It is important to realize that the task frequency is the number of times per year that a task will be performed. This number is used in the determination of annual manhour requirements and SE usage, as well as, provisioning requirements though the determination of an item's Maintenance Replacement Rate (MRR). Appendix 23 details the traceability of the task frequency from the FMECA to the MRR.

Task Narrative: This narrative will be used to some extent to develop the Technical Manuals. It must be determined at the onset the level of detail to be included in this narrative. Appendix 24 gives an example of the requirements to document the tasks as well as selection of Task Codes and Item Category Codes.

Item Category Code: This is a code which identifies a category into which test and support equipment, spares, repair parts, etc. may be divided. (See DED 177, MIL-STD-1388-2B, Appendix F.) This code can be used to group various types of items together in order to produce more meaningful reports. The standard lists a number of categories, these should be looked at and tailored so that only the codes applicable to a program are used in the LSAR. Also if a program has unique categories that are not defined in the standard the program manager may develop new codes or redefine existing codes. (An exception is that the codes 'A', 'B', and 'C' may not be used as they are used to group categories in the report selection process.)

Supply Support: Supply support personnel should identify how the LSAR should be documented in order to provide the appropriate results in the provisioning lists (LSA-036 reports). The LCN structure of the LSAR will define the PLISN assignment so the supply support personnel need to become involved in the LSA process at the very beginning. It would be beneficial if the provisioning conferences could be held in conjunction with the scheduled LSA reviews. In addition, there are many program examples of data items required for the development of the maintenance plan not being included in the data selection sheets because they are part of the provisioning tables (H Tables).

Figure 6.2 LSA concerns across functional areas.

not be produced merely for the sake of generating a report. During the early phases of a program, it may be best to rely on ad hoc reports or scanning the data base to determine the progress being made.

The second issue is the choice of report selection criteria. Even though the reports are called out in the SOW or CDRL, the selection criteria are rarely specified. The general selection criteria for LSAR reports are Service Designator Code, LCN, ALC, and UOC. In addition, each report may have its own unique selection criteria. For example, the LSA-005 (Support Item Utilization) can be run to show either annual hours that a support item is used (useful for support equipment) or to show the annual quantity of a support item used (useful for spares and repair parts). When this report, as well as most others, is run, certain types of support items can be selected out using the Item Category Code (ICC). It is up to the ILS-element manager who will be using or reviewing a report to become familiar with its selection criteria in order to obtain a meaningful report. The guide for these selection criteria is given in MIL-STD-1388-2B. It may be wise for the LSA manager to determine the applicable selection criteria for the reports that will be used and distribute this information to each member of the ILS management team or include it in the program LSA/LSAR procedures guide.

This topic is appropriate for discussion at the Guidance Conference regardless of whether the contractor or government runs the reports. The delivery schedule of the data/reports should be established such that the ILS-element managers have adequate time to review them prior to the LSA review conferences. A good recommendation is to make them available 1 month prior to the review.

6.4.4 Entering the LSAR data

After the LCN structure has been established, required data elements identified, uses of the LSAR data base and reports identified, and sources of the data identified, you can begin entering the LSAR data. Historically, the LSAR has been a major cost driver in the LSA program (see Appendix 17). The task of gathering the data and entering them into a data base to be delivered to the customer often made the logistician more of a data-entry clerk than an analyst. Some reasons for this were that the LSAR was cumbersome to use and understand and many potential users did not see a benefit from its use.

In the 2B standard version the LSAR will become a more accessible and useful tool, and the managers/analysts who realize this will reap benefits from it. One important gain will be the increased integration capability. Many 2A standard third-party software vendors have developed software that integrates data program elements such as reliability to input the FMECA data into the LSAR and maintainability to access the LORA from the LSAR. The 2B standard and other CALS initiatives will make this entire concept of data integration more feasible. As this idea becomes more accepted and CALS linkages are more widely incorporated, the cost of the LSAR will be greatly reduced by eliminating duplicate data development and entry. Many data elements are "owned" and are developed by a design or logistics discipline for uses other than in the LSAR. CALS linkages to the LSAR will permit the data to be entered one time and thus will remove the data-entry-clerk function from the logistician. The design engineers would input the parts requirements to the LSAR from the parts list on the CAD drawings, and the provisioners would be procuring the correct parts on the provisioning lists. The same task analysis would be used to develop both the technical manuals and the training manuals. Task resource requirements would be reflected in the provisioning lists. The same failure data would be in the FMECA as in the LORA and the PPL. This has always been the logic behind the LSAR planning; the 2B standard will go a long way toward making it happen.

The LSAR should not be a collection of independent data developed in isolation to become a final deliverable. Rather, it is information to be built on and used as the program develops. See Appendix 23 for an example of the development of data based on analysis by other disciplines. As the failure modes are documented, one can begin assigning tasks to correct the associated failures. Planning a task-description structure at this point will be beneficial to the LSAR program. Each maintenance task consists of a fault locate/isolate action, the actual repair, and a test/verification. It is usually advantageous to document fault isolate, test/verification (in some cases the verification task may have the same steps as the fault isolate task), and other common tasks (assembly/disassembly) with a task frequency of 0. The subtasks of these tasks will then be included within the appropriate maintenance tasks.

The advantages in this approach are that there is less writing and that the task frequencies of the maintenance task reflect the failure rate and annual operating requirements. The task frequencies of the referenced tasks, if not documented as described above, must reflect the sum of the task frequencies of each referencing task. In other words, each time a new maintenance task is added or its task frequency is changed, each task frequency on the associated referenced tasks must be changed. Another advantage of structuring the tasks in this manner is that the entire time associated with the MTTR is contained in that one maintenance task.

Another associated idea is to establish a task-tree flowchart. This lays out the flow of tasks throughout the system (Fig. 6.3). In this scheme, "remove and replace" (H) tasks identify tasks involving removal and replacement of the LCN item, while "repair" (J) tasks indicate corrective actions involving removal/replacement/repair of subindentured parts. This means that each LCN will have only one H task documented against it.

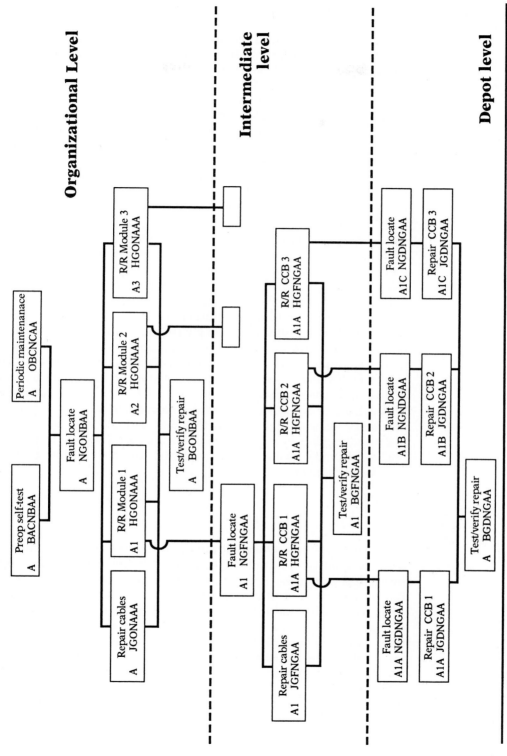

Figure 6.3 Maintenance task tree.

The referencing of task narrative in the 2A and 2B standards is different. In the 2A standard, narrative was referenced by sequence line number. This allowed referencing of a fault isolate task either in its entirety or until a particular fault was identified on one line, i.e., "Fault Isolate IAW LCN A1B TASK NGONBAA." The 2B standard references by subtask. This means that each individual step (subtask) in a fault isolate task must be referenced. This could have major implications on the ease with which narrative data can be translated from the 2A standard to the 2B standard format.

One idea that may be beneficial in planning development of the LSAR data is to envision what that data will be used for, i.e., how will it be rolled up in the various reports and how much detail should be there. With this in mind, the question of documenting like items in the LSAR will be discussed as an example. In many systems there will be identical components. Usually there is an initial inclination to document these items by summing their task frequencies, doubling (in the case of two components) the manhours, or multiplying the use of support items within a single task. This is usually not the recommended approach. If you do some sort of off-line calculation to document an assembly that has repairable subassemblies, you may lose track within the LSAR of what factor is needed to document this subassembly correctly. At the very least, you rapidly lose track of accountability. By only documenting that the assembly is the same as another assembly, manhours and spare/repair parts use are lost. The correct method would be to assign each LSA candidate an LCN and document each separately. Using the LSAR and some electronic data entry, this task can be eased by referencing tasks and copying data from one LCN to another. However this is accomplished, it is important so that the LSAR data base will be a complete and useful database.

6.5 Using the LSAR

Use of the LSAR is somewhat unique in the sense that it can be used both for quality assurance (QA) and as the basis for identifying detailed ILS resources. It is important not only that LSAR ADP data files be consistent within themselves but that they also be consistent with the source data (i.e., FMECA, drawings) and the logistics-element products (i.e., technical manuals, provisioning documentation, support equipment recommendations). To achieve this consistency, it is the responsibility of each logistics-element manager to ensure that his or her logistics data are documented correctly in the LSAR and that the LSAR is used to develop the logistics products. Since the logistics manager is the administrator or manager of the LSAR, the overall responsibility for ensuring that the LSAR data and tables are consistent among themselves and are being used and developed in a manner that most cost-effectively supports the program belongs to him or her. The next two subsections will expand on quality-assurance approaches, use of standard LSAR output reports, and the potential uses for ad hoc reports.

6.5.1 LSAR quality assurance

The logistics analyst must maintain the integrity of the LSAR, conduct logistics analyses, and coordinate the flow of data among the design engineers, R&M engineers, and various logistics elements. The quality assurance (QA) responsibility for the LSAR data base falls on the shoulders of all its contributors and users. This section will identify ways in which this can be accomplished through the use of various LSA output reports. This QA process should be undertaken by the subcontractor prior to data submission to the prime contractor, by the prime contractor, and finally, by the government customer. Reports that may be useful include the following:

LSA-008 (Support Items Validation Summary) and LSA-009 (Support Items List).   These two reports provide useful information in themselves and also can be compared to ensure that the data in the LCN and parts master files are consistent. The LSA-008 lists the quantity per task for the support items; this highlights items whose quantity per task is not equal to 1. In these cases, particularly for spares and repair parts, the tasks should be inspected to ensure that the quantity per task has been calculated correctly. This report also identifies the skill specialty code (SSC) for the task. If different SSCs are being used, a check should be made to ensure that the tasks (task codes) match the SSCs (i.e., has an I-level SSC been assigned to perform an O-level task?). If a task has a Hazardous Maintenance Procedure Code (HMPC) greater than D, the task narrative should be inspected to see that it includes either a warning or a caution.

The LSA-009 can be examined to help determine the status of the H tables: Are the FSCMs correct? Does the LCN match the WUC/TM functional group code (FGC)? Have national stock numbers (NSNs) been entered? Have the quantity per assembly, prices, and source, maintenance, and recoverability (SMR) codes been entered?

The LSA-008 data are obtained primarily from the C tables, while the LSA-009 data come primarily from the H tables. Therefore, these two reports can be compared to ensure consistency between these two files. Does the SMR code support the task code? Repair parts should be replaced (J tasks) in tasks that have the next-higher indenture LCN of the repair part. Are the ICCs correctly entered for the CG table (Support Item) and the CI table (Provisioned Item)? Are the Units of Measure (UMs) the same in the C and H tables? Are support/provisioned items identified in the C tables documented in the H tables?

LSA-019 (Task Analysis Summary). This report should be run in both merged and unmerged format to ensure that tasks have been referenced correctly. It is important when changing any task to update its associated referenced and referencing tasks if necessary. The mean man-minutes and mean elapsed minutes must be added and converted to hours correctly. Remember that the times required to perform referenced steps must be input to the task description where they apply. The support items/spare and repair parts requirements should be used to validate the requirements called out by the narrative.

LSA-027 (Failure/Maintenance Rate Summary). This report is a worksheet that can be used to verify the relationship among failure rates, task frequencies, and the maintenance replacement rates (Table 6.6). It also can be used to identify discrepancies between SMR codes and task codes.

LSA-058 (Reliability and Maintainability Analysis Summary, Part I: Redesign). One of the functions of logistics analysts is to ensure that logistics support has been incorporated into the design of the end item. They document that this has been done, has not been done, or is not applicable for the 13 logistic considerations listed in the BA table of the LSAR. This report summarizes the results of this analysis. It also summarizes any narrative associated with this analysis and any sys-

---

TABLE 6.6   LSAR Interrelationships/Checks

MTBF check:
$$1/\text{MTBF} = 1/\text{MTBF}_{\text{NLA1}} + 1/\text{MTBF}_{\text{NLA2}} + \cdots + 1/\text{MTBF}_{\text{NLAX}}$$

Failure rate checks:
$$\text{FR} = (1/\text{MTBF}) * 10^6$$
$$\text{FR} = \text{FR}_{\text{NHA}} * \text{associated Failure mode ratio}$$
$$\text{FR} = \text{FR}_{\text{NLA1}} + \text{FR}_{\text{NLA2}} + \cdots + \text{FR}_{\text{NLAX}}$$

Task frequency checks:
$$\text{TF} = (\text{AOR/MTBF}) * \text{conversion factor}$$
For an item the sum of the task frequencies of any off-equipment tasks should be equal to or less than the task frequencies of its R/R tasks.

NOTES: NLA is next-lower assembly; NHA is next-higher assembly; the formula for task frequency is a gross check; see MIL-STD-1388-2B for the complete formula.

tem redesign comments. It should be run prior to any LSA reviews so that potential support problems may be added to the agenda.

LSA-058 (Reliability and Maintainability Analysis Summary, Part II: Level of Repair). This report is useful to determine whether corrective maintenance tasks have been assigned to each failure mode. The repair time on this report reflects a gross estimate of the time required to perform a task. It may be compared with the mean elapsed time (using the LSA-019) to identify any major discrepancies.

LSA-054 (FMECA Report, Part III: Failure Mode Analysis Summary). This report can be used to identify any LCNs whose failure mode ratios do not sum to 1.00. It is also a convenient report to use for validating the failure mode criticality calculation.

LSA-080 (Bill of Materials). Part II of this report (error listing) identifies inconsistencies documented against parts and assemblies. Some of these are no SMR code, recoverability code does not agree with SMR code, repairable assemblies identified as parts of nonrepairables, repairable assembly with no parts identified, and wrong or no indenture code.

LSA-151 (Provisioning Parts List Index, PPLI). This is a cross-reference of PLISNs to LCNs and reference numbers. It also can be used to determine the status of key provisioning data elements in the H tables prior to generating an LSA-036 (Provisioning Requirements).

The logistics analyst should be aware of all the data relationships in the LSAR. Validation of the LSAR may require the use of ad hoc reports, particularly in the early phases of LSAR development, when there may not be enough data entered to generate meaningful standard LSAR reports. The logistics analyst is also responsible for ensuring that the logistics-element managers get the reports they require and that they provide the appropriate inputs, validate, and use the LSAR to develop their logistics requirements/products. The analyst should review any deliverable LSA reports to ensure that they have been run properly and that there are no discrepancies.

6.5.2   Uses of the LSAR

Some examples of how the LSAR data may be used by managers and engineers in the various logistics elements, RAM, and design are presented here and discussed in the following paragraphs:

Design interface

Maintenance planning

Manpower and personnel

Supply support

Support equipment

Technical data

Training and training support

Computer resources

Facilities

Packaging, handling, storage, and transportation

Design interface. This is the relationship of logistics-related design parameters, such as R&M data, to readiness and logistics support requirements. The system readiness objectives and logistics constraints are usually specified in the SOW and documented in the LSAR A tables. While the greatest chance to influence design occurs in the earliest phases of development, the LSAR outputs provide a unique opportunity for design influence by identifying design-related shortfalls and targets for additional design study as the program progresses. LSA-001 (Annual Man-Hours, Skill Specialty Code, and Level of Maintenance), LSA-003 (Maintenance Summary), and LSA-006 (Critical Maintenance Task Summary) are useful to identify logistics-related shortfalls in the design. LSA-058 (Reliability and Maintainability Analysis Summary, Part I: Redesign) highlights areas that have been identified as problems during the LSA process. LSA-052 (Criticality Analysis Summary) lists those items which may have the greatest impact on system reliability. It is important that the R&M and design engineers be aware of these outputs, and in those cases where problems have been identified, they should propose solutions.

For these reasons, it is necessary that the "design interface" be established at the very beginning of the program to ensure that logistics-related problems can be solved prior to the final design.

It is also the responsibility of the R&M engineers to ensure that R&M data have been entered correctly into the LSAR. One method of doing this is to validate LSA-056 (Failure Modes, Effects, and Criticality Analysis Report). The first part of this report satisfies the requirements of MIL-STD-1629. Other reports that may be useful in this validation, in addition to those mentioned above, include

1. LSA-023, Maintenance Plan Summary

2. LSA-024, Maintenance Plan

3. LSA-027, Failure/Maintenance Rate Summary

4. LSA-050, Reliability-Centered Maintenance Summary

Maintenance planning. This process is conducted to evolve and establish maintenance concepts and requirements for the life of the end item. Table 6.7 identifies some of the relationships between maintenance planning and the LSAR. The LSAR has many reports that can aid in this process. In particular, LSA-023 (Maintenance Plan Summary) and LSA-024 (Maintenance Plan) identify the maintenance concepts and requirements. These should be examined to ensure that the task codes and the SMR codes reflect the maintenance concept. Also, the item function should adequately describe the purpose of the item.

In the early stages of the program, the initial maintenance concept normally should reflect that given in the system specification or SOW. However, as the program progresses, it may need to be changed as a result of task 302 or the level-of-repair (LOR) analysis. The LSAR data should be compared with the LOR input data and output reports.

LSA-027 (Failure/Maintenance Rate Summary) can be used to verify the MRRI calculations. This data item is a driver in the LSA-024 report to calculate the maintenance factors. The maintenance task distribution (MTD) should be checked for the same reason.

LSA-005 (Support Item Utilization Summary) is useful to identify the annual utilization of the support items at the various maintenance levels.

Manpower and personnel. This logistics element identifies and plans for the availability of the military and civilian personnel with the appropriate skills required to support the end item. LSA-001 (Annual

TABLE 6.7   MPA/LSA/PMS Process Relationships

MP activity	General requirements	Detailed requirements	Data table/form
Maintenance concept	MIL-STD-1388-1A, LSA	MIL-STD-1388-1A	LSAR table BB, RAM characteristics
MPA process FMECA/ FMEA	MIL-STD-785B, reliability program	MIL-STD-1629A, FMECA	LSAR B tables, LSA-056 FMECA Report
RMC logic	MIL-STD-21388-1A	MIL-P-24534A, PMS	PMS form AOPNAV 4780/125 (RCM)
	MIL-STD-1388-1A	MIL-STD-1388-2B, LSAR	LSAR tables BF and BG, LSA-050 RCM summary
Task identification	MIL-STD-1388-1A	MIL-STD-1388-2B, LSAR MIL-STD-2453A	LSAR table BH PMS form OPNAV 4790/84 (MIP)
Task analysis	MIL-STD-1388-1A	MIL-STD-1388-2B	LSAR C tables, LSA-019 Task Analysis summary PMS Form OPNAV 4790/82 (MRC)
LORA	MIL-STD-1388-1A	MIL-STD-1390B, LORA	
Maintenance plan	MIL-STD-1388-1A	MIL-STD-1388-2B, LSAR	LSA-024 Maintenance Plan

Man-Hours by Skill Specialty and Level of Maintenance) and LSA-003 (Maintenance Summary) may be used to monitor manpower and skill-level requirements with respect to system constraints. In addition, LSA-006 (Critical Task Summary) will identify tasks exceeding specified requirements of man-hours per task and annual man-hours.

LSA-075 [LSAR Manpower Personnel Integration (MANPRINT) Summary] lists FMECA data, task summaries, and new/modified skill requirements needed as a baseline for performing hardware-manpower requirements analyses.

Supply support. This includes all management actions, procedures, and techniques used to determine requirements to acquire, catalog, receive, store, transfer, issue, and dispose of principal and secondary items. This includes provisioning for initial support as well as replenishment supply support. The primary products of this logistics element are the provisioning lists. These lists are a direct output of LSA-036 (Provisioning Requirements).

Supply support personnel need to be aware of the relationships described above. In particular, LSA-009 (Support Items List) offers a convenient format to use for the validation of LSA-036 data elements. In addition, supply support personnel should review and maintain a copy of the current LSA-152 (PLISN Assignment/Reassignment) since the LSAR is being used to assign the provisioning list item sequence numbers (PLISNs).

Any additions and/or changes that result from LSA-032 (DLSC Submittals) must be incorporated into the H tables.

The H tables (Packaging and Provisioning Requirements) are the domain of supply support, and these tables should be reviewed for accuracy and correlation with the remainder of the LSAR data base. Of primary concern in this respect is reviewing the LCN structure, which determines the order of the various provisioning lists.

Support equipment. This is all equipment required to support the operation and maintenance of the end item and includes the acquisition logistics support for the support equipment. The LSAR produces various support equipment lists, most of which can be selected on the basis of item category codes. Support equipment personnel should identify to the logistics manager which of these lists they require for review. LSA-005 (Support Item Utilization Summary) and LSA-009 (Support Item List) are good tools for this review.

LSA-070 (Support Equipment Requirements Data) and LSA-02 [Test, Measurement and Diagnostic Equipment (TMDE) Technical Description] identify requirements for new or unique support equipments and their support. Other reports that may be useful or of interest to the support equipment personnel are LSA-071 (Support Item Candidate List) and LSA-076 (Calibration and Measurement Requirements Summary).

Technical data. Technical data are all recorded data of a scientific/technical nature that are related to a program. Technical data constitute one of the ILS cost drivers and are a fertile field for LSAR cost benefits from CALS linkages. This includes drawings, operating and maintenance manuals, test and calibration procedures, and so on. Computer software and financial documentation are not generally considered to be technical data, although software documentation is.

The LSAR is capable of collating extensive data to support the development of technical manuals. LSA-019 (Task Analysis Summary) is the starting point for the maintenance and operation tasks used in technical manuals. The following reports produce outputs in the standard format for technical manuals:

1. LSA-030, Repair Parts and Special Tools List
2. LSA-040, Authorization List Items Summary List

LSA-080 (Bill of Materials) lists items related to or contained in an assembly. This provides a convenient tool for comparing the LSAR with the assembly drawings to ensure that the items required for support of the assembly are contained in the LSAR. LSA-126 (Hardware Generation Breakdown Tree) provides a useful tool for configuration management. Under CALS, an automated link between the engineering bill of materials and the LSAR could be a significant cost saver.

Training and training support. This logistics element identifies and plans for processes, procedures, techniques, training devices, and equipment required to train operators and maintainers. This support is required for both initial and sustained training. LSA-011 (Requirements for Special Training Device), LSA-014 (Training Task List) and LSA-075 (Manpower Personnel and Training Report) identify some of these requirements. LSA-014 identifies tasks for training and provides a basis for that recommendation. Training materials are closely linked to technical manuals. The data for these many times have great commonality. LSAR linkages here can have great cost benefits.

Computer resources support. This includes all computer equipment, software, documentation, personnel, and supplies needed to operate and support all embedded software. Presently, the LSAR system has no output reports to directly support this data element. However, the U tables identify requirements for test program sets (TPS), and the C tables identify test procedures and required support equipment. The computer resource support manager should ensure that the LSAR data are in concurrence with his or her support requirements. The TPS requirements are closely tied with the ATE used to support the unit under test (UUT).

Facilities. Facilities encompass real property assets required for the support of the end item, including both new and modified facilities. The object of facilities planning is to ensure that the required facilities are in place at the time they are needed. Because of the long lead time (approximately 5 years) involved in the acquisition of support facilities, these requirements are identified as early in the program as possible. These requirements and their justification are recorded in the LSAR F tables. LSA-012 (Requirements for Facility) and LSA-077 (Depot Maintenance Interservice Data Summary) are important for this review. The LOR reports also should be examined to evaluate facility requirements.

Packaging, Handling, Storage, and Transportation. This element includes the characteristics, actions, and requirements necessary to ensure the capability to transport, preserve, package, and handle all equipment and support items. LSA-085 (Transportability Summary) identifies transportability engineering requirements for the end item and critical subcomponents in the case that the end item must be sectionalized for transport. These requirements must be approved by the appropriate military service transportation agents.

The packaging, handling, and storage requirements are documented in the H tables. LSA-025 (Packaging Requirements Data) and LSA-026 (Packaging Development Data) are convenient outputs to use when reviewing these data.

The preceding subsections are not in any way meant to be an exhaustive description of the use and verification of the LSAR. These are examples. Each program will likely have unique requirements for reports and data entry. In addition, as the LSAR becomes a more familiar tool, the various disciplines will want to make more use of the data contained therein and to help ensure that the data are accurate.

6.5.3 Ad hoc reports

One feature of the 2B standard LSAR is the ability of a user to use a structured query language (SQL) or SQL-type queries to develop ad hoc reports. For example, the user can produce a list of all the parts having an MTBF of less than 1000 hours with the associated maintenance task or all parts used in organizational level maintenance tasks that cost more than $500. These reports can be useful in reviewing the LSAR, particularly in its early stages, and also in using the data to identify logistics concerns.

6.6 Summary

The LSA data base is the central repository of logistics-related data. As a major component of the LSA data base, it is critical that the LSAR contain valid data, since these data will be used to support the identification and development of the logistics support requirements. If the data are not accurate, the readiness and supportability of the end item once it is fielded can be seriously impaired. It is the responsibility of both government and contractor logistics elements and design personnel to use and validate these data.

MIL-STD-1388-2B is a leap forward in the usefulness of the LSAR. Not only will it make the data more accessible to the logistician, but through ad hoc reporting capability it will also assist in translating those data into information. Its design also will allow greater and easier integration with other data bases and automated programs, such as bills of materiel, reliability analysis, technical manuals, and so on.

This chapter has identified some methods for accomplishing this task. As users become more familiar with the information contained in the LSAR data base, data and/or report formats other than those specified in the 2B standard may be identified. The LSA manager should decide whether it is cost-effective to develop unique new reports from the LSAR data base or if it is possible that the end user can satisfy his or her needs through the use of ad hoc queries.

The objective of the LSAR is to provide a tool for both contractor and government design and logistics elements. This tool can be used universally in all the analysis programs. This will ensure consistent evaluations of the system effectiveness, foster appropriate corrective actions, and validate the correct/optimal design approach. The LSAR can integrate design and logistics data by employing the principle of recording data once and using it many times to achieve cost-effectiveness.

7

Analytical Methods

7.0 Introduction

Use of analytical methods is a given if LSA is to be effective. This chapter discusses various developments in this area in terms of military standards and presents a number of techniques useful in doing LSA. It specifically discusses

General considerations   (7.1)

RCM and LORA developments   (7.2)

System engineering tradeoffs   (7.3)

Readiness modeling methods   (7.4)

Life-cycle costing methods   (7.5)

Summary   (7.6)

7.1 General Considerations

Three basic considerations in using models are

1. The analysis process
2. Input data sources
3. Which model to use

The general analysis approach is well defined in Ref. 4. In summary form, it involves

Defining the problem

Analyzing the goals

Selecting variables, criteria, and quantitative techniques (construction of a model, if necessary)

Running the model, including sensitivity or "what if" analysis when appropriate

Documenting the results

Managing the review/decision, including iteration if necessary

There are many government and commercial models available. Blanchard[4] mentions a number of operations research models available for use, such as linear programming, queuing models, dynamic program models, etc. Many of these now have computer software packages available that can be used to expedite application and use. U.S. Army Pamphlet 700-4, LSA Techniques Guide,[6] lists many DOD models for which software exists. Figure 7.1 shows an example of the data

---

Technique: TIGER Reliability Computer Program (TIGER)

Purpose: To predict system RAM performance

Description: TIGER is a RAM simulation computer program capable of representing complex systems under varying operating conditions. It is applicable to system simulations of all sizes ranging from a microcircuit to a fleet of ships. All that is required is the ability to model the system with reliability block diagrams.

The unique advantage of TIGER is that it can be used in situations where other mathematical methods will not work. For systems that consist of only a few components and operate under steady conditions, the analyst has access to a variety of analytical methods. However, if the system is large, or has complicated operating rules, or is subjected to a scenario of changing conditions and requirements, or logistic considerations are included, then the analyst needs a model of the simulation type, such as TIGER.

In TIGER, all internally generated events are equipment events: failure of an equipment, arrival of a spare for a waiting equipment, and repair of an equipment. The mission begins with all equipments in good condition and all stocks up to allowance. The first failure of every equipment is forecast and placed on the schedule. The program then simulates the mission time-line, phase by phase. Concurrently, TIGER observes the time-line. At the end of each sequence, the program reassesses system state, checks for operating rules, turns equipment on or off if it is entering or leaving the system, and then simulates the next phase sequence in the time-line.

Proponent: Naval Sea Systems Command Code 05MR (Mr. Buckberg), Room 1E60 BLDG NC-2, Washington, D.C., 20360-5101 (202)692-2150	*Current users/POC* Proponent is also user

Technique: TIGER Reliability Computer Program (TIGER)

Inputs: Inputs are structured in four parts. In part 1, the user selects simulation options. Part 2 lays out the mission time-line and accepts print-out options. Part 3 specifies the types of equipments in the system, designates the quantity of equipment of each type, assigns identifying numbers to each item of equipment, and (optionally) sets spares allowances. Part 4 contains a complete system description for each type of phase occurring in the time-line.

Outputs: Outputs include the following: (1) user input echo, (2) simulation progress reports, (3) final figure of merit reports, (4) equipment performance statistics, (5) critical equipment lists, and (6) Restricted Erlang Distribution Model (REDM) report, which provides additional output statistics concerning the reliability characteristics of the system, in restricted Erlang form.

Documentation available: TIGER Reliability Computer Program User's Manual, Version 8.21, Sep 87: TIGER/SENSITIVITY Computer Program User's Manual Version 8.21, Sep 87.

Automation: Available Yes: Language FORTRAN 77: Hardware; Cray, CYBER, IBM, PDP, VAX, UNIVAC, etc. Life Cycle Application: All.

LSA task interface: 203.3,5,7 303.2.1,4,9	*Application:* Numerous ship, aircraft, spacecraft, and transportation

Figure 7.1   LSA technique guide information sheets.

for each model contained in the pamphlet. As computers have become more powerful, many of these models have become usable on PCs. Thus it is common now for these analyses to be performed by logistics analysts rather than LCC experts, LORA experts, etc. However, the reader is cautioned that some models listed in pamphlet 700-4 may not be available because of difficulties in updating the data.

Model input data is frequently a major problem in practice. The type of input data necessary if an analysis is to be done will range from expert opinion to specification data on items in development to actual field data on usage. The analyst must strive to use the most credible data available. It is frequently necessary in practice to use data from one source for part of the analysis and data from a second source for another part. Various shortcuts, assumptions, etc. may be necessary to enable some kind of analysis. Remember, something is better than nothing.

Major sources of data in the U.S. DOD include

Navy Maintenance and Material Management (3-M) System

AFM 66-1, Air Force Maintenance Policy and Procedures

Visibility and Management of O&S Costs (VAMOSC)

Others

7.2   RCM and LORA Standards

As this book is being prepared, the DOD has initiated projects to develop DOD-wide standards on reliability-centered maintenance (RCM)[27] and level of repair analysis (LORA).[39] This is a much-desired effort because the new standards will be in the general form of MIL-STD-1388-1A; i.e., application guidance will be included in the standards. This will result in improved advice on when to apply these standards and their tasks and subtasks. Historically, widely accepted guidance with regard to RCM and LORA has not been available.

7.2.1   Reliability-centered maintenance (RCM)

General.   *Reliability-Centered Maintenance* (RCM) is a systematic approach for identifying preventive maintenance tasks for a system/equipment in accordance with the specified set of procedures and for establishing intervals between maintenance tasks. RCM is a major input to LSA task 301 and the LSAR generated in LSA task 401. Figure 7.2, from Jones,[3] summarizes the general process. While various RCM logic rules differ in detail, Fig. 7.3 illustrates a typical RCM decision logic chart. Depending on data availability, use of a default logic may be necessary. Table 7.1 presents an example of such logic.

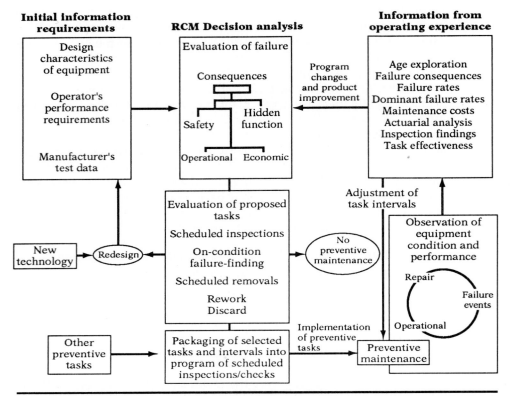

Figure 7.2   Reliability-centered maintenance analysis process. (From James V. Jones, *LSA Handbook*. New York: McGraw-Hill, 1989.)

Draft RCM standard.   The DOD is developing a DOD-wide standard on RCM that will replace MIL-STD-1843 (USAF) and MIL-STD-2173 (AS), NAVAIR. The initial DOD RCM standard will have the following task structure:

Task 100: Influence on design

Task 200: Determination of significant items

Task 300: Decision logic process

Task 400: Implementation

Task 500: AGE exploration

The subtasks and suggested phasing are shown in Table 7.2. Tasks 100 and 500 are not as widely used and widely recognized as tasks 200, 300, and 400. They are, however, included in Appendix 19.

The initial effort will focus primarily on standardizing data charts

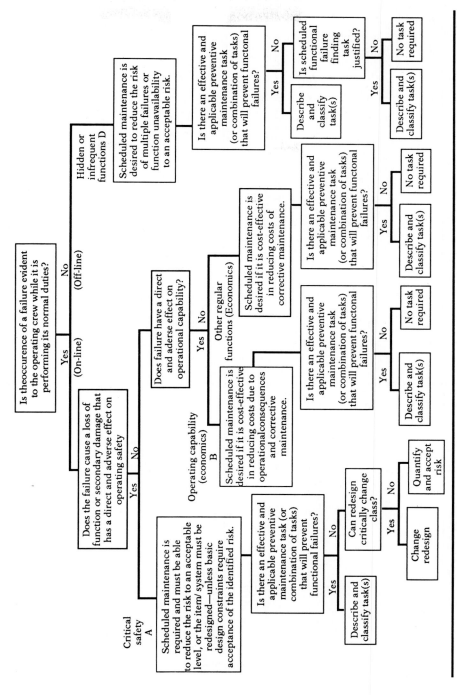

Figure 7.3 RCM Decision Logic Diagram. (From Navy Operational Availability Handbook, June 1986.)

TABLE 7.1 Default Decision Logic Chart

Decision question	Default answer to be used in case of uncertainty	Stage at which question can be answered		Possible adverse consequences of default decision	Default consequences eliminated with subsequent operating information
		Initial program (with default)	Ongoing program (operating data)		
Identification of significant items: Is the item clearly nonsignificant?	No; classify item as significant	X	X	Unnecessary analysis	No
Evaluation of failure consequences: Is the occurrence of a failure evident to the operating crew during performance of normal duties?	No (except for critical secondary damage); classify function as hidden	X	X	Unnecessary inspections that are not cost-effective	Yes
Does the failure cause a loss of function or a secondary damage that could have a direct adverse effect on operating safety?	Yes; classify consequences as critical	X	X	Unnecessary redesign or preventive maintenance that is not cost-effective	No for redesign; yes for preventive maintenance
Does the failure have a direct adverse effect on the operational capability?	Yes; classify consequences as operational	X	X	Preventive maintenance that is not cost-effective	Yes
Evaluation of proposed tasks: Is an on-condition task to detect potential failures applicable?	Yes; include on-condition tasks in program	X	X	Preventive maintenance that is not cost-effective	Yes
If an on-condition task is applicable, is it effective?	Yes; assign inspection intervals short enough to make task effective	X	X	Preventive maintenance that is not cost-effective	Yes
Is a rework task to reduce the failure rate applicable?	No (unless there are real and applicable data); assign item to no preventive maintenance	—	—	Delay in exploring opportunity to reduce costs	Yes
If a rework task is applicable, is it effective?	No (unless there are real and applicable data); assign item to no preventive maintenance	—	X	Unnecessary redesign (safety) or delay in exploiting opportunity to reduce costs	No for redesign; yes for preventive maintenance
Is discard task to avoid failures or reduce the failure rate applicable?	No (except for safe life items); assign item to no preventive maintenance	X (safe life only)	X (economic life)	Delay in exploring opportunity to reduce costs	Yes
If a discard task is applicable, is it effective?	No (except for safe life items); assign item to no preventive maintenance	X (safe life only)	X (economic life)	Delay in exploring opportunity to reduce costs	Yes

NOTE: The default answers to be used in developing an initial preventive maintenance program in the absence of data from actual operating experience.
SOURCE: From Navy Operational Availability Handbook, June 1986.

TABLE 7.2 Applicability of Reliability-Centered Maintenance Tasks to Acquisition Phase

Task/ subtask number	Title	Purpose	Phase			
			Concept	DVAL	EMD	PROD & OPER
100	Influence on design	To ensure RCM philosophies are considered during design	X	X	X	X
200	Significant item selection	To identify components which affect safety, operations, and economics		X	X	X
203.1	Determination of functionally significant items	To identify functionally significant items		X	X	X
203.2	Determination of structurally significant items	To identify structurally significant items		X	X	X
300	Decision logic process	To determine applicable and effective preventive maintenance requirements		X	X	X
303.1	Failure consequence determination	To determine consequences of item failure		X	X	X
303.2	Functionally significant item (FSI) decision logic process	To determine applicable and effective preventive maintenance tasks for functionally significant items		X	X	X
303.3	Structurally significant item (SSI) decision logic process	To determine applicable and effective preventive maintenance tasks for structurally significant items		X	X	X
400	Implementation	To implement the preventive maintenance requirements determined by RCM			X	X
500	Age exploration	To continually update and refine preventive maintenance requirements			X	X

SOURCE: From Draft DOD standard, RCM Requirements for DOD Weapon Systems and Equipments, May 1991.

and guidance. Each of the different service's logic will be included in the standard as an appendix.

7.2.2 Level of Repair Analysis (LORA)

General. *Level of Repair Analysis* (LORA) is the process that analyzes corrective maintenance requirements for a system/equipment leading to the repair decisions for each repairable item. LORA is performed to determine if an item should be repaired or discarded and the maintenance level where repair or discard actions should be accomplished. LORA does not apply to preventive maintenance, discussed in Sec. 7.2.1.

LORAs are classified as economic or noneconomic. *Economic* LORAs use cost algorithms to determine least-cost repair assignments for each significant maintenance item. *Noneconomic* LORAs use "preempting" factors (which override cost considerations) or existing level of repair decisions on similar equipment to determine repair levels. In practice, LORAs can be economic, noneconomic, or a mixture of the two. Various government LORA models have existed for some time. The U.S. DOD is currently developing a standard on DOD LORAs.

Initially, the LORA standards/models contained cost algorithms and criteria but little in the way of application guidance. This issue has been addressed in several good pamphlets:

AMC-P700-27, LORA Procedures Guide[35]

NAVSEA TL081-AB-PRO-010/LORA, NAVSEA LORA Procedures Manual[36]

AFLC/AFSC Pamphlet 800-28, Repair Level Analysis (RLA) Program[37]

Much of the material in this section was derived from Refs. 35 and 36.

All the services consider discard as a viable alternative to repair of an item. The Army stresses the alternative of discard-at-failure in lieu of repair when it can be supported by readiness data and life-cycle cost (LCC) considerations.

The maintenance structure used in conducting LORA varies from service to service. The Air Force considers three levels of maintenance (on-equipment, off-equipment, and depot), the Navy considers three levels of maintenance (organizational, intermediate, and depot), and the Marine Corps considers three levels of maintenance with five echelons [organizational (first and second echelons), intermediate (third and fourth echelons), and depot (fifth echelon)]. The Army utilizes a four-level maintenance concept except for aircraft, for which it utilizes a three-level concept (organizational, intermediate, and depot).

The LORA program is an integral part of the LSA program, as defined in MIL-STD-1388-1A, subtask 303.2.7, repair level analyses. Since LSA serves as the interfacing mechanism between the elements of system engineering (e.g., design, reliability, maintainability, safety, human factors, etc.) and the elements of Integrated Logistics Support (ILS) (e.g., supply support, maintenance planning, technical data, training and training devices, etc.), the LORA environment is also one that integrates design, operations, and logistics support characteristics/constraints to establish the maintenance level at which an item will be removed, replaced, repaired, or discarded.

LORA recommendations identify the repair/discard location, the extent of maintenance, and the resources needed to support the repair process. LORA decisions influence both system effectiveness and the

ILS elements necessary to maintain the operational readiness of the hardware system. The following paragraphs expand on this concept to show how LORA, as part of the LSA program, interfaces with such activities as maintenance planning, design engineering, reliability engineering, maintainability engineering, provisioning, Source Maintenance and Recoverability (SMR) coding, technical manual development, Maintenance Allocation chart (MAC) development, and the LSA Record (LSAR).

1. *Maintenance Planning: Maintenance planning* is the process conducted to evolve and establish maintenance concepts and requirements for a materiel system. This process involves several analyses, in addition to LORA, such as reliability-centered maintenance (RCM) (AMC-P 750-2, Guide to Reliability-Centered Maintenance), LSA Task Analysis (MIL-STD-1388-1A, Task 401), Failure Modes, Effects, and Criticality Analysis (FMECA) (MIL-STD-1629A, Procedures for Performing a Failure Mode Effects and Criticality Analysis), Reliability Program for Systems and Equipment Development and Production (MIL-STD-785B), and Maintainability Program for Systems and Equipment (MIL-STD-470A). The relationship of these analyses, for purposes of maintenance planning, is generally as follows: The FMECA identifies potential design weaknesses through systematic, documented consideration of all likely ways in which a component or equipment can fail, the causes for each failure mode, and the effects of each failure; RCM identifies preventive or scheduled maintenance tasks for an equipment end item in accordance with (IAW) a specified set of procedures and establishes intervals between maintenance tasks; LSA task 401, Task Analysis, analyzes required operations and maintenance tasks for the new system/equipment; the reliability program identifies the frequency of failures; the maintainability program identifies the elapsed time to correct a failure; and LORA identifies the maintenance level and support costs associated with an unscheduled maintenance task.

2. *Reliability Engineering:* Four of the 18 reliability program tasks identified in MIL-STD-785B relate to LORA (i.e., reliability modeling, reliability allocations, reliability predictions, and FMECA). Reliability modeling reorients the functional (schematic) block diagrams into a series-parallel network showing reliability relationships among the various subsystems and components. This reliability block diagram is the first gross breakout of candidate items requiring LORA. The system-level reliability requirement is then allocated down (i.e., in a top-down approach) the reliability block diagram and is levied on the equipment designers. LORA, at this point, can be used as a tool for conducting tradeoff analyses to determine whether a de-

sign a particular item for repair or discard. Reliability predictions are applied as the design progresses to determine whether the reliability allocations are feasible and attainable. Thus LORA, which uses the failure rates determined in reliability prediction, becomes more accurate and detailed. In conducting LORA, the FMECA is used to affect readiness.

3. *Maintainability Engineering:*   With respect to LORA, the maintainability engineering process, to a large extent, tracks with the reliability engineering process. Five of the 12 maintainability program tasks identified in MIL-STD-470A relate to LORA [i.e., maintainability modeling, maintainability allocations, maintainability predictions, failure modes and effects analysis (FMEA), maintainability information, and maintainability analysis]. The maintainability model typically is consistent with the reliability model described above and is used as a tool to perform allocations and predictions. The maintainability allocation and prediction tasks have the same functions as the reliability allocation and prediction tasks. However, in maintainability engineering, terms such as *mean time to repair* (MTTR), not failure rates, are being allocated or predicted. The FMEA is consistent with the FMECA, except that the FMEA is used to ascertain information that relates to fault detection and isolation, which are critical drivers of maintainability at all levels of maintenance. The maintainability analysis task translates data into a detailed maintainability design approach to achieve the system MTTR requirements. LORA plays a vital role in maintainability analysis in that it evaluates the maintainability design alternatives. MTTR data is used by LORA to help determine level of repair and discard decisions for subsystems and components.

4. *Provisioning:*   Provisioning involves the use of several tasks [e.g., cataloging, SMR coding, assignment of failure factors, consumption expenditures, essentiality coding, identification of maintenance-significant items, determination of Maintenance Task Distributions (MTD) and Replacement Task Distributions (RTD) and interfaces with various other efforts (e.g., MAC development, LORA, RAM analyses, FMECA) to ensure the timely availability of minimum initial stocks of support items at the least initial investment cost the system readiness objective (SRO) can achieve. It should be noted the development of failure factors involves the manipulation of failure rates derived in reliability engineering. LORA, relative to provisioning, provides the analytical basis from which the maintenance portion of the SMR code is obtained. In addition, LORA is the analytical basis for the development of a maintenance allocation chart (MAC). The Optimum Supply and Maintenance Model (OSAMM) provides the MTD

and RTD tables as a direct output product when the MAC and SMR codes are input.

5. *LSAR:* The LORA can use data from the LSAR and other sources as other analyses are completed and documented [e.g., mean time between failure (MTBF), MTTR, unit price, deployment/usage]. The LSAR provides a common consistency of data among the various analyses being conducted and a vital source of updated information as the weapon system under acquisition matures through the life-cycle phases. The LORA results are documented in the LSAR through the data elements entitled SMR, MTD, and RTD and the task codes in the LSAR data tables.

After a system has been fielded, LORA can be used as an analytical tool to refine the support system and propose a support structure for Materiel Change Management (MCM) and Engineering Change Proposals (ECP). A LORA on fielded systems evaluates changes in factors (i.e., maintenance policy, costs, utilization rates, capabilities, etc.) that affect the logistics support of a fielded weapon system and which may require adjustment of the support structure to maintain the system operational readiness and cost-effectiveness.

LORA models. LORA can be performed using one or more LORA models/techniques. The model/technique used is dependent on such factors as the availability of data, complexity and type of weapon system being analyzed, life-cycle phase, and the purpose of the analysis (e.g., design tradeoffs, MTD and RTD development, and basis for SMR coding). The U.S. Army generally divides LORA into three general classes of analysis: (1) system/end-item analysis, (2) subsystem/item analysis, and (3) specific aspects of repair analysis. Table 7.3 summarizes the paragraphs that follow.

TABLE 7.3   Class Application of LORA

Life-cycle phase class of LORA	CE/D	CD/V	EMD	PROD	O&S
System/end item		D,N*	D*,N*	D,N,P	D,N,P
Subsystem/item	D,N	D*	D,N shift to above	P*	P*
Specific aspects of repair	D,N GRAPH	D,N GRAPH TPS CEEM	D,N GRAPH or PALMAN TPS CEEM	PALMAN TPS CEEM	PALMAN TPS CEEM

*Mandatory; D = development items; N = NDIs; P = ECP items (may undergo their own life-cycle phases); GRAPH = repair vs. discard alternative; TPS CEEM = test program set determination.

source: From AMC-P700-27, LORA Procedures Guide, February 1991.

1. *System/end-item analysis:* A system/end-item analysis is typically conducted to analyze support alternatives for an entire weapon system and to assign maintenance and supply support costs to a system's maintenance concept. Several computer models have been developed to assist in this type of analysis:

> OSAMM, Optimum Supply and Maintenance Model (Army)
> LOGAM, Logistics Analysis Model (Army)
> NRLA, Network Repair-Level Analysis (AF)
> MCLOR, Marine Corps LORA
> MODIII, Level of Repair Model (Navy)

2. *Subsystem/item analysis:* A subsystem/item analysis can be conducted to analyze specific components of a weapon system to evaluate design tradeoff alternatives or when only limited information is available on selected items of the materiel system. A subsystem/item analysis is normally conducted in the CE/D phase on developmental items and NDIs. It should be done in the D/V phase for developmental items unless a system/end-item analysis is to be conducted. As developmental items and NDIs progress through the acquisition life cycle, a shift can be made from subsystem/item analysis to system/end-item analysis, as shown in Table 7.3. This shift should take place in the D/V phase before the contract has been awarded for the next phase. Subsystem/item analysis is performed in the P/D and O&S phases for MC or ECP items unless a system/end-item analysis is to be conducted. The same models approved for use in conducting system/end-item analysis may be used for subsystem/item analysis. In addition, the Item Repair-Level Analysis (IRLA) model could be used for joint service projects in which the Air Force has the lead responsibility.

3. *Specific aspects of repair analysis:* Specific aspects of repair analysis are conducted to analyze a specific area of repair for a materiel system. Presently, this class of LORA is limited to two select areas: (1) the analysis of test program set (TPS) requirements using the TPS Cost-Effectiveness Evaluation Model (TPS CEEM) and (2) the determination of the break-even cost for repair versus discard of an item using the Interactive Palman Model (IPM) or its derivative, The Graphical Repair/Discard Analysis Procedure Handbook (GRAPH). The GRAPH method can be used during the CE/D, D/V, or EMD phases of the life cycle on developmental items to determine if it is more cost-effective to design an item for repair or for discard. In the EMD phase, a transition should take place from the GRAPH method to the IPM Model for both developmental items and NDIs if a more detailed analysis is needed to determine which items should be repaired or discarded. The IPM model can be used in the P/D or O&S phases of the life cycle for MCM and ECPs or for establishing/

reexamining Depot Maintenance Expenditure Limits (MEL) for specific fielded items. The results of the depot MEL examination may trigger a subsystem/item analysis to reevaluate the classification of the item to consumable. The TPS CEEM can be used during the D/V or EMD phases of the life cycle for both developmental items and NDIs to determine the cost-effectiveness of TPSs, maintenance level at which the TPS will be placed to support the weapon system, and extent of diagnostic capabilities needed (e.g., go/no go, end-to-end) by the TPS. The TPS CEEM also may be used in the P/D or O&S phases of the life cycle if the MC or ECP appears to be a candidate for requiring TPSs.

Since LORAs are performed iteratively, the final analysis may verify the results of earlier analyses or recommend a change based on finalized input data. The LORA and associated results should be used to influence the repair or discard decisions, which will, in turn, provide repair or discard recommendations to the equipment designer. LORA also can provide inputs to assist in reaching the proper ILS decisions.

Repair-level decisions are usually made during the Preliminary Design Review and are documented in the system's allocated baseline. The decisions are reevaluated during the Critical Design Review (CDR) and when revised LORA reports are submitted. The investment decisions made during the D/V and EMD phases have a significant impact on the requirement for and the implementation of any repair-level decisions during the O&S phase. Utilization of LORA results as a part of the tradeoff analysis to influence the investment decisions may eliminate or substantially reduce the need for repair-level decisions/changes in the O&S phase. Conversely, failure to utilize LORA results to influence investment decisions may result in an undesirable O&S cost burden. Therefore, it is important that level of repair decisions be made early in the system design process. The design/support of equipment is influenced by many individuals/disciplines having responsibilities in areas such as performance, reliability, maintainability, safety, and supportability. LORA criteria applied by these individuals/disciplines may not be converted easily into economic quantities. Therefore, each of these individuals/disciplines should assist in the level of repair determination to ensure that any noneconomic factors (e.g., safety, readiness, policy, mission success, limitations of maintenance capabilities, etc.) that affect the repair-level decision are also addressed.

Design features to achieve desired logistics support can be incorporated into preliminary drawings and specifications at lower cost than changes to prototype and production hardware. The potential for greatly improved reliability, particularly in electronics systems, provides the opportunity for designing certain items for discard at failure. During the D/V phase, selected items (e.g., LRUs, SRUs) should be subjected to LORA to isolate those items which should clearly be designed for discard from those which should be designed for repair. Discard at failure should be established as a preferred alternative to repair when supported by readiness data and LCC considerations. Major economic benefits of a discard-at-failure maintenance policy can be realized by the early identification of items not subject to repair. This precludes the development and acquisition of principal support resources and gives the following benefits:

1. Reduced and simplified requirements for base-level support and TMDE
2. Reduction of technical data and manuals
3. Elimination of repair parts (bits and pieces, not spares)
4. Reduced training requirements
5. Reduction in quantity and possibly skill level of maintenance personnel
6. Improved unit mobility and deployment capability
7. Reduced inventory controls and storage requirements
8. Reduced system downtime and turnaround time
9. Improved pipeline response and materiel flow

In repair-level determination, emphasis should be placed on cost, operational availability, operational effectiveness, and any overriding constraints. An example of trading off these factors is cost versus operational availability.

Three individual, but closely linked, LORA evaluations are generally performed: economic, noneconomic, and sensitivity evaluations.

1. *Economic evaluations,* normally using some model, establish the least-cost feasible support alternative for each of the items in the system/equipment.
2. *Noneconomic evaluations* are undertaken to evaluate constraints and intangible factors that may restrict the level at which maintenance can be done. Examples of such factors are doctrine, safety, deployment mobility, technical feasibility of repair, security, transportability factors, human factors, vulnerability, and survivability. The key focus is to eliminate nonfeasible support alternatives prior to cost analysis of alternatives.
3. *Sensitivity evaluation* is an extension of economic evaluation. It

identifies and analyzes the results of variations in key input data that may not be firm to quantify economic risks in making LORA decisions when uncertainty exists.

The effectiveness of the LORA program is largely dependent on the assembly of input data. LORA evaluations require a variety of data, ranging from design features to anticipated logistics factors and constraints. Much of the data needed are system-peculiar data (MTBFs, item costs, etc.), but some of the data needed are common logistics parameters not related to the system (requisition costs, depot labor rates, costs to maintain an NSN, etc.). A library of such data, frequently called *default values,* can be obtained or developed to reduce the burden on the analyst in gathering non-system-peculiar data. Table 7.4 contains examples of such data used for U.S. Navy Air LORA applications. Such factors obviously vary from service to service and command to command.

System/end-item analysis. System/end-item analysis is conducted to determine the maintenance policies for the entire system/equipment and all indenture levels below. It is especially necessary when components and modules are sharing TMDE and repair personnel, and such cost is a significant portion of system cost.

Subsystem/item analysis. Problems may occur when conducting a subsystem/item analysis. Since components and modules may share

TABLE 7.4   Example LORA Default Values

Factor	Default value
Discount rate	10%
Item entry cost	$491.70
Item retention cost	$288.20
Field supply administration cost	$17.33
Training cost	$17.37 per hour $695.00 per week
Personnel attrition rate	0.381 military 00.009 civilian
Space cost	$2.10/ft^2/yr shore hanger $3.08 shore admin. office $102.94 nuclear carrier $117.70 carrier
Required days in stock	90 discard, carrier 30 repair, land base

SOURCE: Naval Air Systems Command, "LORA Default Data Guide," August 1991.

TMDE, a determination must be made as to how much of the cost of the TMDE will be attributed to the respective components and modules. Problems arise in determining exactly how to perform this prorating. Since modules and components are examined on a one-at-a-time basis, contradictory results may occur. For example, a module may be determined to be repaired at general support (GS), while the component that contains the module may be determined to be repaired at the depot or discarded. Item Repair Level Analysis (IRLA) considers three options: repair at intermediate level, repair at depot, and discard. The option with the minimum cost is selected. This is done for each LRU and SRU separately. The model is based on computing the discard marginal cost, intermediate repair marginal cost, and depot repair marginal cost and then selecting the alternative IRLA decision with the lowest marginal cost. Each of the cost elements considers only the differential costs associated with a particular maintenance alternative. In effect, the model evaluates the resources necessary to support a particular maintenance policy. The calculations and equations are outlined in AFLC/AFSC Pamphlet 800-4.

Specific aspects of repair analysis.   Specific aspects of repair analysis include such areas as evaluating TPS cost effectiveness and repair versus discard analysis. The Kasian TPS model, now replaced by the TPS CEEM, can be used for evaluating the cost-effectiveness of TPSs, while the IPM and the GRAPH technique, which is a derivative of the IPM, can be used for determining breakeven cost for repair versus discard.

The TPS CEEM was designed to aid decision makers in determining the economic feasibility of developing a TPS to diagnose and isolate failures to piece parts or development of an end-to-end TPS to do a go/no-go determination of whether the TPS candidate item did or did not fail. The model evaluates six alternative circuit card assembly (CCA) maintenance policies, three of which require the use of a TPS. For each TPS candidate item, the model calculates support cost differences associated with the following alternatives: discard of the CCA when it is replaced due to presumed failure, discard of the CCA after it has been screened at the GS level with an end-to-end TPS, repair the CCA at the GS level using a diagnostic TPS to screen it and diagnose the cause of failure, repair the CCA at the depot level using common TMDE to screen it and diagnose the cause of failure, repair the CCA at the depot level using a diagnostic TPS to screen it and diagnose the cause of failure, and repair the CCA by the contractor. Additional information and documentation can be obtained from the Commander, U.S. Army Communications, Electronics Command, ATTN: AMSEL-PL-SA, Fort Monmouth, NJ 07703-5000.

The PALMAN Repair versus Discard Model calculates the purchase cost based on various input variables over a range of expected deployment densities. If the actual/expected cost of procuring an assembly exceeds the model output, the assembly should be repaired; if less, the assembly should be discarded. Details on the PALMAN model can be found in the PALMAN User's Guide (Commander, U.S. Army Armament, Munitions, and Chemical Command, ATTN: AMSMC-LSP, Rock Island, IL 61299-6000) and the TRA Report of PALMAN (Commander, USAMC Materiel Readiness Support Activity, ATTN: AMXMD-EL, Lexington, KY 40511-5101). The GRAPH technique is a user-friendly off-the-shelf analysis methodology derived from PALMAN. It is discussed in more detail below.

Drivers: Sensitivity analysis.   Performing sensitivity analysis as part of the LORA process is important to both the designer and the user. Early in a program, many of the input factors may be estimates based on expert opinion, similar system data, or best guess. By performing sensitivity analysis, the designer can determine whether the repair level selected is firm or marginal. If a small change in a factor causes the repair level to change, the designer should be prepared for such an event. Sensitivity analysis tests the stability of the system under varying conditions and the effect of uncertain data.

In addition to the designer, users, such as the government, also can benefit from sensitivity analysis. Sensitivity analysis can serve as a means of understanding the risks involved with a particular selection. In general, the most frequently sensitized data elements are MTBF, operating hours, unit cost, TMDE cost (including TPSs), and number of systems to be fielded. The value of sensitivity analysis is enhanced if prior knowledge exists of how an increased or decreased data element will affect the level of repair alternative. The sensitivity analysis trends shown in Table 7.5 provide such insight on the most frequently sensitized data elements.

TABLE 7.5   Frequently Sensitized Data Elements

Sensitized data element	An increase in the sensitized data element will tend to move the level of repair alternative in the direction indicated
MTBF	Organizational > DS > GS > depot > discard
Test equipment cost	Organizational > DS > GS > depot > discard
Operating hours	Discard > depot > GS > DS > organizational
Unit cost	Discard > depot > GS > DS > organizational
Number of systems fielded	Discard > depot > GS > DS > organizational

SOURCE: From AMC-P700-27, LORA Procedures Guide, February 1991.

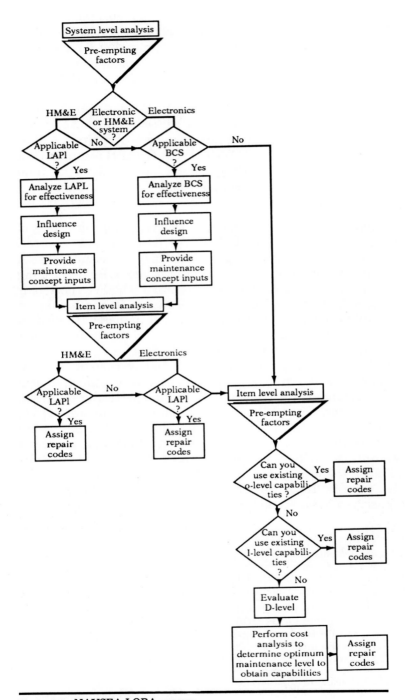

Figure 7.4 NAVSEA LORA process.

Because of the cost of executing LORA computer models, it may be appropriate to vary a number of data elements. Instead of varying the MTBF of one specific LRU, the analyst may want to vary a population of LRU and SRU MTBFs by 200 percent to determine the effect on the repair policy. This may save time and allows the analyst to determine the effects the variation will have on all LRU and SRU policies.

Figure 7.4 illustrates the NAVSEA LORA process, while Table 7.6 cites NAVSEA lead allowance parts list (LAPL) effectiveness parameters for possible assessment and Table 7.7 cites possible negative support drivers to assess through the NAVSEA 3-M data base.

LORA cost estimating. The methodology presented here is a gross method of estimating the cost of LORA. The method is based on the number of LSA candidates, type of acquisition, life-cycle phase of the acquisition, and type of weapon system to determine the man-hours required to perform LORA.

The man-hour estimates shown in Table 7.8 are based on a survey of experience from 50 government and industry experts. This method lumps the expended man-hours for the required skilled personnel (i.e., engineer, engineering technician, clerical, artist, reproduction) in a LORA effort into one aggregate man-hour estimate. This man-hour estimate is the actual time (not calendar time) required to collect input data available from existing sources, conduct the analysis, execute a LORA model, and develop and reproduce a document of the effort. Thus the man-hour estimates shown in Table 7.8 indicate the effort required to develop a LORA report.

TABLE 7.6 **LAPL Effectiveness Analysis Parameters**

The design features (parts, assemblies) that experienced the highest failure rates (major downtime contributors)
The design features that enhance and/or degrade supportability
The design features that affect maintainability (e.g., safety, human factors)
Elements in the life-cycle cost that were higher than projected
Attained system availability: Significant variances between predicted and actual MTBF, MTTR, MLDT
Previously unrecognized requirements for support resources such as additional manning, tools, test equipment, test program sets, facilities, training, or enhanced technical documentation
Continual or unplanned maintenance problems
Frequency of repair actions
Repair turnaround times
Cost of acquiring support resources (e.g., materials or consumable resources used during operation or maintenance) along with projected life-cycle costs
Shortages of resources that were previously in plentiful supply (e.g., diminishing manufacturing sources, material shortages, properly trained instructors and maintenance personnel)

SOURCE: From NAVSEA TL081-AB-PR0-010, LORA Procedures Manual, December 1986.

TABLE 7.7   3-M Support Drivers

Active maintenance time
Causes
Deferrals
Unit price
Failed parts
Man-hours
Man-hours expended
Nondeferrals
Parts
Deferral reasons
Safety codes
Trouble isolation
Work requests
Work required

SOURCE: From NAVSEA TL081-AB-PR0-010, LORA
Procedures Manual, December 1986.

TABLE 7.8   LORA Cost-Estimating Information (Gross)

Number of LSA candidates	Type of acquisition	Life-cycle phase	Type weapon system	Average man-hours estimate*	Worst-case man-hours estimate†
25	NDI	D/V	Electrical or mechanical	96	240
		EMD	Electrical or mechanical	64	200
25	Developmental	CE/D	Electrical	64	280
			Mechanical	92	
		D/V	Electrical	49	240
			Mechanical	85	
		EMD	Electrical	60	200
			Mechanical	80	
300	NDI	D/V	Electrical	120	480
			Mechanical	125	
		EMD	Electrical	132	440
			Mechanical	160	
300	Developmental	CE/D	Electrical	183	560
			Mechanical	152	
		D/V	Electrical	142	480
			Mechanical	155	
		EMD	Electrical	158	440
			Mechanical	180	

* These man-hour estimates are the average man-hour estimates of 50 government and industry experts. No actual historical data is available.
† These man-hour estimates is based on the experiences of MRSA and should be considered as a top end man-hour estimate.
SOURCE: From Draft DOD LORA standard, March 1991.

If a LORA plan is required as a separate deliverable item from the LSAP, the effort required would increase the man-hour estimate shown in the table by one-half. The man-hours required to update a LORA plan or LORA report in a worst-case scenario should only be half those required to develop the initial document. In order to translate these man-hour estimates into dollar amounts, they have to be applied to a contractor labor rate.

Draft DOD LORA standard.   The DOD LORA standard now in development will focus on (1) data element standardization and (2) improved management guidance. Table 7.9 shows the tasks in the draft standard and suggested application by phase.

The standard will incorporate existing LORA models as appendixes. A list of the models in the current draft, dated March 1991, is as follows:

NAVAIR Method 1, Avionics, Model III, Appendix D

NAVAIR Method 2, Support Equipment, Model II, Appendix E

NAVAIR Method 3, Gas Turbine Engine, Appendix F

LOR Techniques for Space and Naval Warfare Systems Command Equipments, Appendix G

NAVSEA LORA, Appendix H

LORA Techniques for Marine Corps Equipment, Appendix I

Army Method 1, Appendix J

Army Method 2, Appendix K

Army Method 3, Appendix L

Air Force Method 1, Network Repair Analysis, Appendix M

Air Force Method 2, Item Repair Level Analysis, Appendix N

Air Force Method 3, Screening, Appendix O (*Note:* Noneconomic)

Federal Aviation Administration Method, Appendix P

TABLE 7.9   LORA Task Applicability

Task number and title	CE	DVAL	EMD	PROD
101 Program strategy and plan	S	G	G	C
102 Program reviews	G(1)	G	G	G(1)
201 Input data compilation	S	G	G	C
301 Evaluation performance, assessment, and documentation	S	G	G	S
401 Utilizing results	S	G	G	G

Code definitions: S = selective application; G = generally applicable; C = generally applicable to design changes only; (1) = selectively applicable for equipment-level applications.
SOURCE: From Draft DOD LORA Standard, March 1991.

"Graphical Repair/Discard Analysis Procedure Handbook." An interesting graphical level of repair technique is contained in MRSA OTSA 84-01, Graphical Repair/Discard Analysis Procedure Handbook.[40] While distribution is limited to U.S. government agencies, it may be obtained under the Freedom of Information Act from the Commander, USAMC, AMCSM-PLD.

The graphical technique in this book can be used in concept development and the DEMVAL phase for items in the range of $100 to $1000 with less than 10 percent error. An example of the nomographs in the book is given in Fig. 7.5.

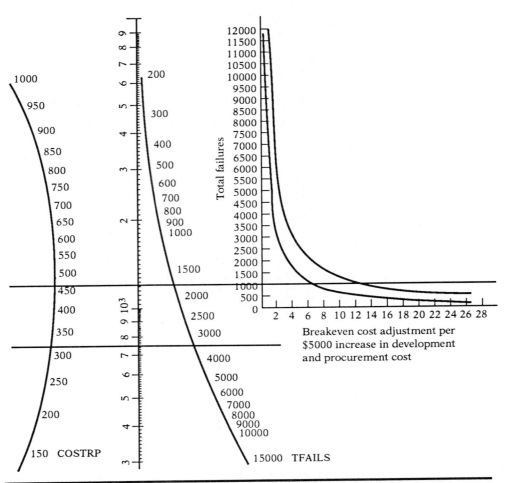

Figure 7.5 Example of Graphical Repair/Discard Chart. (From MRSA OTSA-84-1, Graphical Repair/Discard Analysis Procedure Handbook, 19xx.)

Key input data include

COSTRP, cost of repair parts per repair

COSTFD, development cost of test equipment and facilities (if needed)

COSTFP, procurement cost of test equipment and facilities (if needed)

Other key data codes on the charts are

COSTLH, cost per labor hour

REPNOT, fraction of assemblies not repairable

REPHRS, mean repair time per assembly

This technique, because of the parts cost range involved, could have commercial applications.

7.3 System Engineering Tradeoffs

Tradeoffs are an inevitable part of the design process. The choice between a Philips head or flathead screw, for example, has different implications that must be evaluated to reach a decision. Examples of typical higher-level tradeoffs are given in Table 7.10.

System Engineering Tradeoffs are generally managed by System Engineering (SE) with input from ILS/LSA. The degree of sophistication employed in different projects varies. Different practices observed include:

1. No documented decision approach, ad hoc decision making for each trade study
2. Identification of a list of factors to be employed in all tradeoffs, with no specified mathematical relationship
3. Use of a full mathematical system engineering methodology

TABLE 7.10   Examples of Tradeoffs

Type	Example
Design vs. design	Gas turbine vs. diesel
	Fixed vs. "swing" wing
System vs. system	Aircraft vs. submarine
	New vs. modified design
Operations vs. operations	Centralization vs. decentralization of sea-based aircraft
Design vs. logistics	System redundancy vs. spares
	Reliability vs. maintainability
Logistics vs. logistics	BIT vs. ATE vs. FIM
	ATE vs. skills
	Spares vs. transportation
	Throwaway vs. repair

TABLE 7.11   Generic Parameter Hierarchy

Cost-effectiveness	First order
System effectiveness	Second order
LCC	Third order
System readiness, performance, etc.	
ILS elements, R&M etc.	Fourth order
Accessibility, mounting, etc.	Fifth order

Use of approaches 2 and 3 should be encouraged by ILS/LSA, since their use provides an opportunity to explicitly inject supportability concerns into such decision making. In fact, some of the LSA subtasks could be satisfied by such action. For example, some of the subtasks in tasks 202 and 303 could be incorporated into the SE tradeoff process.

A full SE modeling effort usually involves assigning predetermined weights to the various parameters of interest. Table 7.11 illustrates application of the concept of hierarchical relationships among parameters. Figure 7.6 shows the relative weighting assigned in one major U.S. aircraft project. References 8 and 9 contain good discussions of the tradeoff weighting process. Appendix 20 illustrates application to an engineering judgment tradeoff technique.

Viewing the parameters in this light helps the analyst to determine if a factor has been considered. For example, anyone who has reviewed ILS lessons learned quickly discovers that poor accessibility is a frequent complaint. Anyone who has changed the spark plugs in a car can sympathize with this. However, access time, if not explicitly addressed in the requirements, is contained implicitly if there is an MTTR requirement.

The U.S. DOD now recognizes LSA as part of the System Engineering process. Integration of LSA with System Engineering and the Concurrent Engineering concept should be a major ILS/LSA consideration.

7.4   System Readiness

7.4.1   General

With the advent of more explicit DOD ILS policy in DOD 5000.39, "Acquisition and Management of ILS for Systems and Equipment," dated November 1983, logistics emphasis on system readiness parameters significantly increased. System readiness parameters are useful measures because they represent a useful objective, to military and sometimes commercial management, and because they can be related

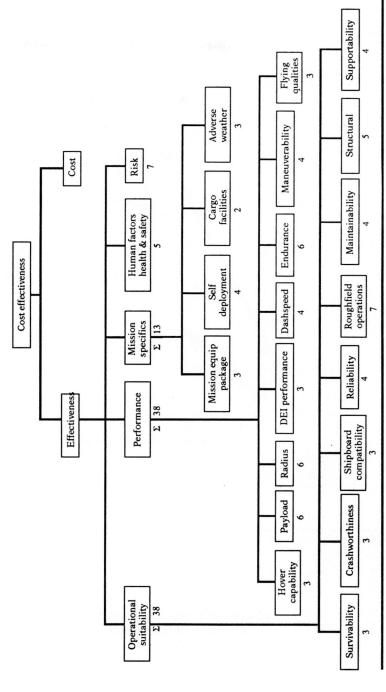

Figure 7.6 Tradeoff study weighting factors.

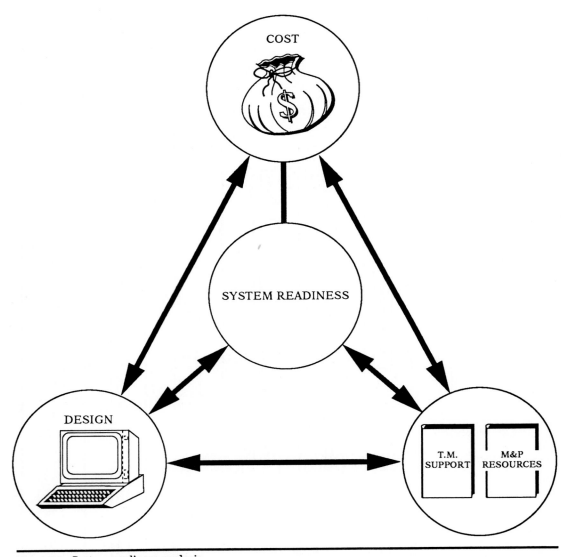

Figure 7.7 System readiness analysis.

mathematically to design, logistics, and cost concerns, as illustrated in Fig. 7.7. Appendix 21 contains a case example of the use of readiness modeling by management.

The system readiness parameters of interest may vary with the type of system, mission, and wartime or peacetime. Appendix 6 contains a highly tailored SOW example for system readiness studies. Opera-

TABLE 7.12 Examples of System Readiness Parameters

		Peacetime/wartime*					
UAV missions	Mission payloads	A_o	FMC	PMC	MC	Sortie rate	% coverage/ day
1. Reconnaissance, surveillance, and target acquisition (RSTA)	EO/IR	X	X	X		X	
2. Defensive electronic warfare	Deceptions repeater (decoy)	X	X			X	X
3. Electronic support measures	ESM set	X	X	X		X	
4. Communication/data relay	Relay transceiver set	X	X			X	
5. Amphibious support	EO/IR	X	X	X		X	
6. Antisubmarine warfare support	Sonobuoy receiver	X	X			X	
7. Search and rescue	EO/IR set	X	X	X		X	
8. Offensive EW	Active/passive set	X	X	X		X	X
9. Mine defense support	EO/IR set	X	X	X		X	X
10. Meteorological Recco.	MET MMP	X	X			X	
11. CBR Recco.	CBR MMP	X	X			X	X

*Specific requirements vary by service; FMC = fully mission capable; PMC = partially mission capable; MC = mission capable; X = applicable; Blank = deemed not applicable.

tional availability is undoubtedly the most widely used military system readiness parameter, but other parameters, such as aircraft sortie rate per day (fighter aircraft) and percent coverage per day (surveillance vehicles), are also useful. More than one readiness parameter may be useful on a system. Table 7.12 illustrates parameters potentially useful on unmanned vehicles. System readiness parameters should be managed in the same manner as other system parameters.

7.4.2 System readiness modeling

System readiness modeling techniques vary in complexity and point of application. Table 7.13 summarizes the major approaches used in practice. Each of these methods will be discussed briefly.

Algorithms: Operational availability

General. Operational availability is undoubtedly the most widely used system readiness parameter. It is frequently used for both wartime and peacetime by the U.S. DOD. It is also increasingly used in the commercial sector as needs similar to those of defense arise more frequently.

Commercial A_o. The most widely used definition of operational availability commercially is

$$A_o = \frac{\text{uptime}}{\text{uptime} + \text{downtime}} = \frac{\text{MTBM} + \text{RT}}{\text{MTBM} + \text{RT} + \text{MDT}} \quad (7.1)$$

TABLE 7.13 Readiness Analysis Approaches

Technique	Description	Advantages	Best use
Simple algorithm studies (A_o, A_1, etc.)	Readiness equations	Limited input data required Not required, a lot of calculations	Broad philosophical questions in conceptual phase
Semi-Markov chain models	Probabilistic based on means; same as simple algorithm if steady state reached	Detailed support systems descriptions not needed; can look at sensitivities fast and economically	Generalized trade studies in advanced and early engineering development availability
Simple trade simulation	Models operation of the support system	Detailed hardware description not needed	More detailed developmental studies Sortie rate
Complex simulation	Detailed example: Air Force LCOM model	Can look at detailed, specific problems	Detailed analysis in late engineering development and production

where MTBM = Mean Time Between Maintenance for both corrective and preventive maintenance actions or $MTTBU = 1/\lambda + f$

λ = sum of the frequency or rate of corrective maintenance

f = frequency or rate of preventive maintenance

MDT = $\bar{M}$ = mean logistic time + mean administrative time

RT = average ready time or uptime (available but not operating)

MDT is discussed in greater depth in the next section.

Navy A_o. The Navy, in Refs. 34 and 41, defines A_o similarly but slightly differently than its most common commercial definition:

$$A_o = \frac{MTBF}{MTBF + MDT} = \frac{MTBCM}{MTBCM + MDT}$$

The MTBF equation is used in development teting; the MTBCM equation is used in operational places. MTBCM is the mean time between corrective maintenance.

$$A_o = \frac{\text{uptime}}{\text{uptime} + \text{downtime}} = \frac{MTBF}{MTBF + MDT} \tag{7.2}$$

$$= \frac{MTBF}{MTBF + MTTR + MCDT}$$

where MDT = MTTR + MCDT

Because different systems are used in different ways, the measurement and interpretation of A_o vary from system to system. For purposes of A_o measurement and analysis, the U.S. Navy divides systems into three classes:

1. *Continuous-use systems:* Systems that are (nearly) always in use during operations of their host platforms. Examples are search radars, radio receivers, and propulsion gas turbines.
2. *Intermittent-use (noncontinuous or on-demand) systems:* Systems that have relatively long periods of standby or inactivity between uses. Examples are fire control radars and radio transmitters.
3. *Impulse (single-shot) systems:* Expendables that are generally used once and not recovered (and so not returned to an operable condition through repair when not recovered). Examples are bombs, missiles, sonobuoys, and torpedoes.

For continuous-use systems, mean calendar time between failure is identical to mean operating time between failure, and the following Equation (7.3) is used. For intermittent-use systems, mean operating time between failure is not equivalent to mean calendar time between failure. Two ways of adjusting the A_o equation are now in use. For aircraft systems, it is common to use

$$A_o = 1 - \frac{MTTR + MLDT}{(K')(MTBF)} \qquad (7.3)$$

where K' is defined as total calendar time over total operating time. For ship systems, a similar equation has been constructed. To do this, K' is redefined to exclude downtime from calendar time. This term, defined as K'', is

$$K'' = K' - \frac{MTBF + MLDT}{MTBF} \qquad (7.4)$$

Equation (7.3) can now be written for intermittent-use systems as

$$A_o = \frac{(K'')(MTBF)}{(K'')MTBF + MTTR + MLDT} \qquad (7.5)$$

Both K' and K'' have been termed the *K factor,* and this has led to confusion. The user should check this factor before using the equations to make sure that the correct "K factor" is used. K' is valid only in Eq. (7.3); K'' is valid only in Eq. (7.5).

Given that Equations (7.3) and (7.5) are identical, either can be employed. Users should choose between them based on current practices in their organization, the way parameters are reported to them, and the ease of computation.

The preceding formulas are not appropriate for impulse systems. Since these systems are generally not recoverable once they are used, the concept of downtime has little significance. As a result, the A_o of impulse systems is quantified as the fraction of attempts at usage (firings, turnons, actuations) that succeed. The formula is

$$A_o = \frac{\text{number of successes}}{\text{number of attempts}} \qquad (7.6)$$

The distinction is that an impulse system spends most of its time in standby, alert, or secured mode; is called on to function for a relatively short time; and is generally not restored to operable condition once it is used.

MLDT, for purposes of A_o, can generally be broken down as follows:

Mean Downtime for Outside Assistance (MDTOA). The average downtime per maintenance action awaiting outside assistance. This is normally caused by the lack of test equipment, tools, or skills beyond those available at the shipboard level. This also occurs when repairs require the use of facilities such as a drydock or floating crane.

Mean Downtime for Documentation (MDTD). The average downtime per maintenance action to obtain documentation needed for fault isolation, maintenance, and checkout. This is normally an insignificant portion of MLDT, but it occurs each time the system/equipment fails and is therefore inversely proportional to the reliability of the system. High MDTD normally is caused by technical publications that are not applicable to the configuration of the system/equipment installed or by errors in the technical documentation.

Mean Downtime for Training (MDTT). The average delay time per maintenance action due to lack of training. This is rarely a factor in the achievement of the A_o threshold and normally occurs when all trained maintenance personnel have been transferred or are otherwise unavailable.

Mean Downtime for Other Reasons (MDTOR). The average downtime per maintenance action for reasons not otherwise identified, including administrative time. It is common practice to provide for 10 hours of MDTOR and include MDTD and MDTT in MDTOR when these are insignificant.

Mean Supply Response Time (MSRT). The average delay time per maintenance action to obtain spare and repair parts from both on and off site. MSRT is the single greatest driver in MLDT. As soon as values are assigned to A_o, MTBF, and MTTR, the value for MLDT can be computed with the equation

$$MLDT = \frac{MTBF}{A_o} - (MTBF + MTTR) \qquad (7.7)$$

The MLDT requirement can be compared with normal supply response times to determine if it is reasonably attainable with the standard supply system. The sample matrix provided in Fig. 7.8 should be tailored to the specific system in order to calculate a close approximation of MLDT. The figure is split into three segments: mean requisition response time (MRRT), MSRT, and MLDT.

Mean requisition response time (MRRT).   Segment I of Fig. 7.8 is a reasonable approximation of supply-system response times whenever a part is not available on the ship and it must be requisitioned from the supply system. The 459 hours of delay awaiting parts from off the ship is based on the following assumptions:

- The wholesale supply system is fully funded to satisfy 85 percent of the requisitions it receives. Therefore, 15 percent of all requisitions that are referred to the wholesale supply system cannot be filled.

- Issue priority group I and II requisitions are considered to be immediate-use, maintenance-related parts requirements. These are readiness-driving parts requirements.

- One-third of the parts requirements are issue priority group I and two-thirds are issue priority group II.

- The 30 days to satisfy issue priority group I and the 75 days to satisfy issue priority group II requisitions, for parts not available in the wholesale supply system, considers cannibalization, reconsignment, active fleet screening, and other extraordinary actions to satisfy the requirement.

- About 30 percent of the units are deployed at any time.

- The times to fill requisitions for deployed and nondeployed units are DOD Uniform Material Movement and Issue Priority System (UMMIPS) time frames.

To the degree that these assumptions are valid for the specific system being analyzed, it takes, on average, about 459 hours to satisfy an off-ship requisition.

Mean supply response time (MSRT).   Segment II of Fig. 7.8 is a logic tree that can be used to compute the weighted average of response times to satisfy repair parts requirements from both shipboard

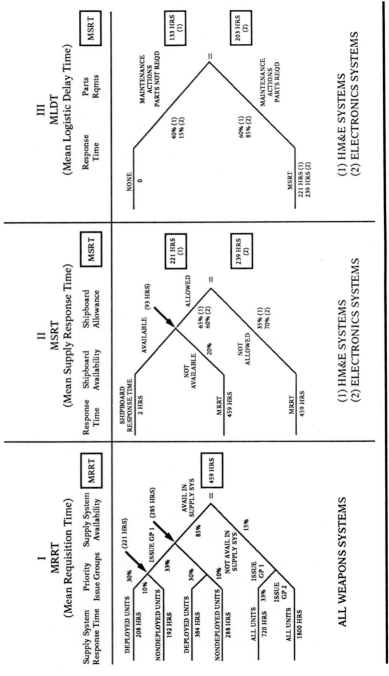

Figure 7.8 Example of MLDT default values. (From TRADOC/AMC Pamphlet 70-11; RAM Rationale Report Handbook, July 1987.)

inventories and the supply system. This computation requires determination of the following variables for the system being analyzed:

- The percentage of maintenance actions requiring parts that are allowed in shipboard stocks
- The percentage of allowed parts requirements that are normally satisfied from shipboard stocks

Depending on the development stage of the system, these data can be obtained by detailed analysis of empirical information on a similar existing system, from 3-M data, from CASREP data, or from APL data. The MRRT value of 459 hours is used to calculate the response time for allowed onboard parts that are not in stock and for not carried parts, both of which require off-ship requisitioning.

Mean Logistics Delay Time (MLDT). Segment III of Fig. 7.8 is the final logic tree by which one can determine MLDT. MLDT is a function of the percentage of maintenance actions that require parts and the mean supply response time to obtain those parts. Generally, a greater percentage of maintenance actions on electronics equipment require parts than maintenance actions on hull, mechanical, and electrical equipment. Ordnance equipments follow the pattern of electronic equipment or hull, mechanical, and electrical equipment depending on the characteristics of the system being analyzed.

Default values. The preceding explanations of each segment of Fig. 7.8 show that the derivation of MLDT is subject to a number of variables that affect MSRT. Some of these variables will differ significantly from one system to another. Although it is possible to actually measure supply response times and the percentage of maintenance actions requiring parts, MLDT information is not available directly from statistical files as a distinct data element. MLDT can be derived through the use of intermediate calculations employing fundamental supply support statistical information. The methodology provided above permits the calculation of MLDT utilizing available data and augmenting those data with reasonable estimates (default values) for selected elements. When estimated values are used for data elements required to calculate MSRT or MLDT and those estimates differ significantly from the default values provided below, the basis for the estimates must be explained as a part of the analysis. The default values that follow reflect current DOD/Navy policies and funding levels, include authorized stock levels at various echelons of supply, supply material availability (SMA) from the supply system, deployment status of fleet units, and supply response time experience for requisitions citing casualty report (CASREP) priorities for critical maintenance requirements.

- Percentage of corrective maintenance actions requiring parts:

  HM&E = 60%   Electronics = 85%

- Percentage of maintenance actions requiring parts that are satisfied from shipboard stocks:

  HM&E = 65%   Electronics = 60%

- Percentage of parts requirements for allowed onboard items that are satisfied from onboard stocks:

  All = 80%

- Time required to obtain a part from onboard stocks:

  All = 2 hrs

Repair parts essentiality coding.   Any repair part with a Military Essentiality Code (MEC) of 1 is potentially a readiness-driving part. An MEC of 1 means that when the part fails, its next-higher assembly fails. If the loss of the next-higher assembly causes failure of the system or equipment, then the part that initiated the chain of failures is a critical readiness-driving repair part. An MEC of 5 means that the part is required for safe operation of the equipment or prevents a personnel hazard. For purposes of sparing, an MEC 5 repair part is treated like an MEC 1 repair part. Any MEC 1 or 5 item that will cause total system failure is capable of being replaced at the MEC 5 level because lack of such a capability is a major deterrent to A_o when the system is deployed.

Source, Maintenance, and Recoverability codes (SM&R).   This is a five-position code that reflects supply and maintenance decisions made during the logistics planning process. The first two positions are the source code that indicates the means of acquiring the item for replacement purposes. The third position is a maintenance code that indicates the lowest level of maintenance authorized by the maintenance plan to remove, replace, or use the item. The fourth position is a maintenance code that indicates whether the item is to be repaired and identifies the lowest level of maintenance authorized by the maintenance plan to return the item to serviceable condition from some or all failure modes. The fifth position is the recoverability code, which indicates the approved condemnation level.

The third position of the SM&R code is the code reviewed for all MEC 1 items to ensure that critical readiness-driving repair parts are replaceable at the o level and that the supply system considers that item as a candidate onboard allowance item when an Allowance Parts List is computed.

Replacement rates. All repair parts have a reliability factor that is provided by the manufacturer at the time of provisioning. This may be expressed initially as mean cycles between failure, failures per million operating hours, mean time between failure, or some other measure of reliability. At the time of provisioning, this is converted to an annual replacement rate so that all parts in the supply system use a common replacement factor. Replacement rates are important because they determine what parts compute for o-level allowances and they identify potential supply support problems.

All U.S. Navy allowance computation models use a replacement factor as a key variable to determine the inventory of spares and repair parts to support organizational-level maintenance. Some allowance models weigh this demand rate more heavily than others. The critical factor for the program manager's consideration is whether the replacement factor continues to reflect the actual replacement rate throughout the life of the system/equipment. If a part is identified as a supply support problem, the first check to make is to compare actual replacements in use against the replacement factor used by the supply system to compute allowances.

In the design phases of system development, the replacement factor for each part should be compared with the SM&R and MEC codes. Any part with a high replacement rate that is an MEC 1 coded item but is not coded for organizational-level removal or replacement requires close scrutiny because this is a key warning of a future readiness problem.

Alternatives other than standard sparing. When the optimal mean response times that the standard supply system can attain and the MLDT are not sufficient to achieve the A_o threshold, given the reliability of the system/equipment, what alternatives are available? Since reliability and MSRT are the two major drivers in the attainment of the A_o threshold, these two variables are the focus of action.

Before Milestone II, design reliability was not frozen and there were many alternatives for increasing the system's reliability to achieve the A_o threshold. After Milestone II, the options available for increasing the system's reliability significantly decreased, and the cost to do so significantly increased.

MSRT can be improved when it has been determined that standard Navy sparing will not provide the response time required to achieve the A_o threshold. OPNAV approval to use an allowance computation model that optimizes the supply support required to achieve A_o must be obtained. The U.S. Navy has approved the use of various sparing to A_o models such as the Availability-Centered Inventory Model (ACIM). When this situation exists, there are procedures to be followed for using a sparing to availability model and for obtaining approval to optimize the APLs that will provide the required onboard stocks to support the system. The time required to complete this process, including approval, the POM and budget cycles, procurement lead times, and positioning the inventory in the using organization, is about 3 years. The timing required dictates that the determination that extraordinary supply support is necessary must be made as soon in the acquisition process as possible.

The objective of supportability design is to ensure that all logistics support elements are consistent with the maintenance plan and support the achievable A_o in the operational environment. A_o is broken down into the components of reliability, maintainability, and supportability to focus management attention on finite segments of uptime and downtime that can be managed. Supportability also can be broken down into finite segments of downtime that can be managed. The goal is for all logistics support elements to be coordinated, compatible, and consistent with the system's operational concepts.

Army A_o. The U.S. Army has different but comparable A_o approaches defined in the "RAM Rationale Report Handbook."[42] This handbook identifies the types of analysis required for different systems (see Table 7.14). The methodology employs comparative analysis techniques (LSA task 203) to identify

Mission reliability

Maintenance manpower

Repair parts cost

The handbook identifies time in the following terms:

TT = total time

OT = operating time

TABLE 7.14   Readiness Analysis Application by Program Category

	Major, new doctrine or technology	Other majors and DAP	Nonmajor development	Nonmajor NDI
Combat developer				
Simulation model	X			
Analytical		X	X	X
Material developer				
Baseline	X	X	X	X
MD proposal	X	X	X	X
State of art	X	X	O	O

Code: X is required; O is optional.
SOURCE: TRADOC/AMC Pamphlet 70-11, "RAM Rationale Report Handbook," July 1987.

ST   = standby time

TCM   = total CM time

TPM   = total PM time

TALDT = total administrative and logistics downtime

The basic A_o formulation is thus

$$A_o = \frac{OT + ST}{OT + ST + TCM + TPM + TALDT} \tag{7.8}$$

The handbook offers a number of other A_o formulations. If the equipment is operating whenever available, simply drop out the ST term in the basic equation.

Operational availability equations for systems in which relocation time considerations vary depending on whether the system is mission-capable while moving to the new location. If it is available during this time, relocation time (RT) can be determined by assuming that TALDT is proportionally distributed between location time and nonlocation time. The resulting equation is

$$A_o = \frac{OT + ST}{OT + ST + RT + TCM + TPM + \left(1 - \dfrac{RT}{TT}\right)TALDT} \tag{7.9}$$

If RT is considered downtime regardless of the system state, the equation drops out the RT term in the numerator and becomes

$$A_o = \frac{OT + ST + RT\left(\dfrac{OT + ST}{TT + RT}\right)}{OT + ST + RT + TCM + TPM + \left(1 - \dfrac{RT}{TT}\right)TALDT} \tag{7.10}$$

If the system is committable during PM, operational availability can be estimated by including TPM in standby time (i.e., combine ST and TPM in the equation).

If the equipment has idle time (e.g., never operated at night), maintenance can be performed at this time, and the equation becomes

$$A_o = \frac{OT + ST}{OT + ST + TCM1 + TALDT\left(1 - \dfrac{IT}{TT}\right)} \tag{7.11}$$

A_o **summary.** Obviously the approach to A_o varies with the specifics of the immediate situation. A_o is very widely used. More and more R&M and support requirements are determined from a top-down analysis early in design rather than from specification of state-of-the-art

R&M requirements and hoping that A_o meets the true need. A_o is now widely used in defense and is gaining use commercially.

Markov chain analysis

General. The application of Markov chain analysis to readiness analysis was summarized in Table 7.13. This section will summarize some key points concerning this analysis method. Excellent discussions of Markov processes are contained in Refs. 43 and 44. Markov chain analysis is easy to do because it involves (1) matrix multiplication and/or (2) solving n variables given $n + 1$ equations that can be reduced to n equations. There are PC programs that address both these situations. Markov chain analysis assumes that the process of interest consists of mutually exclusive states in which the item at hand might be. Figure 7.9 is an example of states in which the Joint Surveillance and Target Attack Radar System (JSTARS) exists.

A key assumption is that only the current state (but not others) affects the next state the system may enter. The probability that a ve-

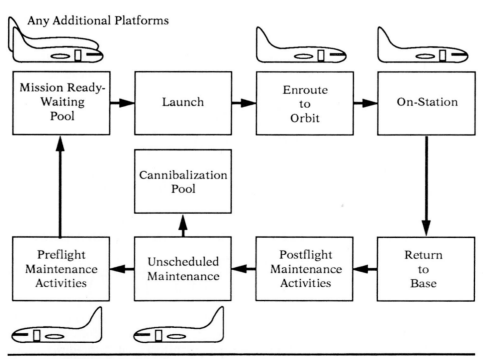

Figure 7.9   JSTAR block diagram.

hicle in one state of going to another is called a *transitional probability*. It is useful to express transition probabilities in a matrix:

$$P_1 = \begin{matrix} P(1,1)\ldots P(1,n) \\ P(2,1)\ldots P(2,n) \\ P(2,2)\ldots P(3,n) \\ \ldots \quad \ldots \quad \ldots \\ \ldots \quad \ldots \quad \ldots \\ \ldots \quad \ldots \quad \ldots \\ P(n,1)\ldots P(n,n) \end{matrix}$$

Since matrix algebra applies,

$$P_n = P_1 P_{n-1} = P_{n-1} P_1 \quad \text{for } n = 2, 3,\ldots, n$$

The conditional probabilities that the system is at state j at time $t + n$ are defined by

$$P_m(i,j) = P(i, k)P_{n-1}(k, j)$$

The unconditional probabilities are defined by

$$P_n(j) = P_{n-1}(k)P(k, j)$$

If we know the initial probabilities that the system is in the various states, we can find the various p_n because

$$p_2 = p_1 P = P p_1$$

$$p_3 = p_2 P = P p_2$$

$$p_n = p_{n-1} P = P p_{n-1}$$

Thus, by using time-line analysis (e.g., time expected in each state), we can calculate the expected results by multiplying matrices as appropriate for the time period (e.g., number of days) of interest.

If we are interested in a steady-state solution, we can solve for the probabilities (translatable to times in state) by solving linear equations, since

$$\lim_{m \to} P_n(i,j) = \pi_j$$

and

$$\pi_j = \pi_k P(k, j) \quad j = 1, 2,\ldots, r$$

and

$$\pi_j > 0 \quad \text{and} \quad \pi j = 1$$

Note that in the steady-state case we, in effect, have algorithms (i.e., firm equations).

Example. As an example, assume that a system can be in one of three states with action as shown in the block in Fig. 7.10. The transition probabilities are

$$P = \begin{bmatrix} 0 & .95 & .05 \\ .2 & 0 & .8 \\ 1.0 & 0 & 0 \end{bmatrix}$$

The stationary (steady-state) probabilities are

$$\pi_1 = 0.2\pi_2 + \pi_3$$

$$\pi_2 = 0.95\pi_1$$

$$\pi_3 = 0.05\pi_1 + 0.8\pi_2$$

The answer is

$$\pi_1 = .36^+$$

$$\pi_2 = .34^+$$

$$\pi_3 = .29^+$$

Solving by substitution gives

$$\pi_1 = 0.36 + \qquad \pi_2 = 0.34 + \qquad \pi_3 = 0.29 +$$

$$= 0.36^+ \qquad\qquad = 0.34^+ \qquad\qquad = 0.29^+$$

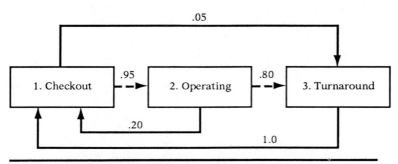

Figure 7.10   System use block diagram.

Simulation models

General. *Simulation* is an analysis approach that imitates the real world in a model that operates over time to allow study of the system of interest. Simulation models have been used effectively for many years by logisticians and engineers. Originally, they needed mainframes, but now they are increasingly worked on PCs, thus allowing logistics analysts to conduct such analyses at their desks. The U.S. defense establishment has a number of simulation models. Some are identified in the "LSA Techniques Guide."[6] Simulation models can be simple or complex depending on the analysis desired and the data available.

Simple simulation models. Simple simulation models may be of two kinds: (1) a simple, high-level model for use early in the program to evaluate system-level parameter variation and (2) a high-level model with significant depth in one or more aspects of interest. Simple high-level models are generally easy to construct in a relatively short time on a PC.

The second kind of simple simulations are sometimes useful for analyzing PIPs and modifications. A simple such simulation was constructed to analyze an aircraft fighter being converted to an electronic warfare (EW) aircraft (see Appendix 21 for illustration). Some 37 line replacement units (LRUs) in the new subsystem were modeled in-depth while the remainder of the aircraft was treated as one (gigantic) LRU. The analysis results matched those of a more complex Logistics Composite Model (LCOM) analysis conducted when comparable.

Complex simulations. There are a number of complex simulation models. Such models generally require large amounts of input data but allow detailed analysis to be performed. Figure A21.3 illustrates some of the results from using such a model. Examples of such models listed in Ref. 6 are

Ammunition Point Simulation (APS) Model

Artillery Battalion Ammunition/Propellant Resupply Simulation (ARTREAM)

Budget/Readiness Analysis Technique (BRAT)

Combat Vehicle Reliability, Availability, and Maintainability (ERAMS)

Force Evaluation Model (FORCEM)

Infantry Division (Light) Simulation Model (ID(LT))

Logistics Composite Model (LCOM)

Maintenance Capabilities Attack Model (MACATAK)

Equipment Repair/Replacement Simulation (REPSIM)

System Effectiveness Level III (SEC III)

TRANSANA Aircraft R&M Simulation, CAA Version II (TARMS-II)

TIGER Reliability Computer Program (TIGER)

Transportation and Supply Activities (TRANSACT)

There is no one "best" readiness (or LORA or LCC) model, since such models are designed with different analysis objectives in mind.

7.5   Life-Cycle Costing

7.5.1   General

Cost is obviously a major consideration. It is generally important to think of life-cycle cost (LCC) as well as acquisition cost, since the operating and support costs are well over 50 percent of the life-cycle cost on some systems/equipments. It is usually therefore important to address operating and support costs in some fashion. The preferred approach is the use of some formal cost-estimating methodology, although it may be necessary to resort to some key O&S cost surrogates such as R&M.

Another key point to keep in mind is that decisions that determine LCC are made *much earlier* than the actual expenditure of the funds. This is illustrated in Fig. 7.11. Because of the need to better manage

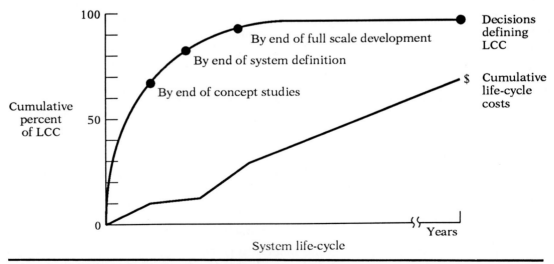

Figure 7.11   LCC fund commitment vs. expenditure.

O&S costs, and because O&S costs were maintained by organizations rather than weapon system/equipment, the U.S. DOD has developed a program called Visibility and Management of Operating Support Cost (VAMOSC) to provide a data base for projecting future O&S costs on new programs.

7.5.2 Uses of LCC analysis

The uses of LCC analysis can be grouped broadly in the following categories:

- *Baseline LCC analysis:* Evaluation of system LCC for a specified configuration and a defined operational and maintenance concept.
- *Sensitivity analysis:* Determination of the sensitivity of cost estimates to variations in key input parameters to identify cost drivers.
- *Tradeoff analysis:* Evaluation of alternative approaches to aid in selection of a preferred candidate.
- *LCC tracking:* Review of LCC estimates as the program progresses to review any variances from previous estimates and to relate variances to causal events.

An example of management decision potential through the use of sensitivity analysis based on T&E results is contained in Appendix 21. Table 7.15 illustrates use by phase.

Trade studies use different analysis techniques depending on the type of study, since no single model is applicable to all such studies. In fact, many trade studies use a variety of models. In addition, trade studies may be time-dependent.

TABLE 7.15 Example LCC Uses by Phase

Phase	Possible LCC utilization
Preconcept	Establish baseline LCC for each alternative; identify affordability constraints
Concept exploration	Establish baseline LCC for selected alternative; determine DTC objectives; identify logistics drivers; demonstrate affordability
Demonstration/validation	Set DTC goals; support tradeoff analyses; analyze cost risks; identify logistics drivers; verify affordability
Engineering manufacturing development	Update LCC; support tradeoff analysis; analyze logistics alternatives; verify affordability
Production deployment	Estimate LCC of product improvement plans and ECPs; use LCC to monitor contract; monitor cost performance in the field

7.5.3 Cost-estimating approaches: types of models

There are basically four methods of estimating costs:

1. Analogy
2. Parametric
3. Engineering estimate
4. Projection of actuals

Analogy is akin to scaling by comparison. This is usually done linearly; that is

$$C_{new} = BC_{old}$$

where b is a selected constant or scalar. This type of O&S estimating has gained popularity as the implementation of VAMOSC has proceeded.

Parametric cost estimating has been used for years in the United States, particularly by independent cost-estimating groups established to provide management with an unbiased second opinion. Parametric estimating uses mathematical relationships developed by minimizing the statistical variance of a line/curve from actual existing data points for similar items. Figure 7.12 shows the concept. The equation may be simple:

$$C_{new} = a + bX_i$$

or

$$C_{new} = aX_i + j$$

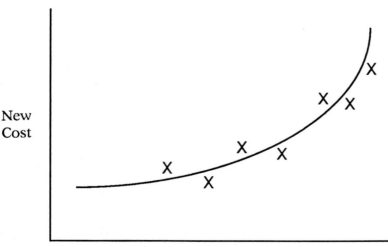

New
Cost

Existing Systems

Figure 7.12 Correlation analysis example.

where a, b, and j are constants and X_i is the value for the variable.

An actual example is one estimate for aircraft airframe production costs:

$$\text{cumulative labor} = 28.9A^{0.74}S^{0.54}Q^{0.52}L$$

$$\text{cumulative material} = 37.6W^{0.69}S^{0.62}Q^{0.79}$$

where W = ampr weight
S = maximum speed at altitude
Q = cumulative quantity produced
L = composite labor rate

Parametric estimating is particularly useful early in a program when detailed data are not available.

Detailed engineering analysis is useful when more data are available. It essentially adds up a series of estimates for various aspects of the item. Such models are normally called *accounting models* because of the additive feature. One example of this approach is presented in Table 7.16.

The last method, *use of actuals*, is similar to the detailed engineering analysis approach, but it takes advantage of actual data. This means that such estimates may only be possible after some production has occurred.

Fundamentally, cost is usually one major criterion in any decision with different options. Cost estimating is important in many trade studies. Blanchard[4] has an excellent discussion of various types of specific cost analysis techniques. Jones[3] has an excellent discussion of accounting models.

TABLE 7.16   Engineering Cost Analysis Example (Aircraft On-Equipment Maintenance Costs)

$$\sum_{i=1}^{N} \left[\frac{(\text{TFFH})(\text{QPA}_i)(\text{BLR})}{\text{MFTBMA}_i} \right] [(\text{RIP}_i)(\text{IMH}_i) + (1 - \text{RIP}_i)(\text{RMH}_i)](K_{\text{utilization}}) + \frac{(\text{TFFH})(\text{SMH})(\text{BLR})}{(\text{SMI})}$$

where i = item i
TFFH = Total fleet flying hours
QPA_i = Quantity of item i on aircraft
MFTBMA_i = Mean flying time between maintenance item i
BLR = base labor rate
RIP_i = percent of item i failures repaired in place
IMH_i = Average man-hours to repair item i in place
RMH_i = Average man-hours to remove and replace
SMH = average scheduled maintenance hours
SMI = scheduled maintenance interval

7.5.4   Models available

There are many LCC models currently in existence. Table 7.17 lists some of these by type. Additional models are listed in AMC-P700-4.[6]

The CASA model.   While there are too many cost models available for a discussion of each, two models will be discussed. The first is the Cost Analysis Strategy Assessment (CASA) model. CASA was developed and is maintained by the U.S. DOD Defense Systems Management College. CASA is based on Honeywell's Total Resources and Cost Evaluation (TRACE) family of logistics and LCC models. TRACE has been used on a number of DOD programs. CASA can operate on PCs and is available from the Defense Systems Management College.

The CASA model can be used for a number of tasks, such as

LCC estimates

Tradeoff analyses

Repair-level analyses

Production rate and quantity analyses

Warranty analyses

Spares provisioning

Resource projections (e.g., manpower, support equipment)

Risk and uncertainty analyses

Cost-driver sensitivity analyses

Reliability growth analyses

Operational availability analyses

Spares optimization

With these capabilities, the CASA models also can be used in design to life-cycle cost (DTLCC) studies.

TABLE 7.17   Sample Available O&SD/LCC Models

Model	Type	Source	Application	O&S/LCC
GE Price HL	Parametric	GE	Hardware	LCC
GE Price SL	Parametric	GE	Software	LCC
AF LSC	Accounting	AF	Avionics/ground	O&S
LCCH	Accounting	AF	Avionics/ground	LCC
CASA	Accounting	DSMC	Avionics/ground	LCC
Night vision LAB	Accounting	Army	Avionics/ground	LCC
NAVMAT FLEX	Accounting	Navy	Avionics/ground	LCC
EDCAS	Accounting	System exchange	Avionics/ground	LCC
Munitions DSN/ TRADE	Accounting	AF	Missiles/ munitions	O&S

FEA/FECA model.   While CASA and other models have been very useful, many require considerable input data (CASA has over 400 inputs, although all are not needed to do early analysis). Some members of the logistics community have therefore advocated development of a simpler cost-estimating tool. Such a tool is now in development in the front-end cost analysis (FECA)/front-end analysis (FEA) models. The term *front-end analysis* is used for the model developed and published by MRSA. The joint project Office for Unmanned Aerial Vehicles has adopted the FEA approach and called the resulting model *front-end cost analysis.*

FEA/FECA has a little over 200 data inputs. It can be applied using simple spread-sheet techniques. It is intended as a "quick and dirty" technique usable by the LSA analyst.

Although FEA/FECA has limitations, the concepts can be adopted to a variety of circumstances, including commercial applications. Appendix 26 has a more complete discussion of FECA and its algorithms to give readers a departure point for development and use of a comparable technique for their own work environment. FECA/FEA also can include simple algorithms for other measures of merit, such as A_o.

7.5.5   Cost-risk analysis

Cost-risk analysis is an analytical process that quantifies uncertainty associated with point estimates. In single point estimates, all input should be known with certainty. This is seldom the case, however. Single point estimates convey no information about the potential variance in the actual outcome, thus creating false confidence in the results.

Cost-risk analysis assumes that key variables, such as unit cost and reliability, are random variables. Various probability distributions, such as normal, exponential, triangular, uniform, etc., are assigned to the key variables. A number of iterations are then performed, selecting random values from the various distributions as inputs. Frequency tables and cumulative probability graphs can be constructed to portray the results, which can be tested for statistical goodness of fit using a beta distribution approach. Table 7.18 illustrates the process.

7.5.6   Software LCC estimating

The rise in software usage, complexity, and cost and the differences in the hardware and software processes have resulted in software modules in LCC models and in separate software LCC models. Note the Price software model in Table 7.16.

Software maintenance may be many times the original software development cost and a major portion of the LCC of the weapon system

TABLE 7.18   CASA Cost-Risk Analysis

- The following system parameters can be considered as random variables
  System cost
  Item unit cost
  Item MTCF
  Item MTTR
- Allowed probability distributions
  Normal (mean, standard deviation)
  Triangular (low, most likely, high)
  Uniform (minimum, maximum)
  Constant
- Iterations: minimum of 50 to maximum of 1000
- Frequency table and cumulative probability plot provided: performs goodness-of-fit
  test vs. a beta distribution
- Dependencies between parameters not considered

---

itself. Maintenance here again includes software functional changes as well as software repair. Software maintenance factors include

- Quality of software delivered
- Length of the operations phase
- Maturity of the product
- Number of users
- Type of software

Software models include COCOMO, SLIM, and GE Price. Most models use algorithms that are a function of lines of codes and development cost or man-months.

7.6   Summary

This chapter has discussed various analysis methods used in performing LSA. The use of such methods makes LSA results more credible and marketable. Use of such methods is one key factor in making LSA efficient and productive.

MRSA Electronic Bulletin Board

The Material Readiness Support Agency (MRSA) operates an electronic bulletin board dedicated to LSA/LSAR issues. Some of the topics on this board include 2A–2B, data map, 2B, report specifications, current events, CALS/LSA model documentation, LSAR ADP system line changes, LSA bulletin articles, and user comments. The telephone number to access this board is (606) 293-3700/3709. It is a 2400-Bd 8-*N*-1 configuration using a VT100 emulation. At the prompt "login:", type "lsa" (lowercase). Below is the listing from February 1992.

The remainder of this appendix contains examples of the contents of the bulletin board. Items that are printed in this font are taken directly from the bulletin board.

A1.1 Main Menu

Below is the main menu that appears after you have logged on.

```
          <   <   < MAIN MENU >   >   >

A LSAR ADP System Updates                    [Apr 4, 91]
    (Program Line Changes)
B You Might Want To Know...                   [Feb 14, 92]
C CALS/Data Model Documentation
D MIL-STD-1388-2A to 2B Data Map             [Mar 13, 91]
    (Cross Reference Between 2A and 2B)
E MIL-STD-1388-2B (Excerpts and Supplements) [Oct 19, 91]
F Functional Support and Publications         [Aug 1, 91]
G File Download


.X Exit BBS
? HELP
```

A1.2   Main Menu Selection A

This selection contains line changes to the MIL-STD-1388-2A LSAR ADP program.

A1.3   Main Menu Selection B

This selection primarily contains information on upcoming events, such as courses and seminars.

A1.4   Main Menu Selection C

This selection contains the data model that was used to develop MIL-STD-1388-2B.

A1.5   Main Menu Selection D

This selection contains the crosswalk from the 2A data records to the 2B tables. It is very detailed, and most logisticians will have no need for this information. A sample is printed here.

First screen.   User can view data mapped from either 2A to 2B or from 2B to 2A:

```
                        LSAR Data Map
             There are two ways of viewing the Data Map

                 1 2A Data Record Sequence [Mar 13]
                 2 2B Table Sequence       [Mar 13]
```

Second screen.   User selects which record or table he or she wishes to examine:

```
                          Data Map
                   2A Data Record Sequence

                 1 Explanation of Map
                 2 'A'  Record - Data Map [Mar 13]
                 3 'B'  Record - Data Map [Mar 13]
                 4 'B1' Record - Data Map [Mar 13]
                 5 'B2' Record - Data Map [Mar 13]
                 6 'C'  Record - Data Map [Mar 13]
                 7 'D'  Record - Data Map [Mar 13]
                 8 'D1' Record - Data Map [Mar 13]
                 9 'E'  Record - Data Map [Mar 13]
                10 'E1' Record - Data Map [Mar 13]
                11 'E2' Record - Data Map [Mar 13]
                12 'F'  Record - Data Map [Mar 13]
                13 'G'  Record - Data Map [Mar 13]
                14 'H'  Record - Data Map [Mar 13]
                15 'H1' Record - Data Map [Mar 13]
                16 'J'  Record - Data Map [Mar 13]
```

Third screen. This section begins with a preface that identifies and defines the format of the crosswalk from tables to records and then continues with the record or table that was selected. In this case, it is a portion of the 'B' record as an example:

```
PREFACE

A. General: This MIL-STD-1388-2A to MIL-STD-1388-2B cross reference
is not intended to stand alone when converting files. It would be
impossible to perform conversion correctly without the 2A and 2B
documents and the conversion guide developed by this activity. THESE
TABLES WILL BE UPDATED AS REQUIRED. THE DATE OF THE VERSION IS IN
THE UPPER LEFT CORNER OF THE DISPLAY. This cross reference supplies
the source of the 2B elements. The logic for creation of the 2B
tables may be acquired from the documents. For example; This cross
reference lists the A01 record as the 2A source for LCN and ALC
which are the major keys for most of the RAM, FMECA, and RCM tables
(B tables). In fact, the B table entries (rows) will not be created
during conversion unless the B, B1, B2 records are present.
MIL-STD-1388-2A logic demands presence of an A01 master record when
establishing B, B1, or B2 data—hence the A01 source for the LCN and
ALC.


B. The following is a list of columns displayed on the maptable
lists and their meaning.


Column                    Definition
arec                      The 1388-2A record-card source of this data.
ablk                      The 1388-2A record block source of this data.
aname                     The short name of the 1388-2A source of this
                          data.
aded                      This is the 1388-2A DED number.
bded                      This is the 1388-2B DED number.
bcode                     This is the 8 position element code as in
                          appendix A.
btab                      This is the 2 position table code as in appendix
                          A.
conv_remarks              Comments on elements which require special
                          handling during conversion.


C. The lines in the 2B tables listings which are blank in arec,
ablk, and aname are either new data elements to 2B or 2A elements
which cannot be referenced in the 2A format. In either case, these
elements are not convertible.


D. The lines which have ''***'' in the arec column are elements
which must be manually supplied during the conversion process.


E. 999 in bded indicates this column is composed of several DED's.
Consult Appendix A of 2B for complete description. F. ''NEW'' in the
aded column indicates this is a new data element and is not in
1388-2A.
```

B TABLES

AREC	ABLK	ANAME	ADED	BDED	BCODE	BTAB	CONVREMARKS
***		EIAC	106	096	EIACODXA	BA	data from conversion record type ''4''
A01	01	LCN	197	199	LSACONXB	BA	expand to 18 pos based on conv record type 5
A01	02	ALC	023	019	ALTLCNXB	BA	expanded to 2 positions-zero fill if blank
			NEW	203	LCNTYPXB	BA	''P'' (physical) assigned in conversion
			NEW	243	MEQLINBA	BA	
A03	05	CONV FAC	069	059	CONVFABA	BA	
B06	04B	FAULT	040	143	FIAMBABA	BA	-2A DED 040 is divided into multiple DEDs
B06	04B	PERCENT	040	143	FIPFGABA	BA	-2A DED 040 is divided into multiple DEDs
B06	04A	DETECT	040	032	BDLPGABA	GA	-2A DED 040 is divided into multiple DEDs
B06	05B	FAULT	040	143	FIAMBBBA	BA	-2A DED 040 is divided into multiple DEDs
B06	05B	PERCENT	040	143	FIPFGBBA	BA	-2A DED 040 is divided into multiple DEDs
B06	05A	DETECT	040	032	BDLPGBBA	BA	-2A DED 040 is divided into multiple DEDs
B06	06	BIT CND	041	031	BITNDPBA	BA	
B06	07	BIT RTOK	042	033	BITROPBA	BA	
B16	05	FAIL SOR	133	141	FRDATABA	BA	
B06	08	PIL R/O	321	292	PREOVCBA	BA	
			399	369	SECCLEBA	BA	not convertible from 1388-2A
B06	09	ICS	165	410	SUPCONBA	BA	
B06	14	WEAROUT	540	505	WEOULIBA	BA	
B06	15	WEAROUT MB	244	238	WOLIMBBA	BA	
B06	03B	LOG CONSID	195	196	LOGACCBA	BA	
B06	03H	LOG CONSID	195	196	LOGCONBA	BA	
B06	03M	LOG CONSID	195	196	LOGCRCBA	BA	
B06	03L	LOG CONSID	195	196	LOGDSPBA	BA	
B06	03J	LOG CONSID	195	196	LOGFLOBA	BA	
B06	03K	LOG CONSID	195	196	LOGLABBA	BA	
B06	03C	LOG CONSID	195	196	LOGMAIBA	BA	
B06	03I	LOG CONSID	195	196	LOGPATBA	BA	
B06	03D	LOG CONSID	195	196	LOGSAFBA	BA	
B06	03F	LOG CONSID	195	196	LOGSKIBA	BA	
B06	03A	LOG CONSID	195	196	LOGSTABA	BA	
B06	03E	LOG CONSID	195	196	LOGTEPBA	BA	
B06	03G	LOG CONSID	195	196	LOGTRABA	BA	
***		EIAC	106	096	EIACODXA	BB	data from conversion record type ''4''
A01	01	LCN	197	199	LSACONXB	BB	expand to 18 pos based on conv record type 5
A01	02	ALC	023	019	ALTLCNXB	BB	expanded to 2 positions-zero fill if blank
			NEW	203	LCNTYPXB	BB	''P'' (physical) assigned in conversion
			NEW	341	RAMCNABB	BB	RAMCNABB generated during conversion based
				450	TEXSEQBB	BB	on B08-4 = A, B10-4 = B, B09-4 = D, C cannot
				—	RAMNARBB	BB	be converted. TEXSEQBB is generated from CSC.
				096	EIACODXA	BC	BC table is not convertible from 1388-2A
				199	LSACONXB	BC	BC table is not convertible from 1388-2A
				019	ALTLCNXB	BC	BC table is not convertible from 1388-2A
			NEW	203	LCNTYPXB	BC	BC table is not convertible from 1388-2A
			NEW	425	LOCOCOBC	BC	BC table is not convertible from 1388-2A
				450	TEXSEQBC	BC	BC table is not convertible from 1388-2A
				426	LOGNARBC	BC	BC table is not convertible from 1388-2A
***		EIAC	106	096	EIACODXA	BD	data from conversion record type ''4''
A01	01	LCN	197	199	LSACONXB	BD	expand to 18 pos based on conv record type 5
A01	02	ALC	023	019	ALTLCNXB	BD	expanded to 2 positions-zero fill if blank
			NEW	203	LCNTYPXB	BD	''P'' (physical) assigned in conversion
B07	03	R/M IC	379	347	RAMINDBD	BD	
B06	12	AA	003	001	ACHAVABD	BD	
B06	11	AI	158	164	INHAVABD	BD	expanded from 6 N R 4 to 8 N R 6

A1.6   Main Menu Selection E

This selection has a great deal of information. The first two selections are electronic copies of Appendixes A and E of MIL-STD-1388-2B. The next two items on this menu contain the DIDs and report specifications associated with MIL-STD-1388-2B. Finally, the user may choose to look at the new DEDs that have been added to the data-element dictionary. Here we will look at two DID examples (LSAR and FMECA) and one report specification for the FMECA.

First screen.   Here the user can select what type of information he or she wishes to examine. The entries for selections 1, 2, and 5 are digital copies of the information that is in MIL-STD-1388-2B and will not be looked at any further here.

```
                 MIL-STD-1388-2B
               Excerpts and Supplements

     1  Appendix A...LSAR Relational Tables     [Mar 13]
     2  Appendix E...Data Element Dictionary    [Mar 13]
     3  New Data Item Descriptions (DID)        [Jun 13]
     4  LSAR Report Specifications              [Oct 19]
     5  New Date Element Descriptions (DED)     [Oct 19]
```

Second screen.   Having chosen selection '4' from the previous menu, the user may now select the range of report specifications that he or she wishes to examine:

```
                   LSAR Report Specs
       Select range in which specification of interest
                         falls.

     1          LSA-001—LSA-009            [Aug 2]
     2          LSA-010—LSA-023            [Aug 2]
     3          LSA-024—LSA-040            [Aug 2]
     4          LSA-046—LSA-075            [Aug 2]
     5          LSA-076—LSA-155            [Aug 2]
```

Note: The LSA report descriptions are in MIL-STD-1388-2B. The report specifications are a much more detailed explanation of the reports. LSA-056 is shown as an example. It is followed by its DID.

Fourth screen.   The specification for the FMECA is partially listed below as an example:

```
          LSA-056 FAILURE MODE, EFFECTS,
        AND CRITICALITY ANALYSIS (FMECA) REPORT

               REPORT SPECIFICATION

1. SELECTION. The LSA-056 summary is selectable by mandatory EIAC,
Start LCN, ALC, TYPE, UOC and SERV DES; and optional Stop LCN. The
report may also have selections for SHSC and RPT RT. The report
selections consist of:
     RPT PT:
     a. Part 1    1. Enter a (y) if part 1 of the 056 report is
                  required, otherwise, leave blank.
```

2. If part 1 has been selected then enter only those SHSC's (1, 2, 3,4) of the failure modes which are of interest. If the SHSC field has been left blank, data involving only SHSC's of 1 and 2 will be reported.

b. Part 2   1. Enter a (y) if part 2 of the 056 report is required, otherwise, leave blank.

2. If part 1 has been selected then enter only those SHSC's (1, 2, 3,4) of the failure modes which are of interest. If the SHSC field has been left blank, data involving only SHSC's of 1 and 2 will be reported.

3. Enter the lowest values of a failure mode criticality number to be included in the report. If an entry is made in b.4, Failure Probability Level then leave this field blank.

4. Enter the lowest value of a Failure Probability Level to be included in the report. An entry must be made in either b.3 or in this field. If an entry is made in both blocks; failure probability level is disregarded.

c. Part 3   1. Enter a (y) if part 3 of the 056 report is required, otherwise, leave blank.

2. If part 1 has been selected then enter only those SHSC's (1, 2,3,4) of the failure modes which are of interest. If the SHSC field has been left blank, data involving only SHSC's of 1 and 2 will be reported.

2. PROCESSING.   a. Part 1. The report will pull the information from the appropriate tables, except for:

Calculated Item Criticality Number, where

$$Cr = (Cm)n \quad n = 1,2,3 \ldots j$$
$$Cr = \text{Criticality Number for the item}$$

Cm = Failure Mode Criticality Number
n = The failure modes in the items that fail under a particular severity classification/mission phase combination
j = Last failure mode in the item under the severity classification/mission phase combination

Calculated Failure Mode   Criticality Number, where

$$Cm = (B*a*FR*T) \ (1,000,000.)$$
B = Failure Effect Probability
a = Failure Mode Ratio
FR = Failure Rate
t = Operating Time

NOTE: FM Criticality Number is computed for each LCN/ALC and Failure Mode combination. This report will print a part 1 for each LCN that Falls within the LCN select range. The failure rates are depicted by comparative analyzed (C), allocated (A), predicted (P), and measured (M) (Where a measured value has not been entered, the report will default to the predicted, then allocated, and finally comparative analysis) (these come from RAM Indicator Code). If Type is (F) the Reference Number and CAGE may not appear on report.

b. Part 2 of the report will pull the required information from the appropriate tables. Reference Number is wrapped at 16 positions.

c. Part 3 of the report will pull the required information from the appropriate tables, except for:

Calculated Failure Mode Criticality Number,

$$Cm = (B*a*FR*T) \quad (1,000,000.)$$

where B = Failure Effect Probability
a = Failure Mode Ratio
FR = Failure Rate
t = Operating Time

NOTE: FM Criticality Number is computed for each LCN/ALC and Failure Mode combination. The Failure mode ratio(s) for an LCN should be added up (TOTAL) and placed on the report as specified in attachments 1 and 2. An edit check should be made against this total to make sure it never exceeds 1.00. If the totals of the Failure Mode are greater than 1.00 then an (***) should appear.

$$Total = E \ FMR$$

where   FMR = Failure Mode Ratio
n   = The failure modes indicator that falls
under a particular LCN and ALC combination
j   = Last failure mode indicator in the LCN and
ALC combination

The failure rates are depicted by comparative analyzed (C), allocated (A), predicted (P), and measured (M) (Where a measured value has not been entered, the report will default to the predicted, then allocated, and finally comparative analysis)

d. Each part of the 056 report should start at the top of a new page.

3. REPORT FORMAT. See attachment 1.

4. REPORT SEQUENCING.

a. Part 1, The report is sequenced by ascending values of LCN then FMI then MPC then SHSC. Part 1 will be produced once for each LCN ALC combination which falls within the select range.

b. Part 2, The report is sequenced by ascending values of FPL then LCN.

c. Part 3, The report is sequenced by ascending values of LCN.

5. DATA SOURCE.

Report Header	Table Codes	Ded	Element Code
EIAC	XB	096	EIACODXA
LCN NOMENCLATURE	XB	201	LCNAMEXB
START LCN	XB	199	LSACONXB
ALC	XB	019	ALTLCNXB
TYPE	XB	203	LCNTYPXB
STOP LCN	XB	199	LSACONXB
UOC	XC	501	UOCSEIXC
SERVICE DESIGNATOR	AA	275	SERDESAA
RPT PR (SELECTED)			
SHSC (SELECTED)			

PART I

—FMECA WORKSHEET SUMMARY—

	Table Codes	Ded	Element Code
LCN	BA	199	LSACONXB
ALC	BA	019	ALTLCNXB
LCN-TYPE	BA	203	LCNTYPXB
LCN NOMENCLATURE	XB	201	LCNAMEXB
FAILURE RATE	BD	140	FAILRTBD
FR MB	BD	238	FARAMBBD
RAM I C	BD	347	RAMINDBD
FAIL RATE SOURCE	BA	141	FRDATABA
DRAWING NUMBER (HEADER)			
REFERENCE NUMBER	HA	345	REFNUMHA
ADDITIONAL REFERENCE NUMBER	HB	345	REFNUMHB

To retrieve a drawing number you will need to check first in the HA table to see if the reference number documented in that area is a drawing number. To make this check look at the Reference number category Code and see if the code of ''D'' has been entered. If a ''D'' has not been entered check in the HB table under additional reference number and go through the same check of the ARNCC.

This specification continues and identifies the LSAR data source for each item on the FMECA report.

Second screen. Having chosen selection '3' from the previous menu, the user may now select the range of data item descriptions that he or she wishes to examine:

NEW DIDs
Select range in which DID of interest falls.

```
1 LSA-001–LSA-009 [Jun 13]
2 LSA-010–LSA-023 [Jun 13]
3 LSA-024–LSA-040 [Jun 13]
4 LSA-046–LSA-075 [Jun 13]
5 LSA-076–LSA-155 [Jun 13]
6 MISCELLANEOUS    [Jun 13]
```

Third screen. Here the user may select the DID that is of interest. Two different DIDs are shown here. The FMECA DID (DI-ILSS-81164) is shown so that the reader may relate it to the FMECA report specification above. Also, the LSAR DID (DI-ILSS-81173) is displayed immediately following the FMECA DID for its interest to ILS managers.

New DIDs

```
1 DI-ILSS-81161,   LSA-046 NUCLEAR HARDNESS CRIT ITEM SUMMARY   [Jun 13]
2 DI-ILSS-81162,   LSA-050 RCM SUMMARY                         [Jun 13]
3 DI-ILSS-81163,   LSA-056 FMECA REPORT                        [Jun 13]
4 DI-ILSS-81164,   LSA-058 RELIABILITY & MAINTBLTY ANALYSIS    [Jun 13]
5 DI-ILSS-81165,   LSA-065 MANPOWER REQUIREMENTS CRITERIA      [Jun 13]
6 DI-ILSS-81166,   LSA-071 SUPPORT EQUIPMENT CANDIDATE LIST    [Jun 13]
7 DI-ILSS-80288A,  LSA-072 TMDE REQUIREMENTS SUMMARY           [Jun 13]
8 DI-ILSS-80289A,  LSA-074 SUPPORT EQUIPMENT TOOL LIST         [Jun 13]
9 DI-ILSS-80290A,  LSA-075 CONSOLIDATED MANPRINT REPORT        [Jun 13]
```

Fourth screen. The DID for the FMECA is listed below:

LSA-056, FAILURE MODES, EFFECTS, AND CRITICALITY   DI-ILSS-81163
    ANALYSIS (FMECA) REPORT

3.1 This report contains failure mode and effect analysis, criticality analysis, maintainability information, damage mode and effect analysis, and minimum equipment listing information. This report should be used to identify candidates for RCM analysis or design reviews, identify failure modes which impact item criticality number and SHSC assignment.

TM

7.1 This Data Item Description (DID) contains the format and content preparation instructions for the Logistic Support Analysis Report report required by appendix B, paragraph 30.31 of MIL-STD-1388-2B.

7.2 This DID is applicable to the acquisition of military systems and equipment.

7.3 This DID supersedes DI-L-7176 and DI-L-7178

10.1 Reference documents. The applicable issue of the documents cited herein, including their approval dates and the dates of any applicable amendments, notices, and revision, shall be specified in the contract.

10.2 Format. The report shall be prepared in the format of MIL-STD-1388-2B, appendix B, Figure 45, using one or more of the options specified in paragraph 30.31.

10.3 Content. The source of data appearing on the LSA-056 summary is contained in appendix B, Figure 14, LSAR Data Tables to Report Matrix. Data displayed on the report is not required if the data is not specified on DD Form 1949-1, LSAR Data Selection Sheet, contained in the statement of work.

10.3.1 Calculations of Item Criticality Number, Failure Mode Criticality Number, and Total Failure Mode Ratio shall be performed in accordance with the

10.3 Content. The source of data appearing on the LSA-056 summary is contained in appendix B, Figure 14, LSAR Data Tables to Report Matrix. Data displayed on the report is not required if the data is not specified on DD Form 1949-1, LSAR Data Selection Sheet, contained in the statement of work.

10.3.1 Calculations of Item Criticality Number, Failure Mode Criticality Number, and Total Failure Mode Ratio shall be performed in accordance with the formula contained in appendix B, paragraph 30.31.2.

10.3.2 The FMECA Report consists of the following:

    a. Part I, FMECA Worksheet Summary.
    b. Part II, Criticality Analysis Summary.
    c. Part III, Failure Mode Analysis Summary.

DISTRIBUTION STATEMENT A: Approved for public release; distribution is unlimited

Fourth screen.   This is the DID that describes the LSAR data delivery.

LOGISTIC SUPPORT ANALYSIS RECORD (LSAR)   DI-ILSS-81173
  DATA TABLE EXCHANGE/DELIVERY

3.1 This Data Item Description (DID) is used for automated exchange/delivery of LSAR data to the requiring authority per MIL-STD-1388-2B. Exchange/delivery of the LSAR data may take the form of full file replacement (all MIL-STD-1388-2B Appendix A data tables are delivered), or ''change only'' data (changes to the data tables since the previous submittal of LSAR data are delivered).

TM

7.1 This DID contains the format and content preparation instructions for the LSAR data exchange/delivery required by appendix A, paragraph 30.3 of MIL-STD-1388-2B.

7.2 This DID is applicable to the acquisition of military systems and equipment.

7.3 This DID supersedes DI-ILSS-80115, DI-ILSS-80116 and DI-ILSS-80117.

10.1 Reference documents. The applicable issue of the documents cited herein, including their approval dates and the dates of any applicable amendments, notices, and revision, shall be specified in the contract.

10.2 Format. This DID shall be delivered in variable length ASCII file format, with all data elements positioned at their respective offsets in their table row field, as specified in paragraph 30.3, appendix A of MIL-STD-1388-2B. Also, this DID shall be delivered in the media specified by the requiring authority in the Contract Data Requirements List, DD Form 1423.

10.3 Content. The source of data appearing on this DID is all data tables (X - J tables) of Appendix A, MIL-STD-1388-2B. Data elements from these tables are not required if the data has not been specified on DD Form 1949-1, LSAR Data Selection Sheet, contained in the statement of work. notices, and revision, shall be specified in the contract.

10.2 Format. This DID shall be delivered in variable length ASCII file format, with all data elements positioned at their respective offsets in their table row field, as specified in paragraph 30.3, appendix A of MIL-STD-1388-2B. Also, this DID shall be delivered in the media specified by the requiring authority in the Contract Data Requirements List, DD Form 1423.

10.3 Content. The source of data appearing on this DID is all data tables (X - J tables) of Appendix A, MIL-STD-1388-2B. Data elements from these tables are not required if the data has not been specified on DD Form 1949-1, LSAR Data Selection Sheet, contained in the statement of work.

```
10.3.1 LSAR data table exchange/delivery consists of two options:

a. Full file replacement
b. Change only data.

DISTRIBUTION STATEMENT A: Approved for public release; distribution
is unlimited.
```

A1.7 Selection F

This section has a directory of points of contact for various ILS/LSA/ LSAR topics.

This list was current as of February 1992.

FUNCTIONAL AREA FOR ILS	POINT OF CONTACT (POC)	PHONE EXTENSION (EXT)
ILS GENERAL	TED SCHMIDT	3340
ILS LESSONS LEARNED	LANGSTON THOMAS	3393
ILS POLICY	RAY CRONK	3393
ILS STANDARDIZATION	TED SCHMIDT	3340
LOGPARS	DAMITA BUMPIOUS	3393

```
ADDRESS: COMMANDER
         USAMC MATERIEL READINESS SUPPORT ACTIVITY
         ATTN: AMXMD-EI (POC) *Substitute one of above names for POC
         LEXINGTON, KY 40511

PHONE: (606) 293-EXT (COMMERCIAL)                    745-EXT (DSN)
```

LSA	POC	EXT
ARMY SUPPORT	MALINDA SCHMIDT	3985
DOD SUPPORT	ELLIS ATKINSON	3962
CELSA	JAY LASHER	3962
IR&D	BUD ADKINS	3985
LIFE CYCLE COSTS	CAROLYN BELL-ROUNDTREE	3985
LOG PARAMETERS LIBRARY	LI PI SU	3985
LORA	LES KARENBAUER	3985
LSA TECHNIQUES	NICK GIORDANO	3985
LSAR ADP SYSTEM VALIDATION	LOUIS SCIARONI	3962
MANPRINT	GREG TARVER	3985

```
ADDRESS: COMMANDER
         USAMC MATERIEL READINESS SUPPORT ACTIVITY
         ATTN: AMXMD-EL (POC) *Substitute one of above names for
         POC
         LEXINGTON, KY 40511

PHONE: (606) 293-EXT (COMMERCIAL)                    745-EXT (DSN)
```

Pub Number/ Notice/Date	Pub Title	POC	Phone
MIL-STD-1388-1A, Notice 3, 28 Mar 91	Logistic Support Analysis	Lou Sciaroni	3962
MIL-STD-1388-2A, Notice 2, 15 Jan 87	DOD Requirements For a Logistic Support Analysis Record	Lou Sciaroni	3962
MIL-STD-1388-2B 28 Mar 91	DOD Requirements For a Logistic Support Analysis Record	Lou Sciaroni	3962
MRSAP 700-11, 1 Dec 88	Cost Estimating Methodology for Logistics Support Analysis (CELSA) Guide	Jay Lasher	3962
AMC-P 700-4, 20 Feb 91	Logistic Support Analysis Techniques Guide	Nick Giordano	3963

Joint Service LSAR ADP System:

Release 3.6	Functional Operating Instructions	Lou Sciaroni	3962
Release 3.5	Computer Operating Instructions	Jay Lasher	3962
Release 3.5	Installation and Testing Guide	Jay Lasher	3962
Release 3.4	Provisioning Guide	Sean Connors	3962
Release 3.5	Automated LSAR Utility Routines	Joe Ketron	3962

Validation of Independently Developed LSAR ADP Systems:

Mar 91	List of Validated Systems	Lou Sciaroni	3962
Jan 87	Validation Guide, LSAR ADP Systems	Lou Sciaroni	3962
Nov 89	LSAR Independently Developed ADP Systems	Lou Sciaroni	3962
AMC-P 700-11, Apr 86	Logistic Support Analysis Review Team Guide	Bob Orendas	3964
AMC-P 700-12, 10 Apr 89	ILS Funding Guide	Carolyn Bell-Roundtree	3963
AMC-P 700-13,	Independent Research and Draft Development (IR&D)	Jerry Muszik	3964
AMC-P 700-22, Sep 88	LSA Primer	Bob Orendas	3964
AMC-P 700-27, 20 Feb 91	Level of Repair Analysis (LORA) Procedures Guide	Nick Giordano	3963
AMC-R 700-27, 20 Feb 91	Level of Repair Analysis (LORA) Program	Nick Giordano	3963

Department of Defense Acquisition Policy

A2.1 Purpose

The community that deals with LSA/ILS is, as of 1992, predominately influenced by U.S. DOD policies. Since *Acquisition* is DOD speak for "buying things," the policy is central to doing business in that arena. Detailed knowledge of that policy is not essential to performing LSA; however, it is essential that professionals have some basic understanding of how the policy works, common terms, and its influence on the LSA world. This appendix is intended to provide a brief overview of the basic elements in the new DOD acquisition policy.

More important, this appendix is intended to give the ILS/LSA professional a road map or summary of where and how the new policy and procedures address support and supportability. This is vital, since support concerns are spread *throughout* the acquisition documents. Two detailed tables are included to provide statements affecting support, their ILS/LSA ramifications, and their locations in the acquisition documents. The ILS professional needs to be aware of these many scattered provisions. Table A2.1 cross-references DODD 5000.1 significant statements which affect ILS; Table A2.2 presents the cross-references that apply to DODI 5000.2. There are many changes, or modifications, throughout the documents which have special significance for the ILS/LSA community. Many emphasize the integration of the disciplines into the "total system" context. This is central to the concept of the *integrated management* that pervades the document but has not been the practice in the past. The following list is of key concepts embodied in the new documents that are of major importance to the ILS/LSA disciplines:

1. *Performance definition:* Whenever the word is used, the meaning is "those operational *and support* characteristics of the system."

2. *Total system* includes support and infrastructure.

3. *Life-Cycle Cost* (LCC) is required from the inception of the process. Since LCC must, by definition, include operating and support costs,

TABLE A2.1   DODD 5000.1 Statements with Logistics Impacts

Page; paragraph	Statement	ILS impact
1-A2	Replaces 5000.1, dated September 1,1987 Canceled; see enclosures 4 and 5.	Major reorientation: ILS is integrated within the process and is no longer a stand-alone reference (e.g., 5000.39).
3-E	"Integrated Management Framework Requirements Definition, Planning Programming and Budgeting"	ILS integration and input opportunities: Contributions are essential to defining and procuring necessary support resources.
3-G	"Supplementation"	Service-peculiar instructions severely limited. Affords more opportunities for cost-effective industry and government universal ILS.
Part 1: 1-1; A-1	"An Integrated Management framework shall be used...for acquisition programs that meet users needs and can be sustained given projected resource constraints."	Resource constraints require ILS input for both definition and cost.
1-2; 3b.	"Threat projections, *life-cycle costs*, cost-performance, schedule tradeoffs...shall be major considerations at each milestone beginning with the new start decision milestone."	ILS is integral to LCC. This directs the inclusion of supportability and support resources from the inception of the process. There is a renewed emphasis on LCC.
1-4; 5, C1.a	"Acquisition strategies shall be event driven and explicitly link...commitments and milestone decisions to demonstrated accomplishments."	There is specific linkage between setting objectives and demonstrating achievement by testing. ILS must prepare appropriate input data.
1-4; C1.b	"Program plans must provide for...simultaneous design of the product and its associated manufacturing, test *and support* processes. This Concurrent Engineering...."	Concurrent Engineering must include supportability (and associated support) considerations. The ILS community must participate in the CE effort.
1-4; C2.a	"Risk Management...a. Critical parameters that are design cost drivers or have a significant impact on *readiness...and life-cycle costs.*"	ILS contributions to readiness and LCC must be examined and appropriate management demonstrated.
1-5; C2.d	Solicitation documents shall require contractors to identify risks and specific plans to assess and eliminate...."	Contractor ILS should prepare to respond with evaluations and plans for the appropriate areas.
1-5; C2.e	"Risk areas to be assessed at milestone decision points shall include: (1) Threat, technology...*support.*"	Risk management is an area which is emphasized repeatedly and is one which ILS must consider for all the elements.
1-5; C4.b	"Performance objectives must satisfy identified operational needs....They must include critical *supportability* factors such as reliability, availability, and maintainability."	The statement is significant and includes ILS within "performance." This is further underscored in 5000.2, "Definitions" and means ILS is included whenever the word *performance* is used.
1-6; C6.c	"Contractors' past performance and current capability (technical,...) shall be considered in source selection."	No further emphasis need be added to the "shall be." This is an additional incentive for ensuring ILS participation in all contract obligations.
1-9; D6.c	"The objectives...shall be to establish a...process...and to avoid the proliferation of documents and guidance."	The opportunity for cost savings by using common approaches is an ILS opportunity. It also requires rigorous review of existing approaches and methods to ensure compliance.

TABLE A2.2   DODI 5000.2 Statements with Logistics Impacts

Page; paragraph	Statement	ILS impact
Section C:	Definitions	
2; C7	"Performance. Those operational and support characteristics of the system that allow it to effectively and efficiently perform its assigned mission over time. The support characteristics of the system include both supportability aspects of the design and the support elements necessary for system operation."	*Major significance in that the word *performance,* wherever used, includes support by definition.
4; G	"Supplementation and Implementation...policies and procedures set out in this instruction shall not be supplemented without the prior approval of the Under Secretary of Defense for Acquisition."	Old supplements are obsolete, by definition, for new contracts. Review required by both commercial and service logistics organizations to ensure only compliant documents are resident in ILS internal files for applications.
Part 2: General Policies and Procedures		
2-1; B1	"Acquisition Process Acquisition Milestones and Phases: Milestones 0 through 4 and associated phases."	*New titles and descriptions
2-3; table	"Acquisition Categories and Review Authorities"	ACAT definitions under new policy.
2-4; B3	"Acquisition Strategies, Exit Criteria and Risk Management. 'Acquisition Strategy' must explicitly link milestone decision reviews...to...demonstrated accomplishments."	*This concept is critical to the new directive and policy. The progression through each milestone is linked to demonstrated achievement. ILS must be prepared to integrate support requirements and demonstrate achievements appropriate to the developing system.
2-5; B4a	"Total System Acquisition...Optimize total system performance."	By the definition supplied in this reference, both 4a (3) the logistic Support structure and 4A(4) the other elements of the operational support infrastructure are included. ILS elements are always considered when "total system" is discussed.
2-7; C1	"Milestone Review Documentation Review Concept. Stand-alone Documents: Manpower Estimate Report. Integrated Program Summary includes Cost Drivers and Major Tradeoffs. Annexes include Program LCC estimate summary."	*A review of the documentation requirements illustrated in these references illustrates the range of ILS input. All contract documentation should be coordinated for both analysis and data inputs from ILS.
Part 3: Acquisition Process and Procedures		
3-3; 2d(3),(b)	"The purpose of the subcommittee review is to...recommend study efforts to the board at milestone 0."	Study efforts require analysis, and analysis equates to LSA.
3-8; 3b(1)	"Early *LCC* estimates will be analyzed."	Note *early* and *will be,* which require ILS input at inception of process.
3-21; table	"Minimum Required Accomplishments (Phase II) Programming of adequate resources to support production, deployment and *support.*"	The resource requirements which will be required for support are an ILS task. There is a significantly increased emphasis on ILS early input to budgets.
3-24; table	"Decision Criteria...the design is *logistically* supportable. Milestone 3."	Analysis and evaluation required from ILS.
3-28;3i(2)(c)	...a major modification may be brought about by one or more...factors...(c) an opportunity to reduce *cost of ownership.*"	Clear indication for the continuing documentation of postproduction support costs. The logistics responsibility may reside in either service or contractor sector.

TABLE A2.2   DODI 5000.2 Statements with Logistics Impacts (Continued)

Page; paragraph	Statement	ILS impact
Part 4: Requirements Evolution and Affordability		
Section C: Critical System Characteristics		
4-c-1; 2(a)2	"Critical system characteristics...shall be identified early and specifically addressed....They include...*support* infrastructure."	Support is included in critical characteristics.
Section D: Affordability		
4-D-1;2d	"A program shall not be approved to enter the next acquisition phase...unless sufficient resources...including manpower are or will be programmed...to support projected development and *support* requirements."	The repeated emphasis on the need for ILS to quantify resource requirements, participate in the budget process, and ensure that the resources will be available.
Part 5: Acquisition Planning and Risk Management		
Section C: Technology Development and Risk Management		
5-c-1; 2b	"The DOD components shall establish technology development programs...including *Logistics* Research and Development."	Service and commercial objectives may be similar and present an opportunity for mutually reinforcing programs.
Part 6: Engineering and Manufacturing		
Section A: Systems Engineering†		
6-a-1; 2a(2)	"Integrate the technical inputs...(including the concurrent engineering of manufacturing, *logistics*...)."	Systems engineering management must provide for the integration of ILS, and specifically included are all facets, e.g, CE.
6-A-2; 3a(1)	"An effective systems engineering management program will be implemented. MIL-STD-1388...are major elements and will be integrated."	*LSA and LSAR are specifically included in the procedures section of the systems engineering section.
Section C: Reliability and Maintainability		
6-C-2; 3a(3)	"Reliability objectives will address both mission reliability...and *logistic* reliability (e.g., demand for maintenance...)."	This aspect of reliability is frequently slighted.
6-C-4; 3b(6)(b)	"Maintainability Analysis will be conducted under logistic support analysis (LSA) process."	Another specific inclusion of 1388-1A.
Section E: Transportability		
6-E-1; 2b(3)	"Transportability shall be a major consideration in: Developing integrated logistic support for systems and equipment."	The inclusion is obvious and usual.
Section F: Survivability		
6-F-4; 3f	"The...ILSP for systems with critical survivability characteristics will define a program to ensure those characteristics are not compromised through loss of configuration control...etc."	Note the close interrelationship, specified here, between ILS and configuration control.

TABLE A2.2   DODI 5000.2 Statements with Logistics Impacts (Continued)

Page; paragraph	Statement	ILS impact
Part 6: Engineering and Manufacturing		
Section L: Nondevelopmental Items		
6-L-3; 4c(1)	"Programs using commercial items...should make maximum use of existing commercial logistic support and data. Development of new organic elements will be based on critical mission need or substantial cost savings."	NDIs offer possible savings when analysis and data requirements are examined in light of this guidance. This offers a window of opportunity.
6-L-3; 4d(2)	"Accepting...commercial...*logistics support,* consistent with user's operational needs."	See above comment.
Section O, Attachment 1: Production Readiness Considerations		
6-O-1-4; 6	"Logistics"	All elements of ILS must be programmed to ensure that support resources are available for the anticipated deliveries.
Part 7: Logistics and Other Infrastructure		
Section A: Integrated Logistics Support		
7-a-1; 1	"Purpose"	
7-A-1; 1b(1)	"Support considerations are effectively integrated into the system design."	Restatement of the intent. The reference serves to reinforce the statements in all the preceding material that ILS must be a full participant from the inception.
7-A-1; 1b(2)	"Required support structure elements are acquired concurrently."	The acquisition of these elements can only be accomplished if they are planned and estimated consistent with the program and budget cycle requirements.
7-A-4; K(1)	"The contract for EMD will require the contractor to include post-production support...in early tradeoff studies prescribed by MIL-STD-1388."	Postproduction support is emphasized early in the planning. Use of LSA is the tool, and since the plan must be developed for the EMD phase, ILS is an integral part of the early design development.
Part 9: Configuration and Data Management		
Section B: Technical Data Management		
9-B-3; 3a(2)(b)	"Production contracts must include product drawings and associated lists for items that will be reprocured or manufactured in-house."	*ILS review should include identification of the many requirements for drawing submissions so that while each submission is satisfied, the task is one of simple duplication. It may be possible to combine submissions or implement other cost-saving steps.
9-B-4; 3a(5)(6)	"Logistics Support Analysis data will be used to the maximum extent to define and develop source data for technical manuals."	*Management challenge to integrate the disciplines for cost-effective data development. Management information systems provide an efficient base tool.

*Major impacts.
†This entire section is of paramount importance to ILS. Specific paragraphs are cited for emphasis.

its mandated inclusion involves the ILS/LSA communities from the inception of the program.

4. *Analysis (including LSA)* is required, emphasized, and introduced from Milestone 0.

5. *Risk analysis* includes support elements, and ILS is specifically directed to evaluate support and individual element risks.

6. *Concurrent engineering* must include support and supportability and thus ILS and LSA.

7. *ILS and LSA* are integrated throughout the process. A number of mandatory documents and evaluations are required as the program moves through the Milestones. These should be thoroughly understood by the support community, and appropriate inputs should be prepared.

8. *Integrate ILS elements:* There are illustrations throughout which emphasize the cost savings expected by nonduplicating efforts (e.g., developing technical manuals and using common source data for other ILS products).

9. *Cost estimates for support* are required from the inception of the program. This is a variation of the LCC input but identifies specific new (or additional) investments associated with the program.

A2.1.1 Acquisition policy and the instruction document

The policy is designed to move a requirement through a process which will result in the satisfaction of a verified requirement. The instruction addresses the details, "ilities," engineering, and other concerns involved in making the policy into a smooth reality.

The first page of the new policy, DODD 5000.1, after the title, cites "References," and after the (e) it says through (bbbb), see enclosures 4 and 5. Enclosure 4 is a standard reference list, but 5 with the heading *"CANCELLATIONS,"* continues for two more pages. Those with continuing contracts will, of course, maintain the "canceled" documents if they are still applicable under that agreement. A word of caution: It would be a wise investment for a manager to obtain a large rubber stamp: "Superseded by 5000.1—dated Feb. 23, 1991." This and a copy of enclosure 5 should go to every member of the ILS/LSA staff, starting with the most senior member. The shelves of documents stored constituted a security blanket—doesn't everyone want his or her very own copy of everything? As new RFPs are processed, the new umbrella documents will become more familiar, but until this happens, an ounce of prevention can't hurt.

There are many minor changes that affect logisticians. In many cases it is a question of the old procedures being folded into the new instruction document. DOD Directive 5000.39, "Acquisition and Management of Integrated Logistic Support for Systems and Equipment," dated November 17, 1983, is canceled. This does not mean that there is no longer an ILS concern. To the contrary, the concern is throughout the document.

The specific integrated logistics support discussion is part 7, section A of the instruction. It is important to emphasize that MIL-STD-1388-1A is a significant inclusion in the instruction. In directing LSA, the instructions states "(LSA)...*will be used iteratively throughout the acquisition program as an integral part of the systems engineering process*" (emphasis mine). In directing what essential tasks will have been accomplished by Milestone 2, Development Approval, the instruction states, "A logistic support analysis program has been initiated to serve as the single data base for Integrated Logistic Support documentation."

A2.2   The Acquisition Process

The acquisition process for the Department of Defense is defined in DOD Directive 5000.1, dated February 1991. This is a revision of the old procedures and service-peculiar implementation supplements. Issued in February 1991, the objective is stated in the directive as establishing a disciplined, uniform policy for acquisition and to avoid the "proliferation of documents and guidance." Acquisition policy, for contractors seeking DOD contracts, is obviously the starting point for determining the procedures to be followed and judging the compliance capabilities of their internal organization. The entire policy stresses the interrelationship of these systems:

- Requirements generation
- Acquisition management
- Planning, Programming, and Budgeting

The superseded DOD 5000.1, "Major and Non-Major Defense Acquisition Programs," has been the subject of discussion for several years. In July 1989, the secretary of defense report, "Defense Management Report to the President," was submitted in response to a growing concern about the entire procedure. The 1987 document had been variously implemented, supplemented, and affected by both previously issued directives/instructions which were additional guidance (e.g., "Post-Production Support DOD Instruction 4000.26, dated August 19,

1986) and by subsequent documents (e.g., DOD Directive 5010.19, "DOD Configuration Management Program," dated October 1987). The new directive has several pages of cancellations which document these and similar documents that are no longer in force.

The new policy does not imply that, for example, postproduction support is no longer a concern. It does integrate postproduction support as part of an umbrella policy that seeks to correct the fragmented approach implicit in the multitude of documents that concerned "acquisition."

Problems that had been cited by numerous critics, in addition to the multitude of layered documents, included concern with the requirements-generation process, the mismatch between program estimated cost at inception and actual costs, the timing of the acquisition and budget processes, and the numerous layers of review authority. In addition, the concern for quality and testing, while not new, was repeated throughout many reviews. While many lumped all the concerns under the "management" umbrella, because of the complexity of the entire process, the deficiencies were addressed in the changes inherent in the three interrelated systems.

The *Requirements-Generation System* is where the "requirements" are first described in broad operational terms by the initiator. The procurements are categorized by dollar value, and not surprisingly called "Acquisition Categories (ACAT)," with ACAT 1 the highest priority.

Requirements are first described in broad operational terms by the initiator. Through a series of evaluations, these become more specific and are submitted to an assessment group. At this point, the objective is obviously to determine validity of the need, but in addition to identify what appropriate steps are needed for correction. The choices are nonmaterial fixes, fix or upgrade existing material, or introduce new capabilities.

Nonmaterial fixes are those which include a change in doctrine, tactics, training, or organization. Deficiencies that can be remedied in this fashion are sent to the appropriate military department for action. *Upgrades* are equally familiar and need no clarification. Where *new procurement* is the recommended solution, a Mission Need Statement (MNS) defines these projected needs, again in broad operational terms. If the projected buy could result in a major defense acquisition program, the MNS is reviewed by the Joint Requirements Oversight Council, where it is disapproved or approved with a priority assignment. For less than major acquisitions, a review is held at a lower level, but the same decision process holds. One of the problems targeted for resolution by the reforms was the layers of approval programs accumulated during their progression through the system. This streamlining is identified as "short, clear lines of Authority and Accountability" and applies to major and nonmajor programs.

The *Planning, Budgeting, and Programming System* is the fiscal umbrella. The *Planning phase* results in the development of the long-range investment plan. This considers all major modernization, as well as major acquisitions on the horizon. There are obviously major considerations which set the affordability envelope, the size of the total dollars, and the distribution of funding. A full discussion of these complexities is beyond the scope of this appendix, but the fact that the long-range plan exists is relevant. Long-range planning is set forth in a defense planning guidance document, published by October 1 of every other calendar year.

The *programming phase* of the system results in a 6-year defense program. The 6-year program proposals of each DOD component with programming responsibilities is described in a document called a "Program Objectives Memorandum (POM)." These documents are submitted in April of every other calendar year. After a review of the POMs, decisions which provide the basis for the financial planning during the budgeting phase are issued by the deputy secretary of defense as a "Program Decision Memorandum."

The *budgeting phase* is one with which the general public is most familiar, since the DOD portion is, after presidential approval, submitted to the congress by the president as part of the federal budget.

The timing of the budgeting, the long planning horizon, the cost estimating and control for program planning, and the identification of requirements all must be orchestrated. That is, discussions of the DOD acquisition process refer to "categories" of procurement, as well as to events and activities. The full designation is "Acquisition Category (ACAT)," as was indicated earlier, and the range is from ACAT 1 to ACAT 4. Categories set the level of authority needed to move through each acquisition phase. The category level is established at Milestone 1. "Milestones" are major decision points that separate "phases," or periods of activity which occur between Milestones, of an acquisition program. The Acquisition milestones and Phases are illustrated in DODI 5000.2, Part 2. The thrust of the new directive and instruction can be illustrated in an examination of the activities during phase 0, concept exploration and definition.

The MNS approval, passing the gate of milestone 0 and entering phase 0, carries specific conditions. Since the phase, Concept Exploration and Definition, could carry with it the inference of extensive studies, the approval identifies not only the start of the phase but also the funding available for this exploration. At this point the exit criteria for passing the next gate, Milestone 1, are understood. The intent is to tie the progression of a program from phase to phase to objective assessments of achievements and demonstrated accomplishments. In addition, there is requirement that risk assessment be realistic and risk management be evident, all this again in concert with the planning, programming, and budgeting system.

The framework established in the directive and the instruction sets forth the systems, the integration, and the emphasis on quantitative/objective assessments for moving through the process.

The definitions of category requirements is listed in the document with extensive clarifications. In reviewing the programs from ACAT 1 through ACAT 4, the significant changes are, as noted previously, in the approval levels required to move through the gates, or Milestones. Tailoring permits the process to fit the program within specific constraints; approval authority which is compatible with program needs brings the process to the level closest to the original requestor and the ultimate user.

A2.2.1 Summary

The new policy not only changes the management of the progress of acquisition but also makes a concerted effort to reduce the number of duplicative or conflicting instructions. These were both DOD and individual service originated. The documents which survive appear to be those which are discipline oriented (e.g., MIL-STD-470, "Maintainability Program for Systems and Equipment").

The changes in the process are not a reshuffling for the purpose of a new "spin" on acquisition. The intent is, as before, to acquire quality products that do the job, give reliable service, and achieve both initial procurement and subsequent support within an affordable budget. That said, the changes are an attempt to incorporate the diverse disciplines which have always been required, match the maturing program requirements within the budgeting requirements, and shorten the entire process.

Whatever the initiating activity (e.g., Army), the system is uniform. There is, in this new directive, a clear message that "supplementation" or documents issued to "add to, restrict, or otherwise modify" are not encouraged.

A2.3 Defense Acquisition Management Policies and Procedures

DOD Instruction 5000.2 establishes the framework for translating the need into a working program and details the event-oriented management program to be employed. When the words in the "Applicability and Precedence" portion of the instruction are examined, it is clear that this instruction applies from the office of the Secretary of Defense through all Services, which have always been known under the catch-all, "DOD Components."

The instruction, for brevity the "DODI," repeats the objective of reducing the maze of guidance documents that governed acquisition and management. It establishes a "core" of policies and procedures. These are organized under functional and organizational lines. A word of caution may be useful here: The DODI not only addresses acquisition policy and procedures but also incorporates how the individual disciplines are integrated, and each section includes references to canceled documents. A general review of these canceled documents is a useful first step. Under the pressure of time, there may be the tendency to get right to that portion of the material which is of immediate application, thus overlooking the essential housekeeping. The examples cited earlier serve to illustrate the point: DOD Instruction 4000.26, "Post-Production Support," is canceled, but the new 5000.2 incorporates Post-Production Support as part of this document.

A2.3.1 Impact on ILS/LSA

Many statements of concern with the earlier and now revised acquisition process cited the mismatch between the estimated costs of a program and the budget requirements that emerged as the program progressed. Other heated dialogue centered on what the capabilities purported to be incorporated in the system were and which were actually delivered. The mismatch between budget and real cost is only reconciled with thorough analyses, of which LSA is a significant part. The mismatch between promised and real capabilities is analyzed sequentially, at each milestone. A significant improvement directed in the policy and DODI is that before approval is granted to enter the next phase, a series of measurements must be established. These will be applied at the end of the phase before the next gate (Milestone) is passed. LSA/ILS is a significant contributor, in concert with systems engineering (Tables A2.1, A2.2). Risk analysis is important. Realistic goals and realistic test criteria are also important. The whole package of contributing disciplines is going to be integrated or the goal measurement will not be met.

There is another change, driven by budget constraints, which, if implemented, will elevate LSA/LSAR to a central role. The change being discussed is that more programs will stop short of full-scale production, and many will be fielded in a limited fashion. During this limited fielding, there will be extensive documentation of "lessons learned." During that interval, of whatever duration, there must be some documented baseline against which performance is measured. The documented controlled baseline for all the decisions made is LSA, with some elements retained in the LSAR. This measurement and feedback procedure will provide the justification for modifications to the system and is an indispensable contributor to full production when it is required. Organization of the appendix topics will follow, as far as feasible, the outline of the DODI. The range of topics in the DODI is too broad, however, to be included individually in this book.

There are 16 enclosures to the instruction, called "Parts." These start with part 1, reasonably enough called "Document Background and Table of Contents," and rapidly get down to the essence of the instruction. Part 2, "General Policies and Procedures," defines the Milestones and Phases.

The activity which precedes the first milestone, called milestone 0, "Concept Studies Approval," is Determination of Mission Need. The evolution of a "want" into a documented "need" has been discussed, and this portion of the instruction documents the process of fleshing out that need into a delivered program. Milestones proceed through 1, Concept Demonstration Approval, Milestone 2, Development Approval, and Milestone 3, Production Approval. Milestone 4, as required, is Major Modification approval. The phases, or periods of activity, which follow each milestone are phase 0, Concept Exploration and Definition, phase 1, Demonstration and Validation, phase 2, Engineering and Manufacturing Development, and Phase 3, Production/Deployment.

Earlier in these discussions the simile of "milestone" to "gate" was used, and it is appropriate now. Equally applicable is an understanding of the purpose of each phase, which is to produce the results which open that "gate." The thrust of the policy is to ensure that at the beginning of the phase the objectives to be achieved are defined so that going through the next "gate" is not an exercise in shooting a moving target. Additionally, the "risks" associated with each progression through the phases are to be detailed. Significantly, there is an emphasis on realism, and the surest way to achieve the goals of the policy, in our opinion, is a disciplined analytical approach. This will increase the use and professionalism of the LSA products.

A2.4   Part 3:   Acquisition Process and Procedures

The purpose of the section or "part" is to highlight the key features and characteristics of the acquisition process. Particularly helpful are the displays which describe the objectives and decision criteria at each milestone.

Reading these displays keeps the objectives in mind and focuses attention on the important facets of the phase activity. For example, "Phase 0" in the display "Minimum Required Accomplishments" states "Proposes program-specific exit criteria that must be accomplished during phase I." The program must meet the objectives of phase 0 itself to get through the "gate" and must establish the exit criteria for the next phase to be successful. To restate the obvious, this as a familiar principle: Say what you're going to do, do it, and prove you've done it. The benefits of reading and rereading part 3 are significant. It supplies an understanding of the objectives and focuses the attention on the important aspects of the activity to be performed. In tailoring both the requirements and execution of LSA tasks, the guidance cannot be specific, but a grasp of the philosophy does keep the focus on what are the important questions to be asked and how detailed the answers must be. There is a consistent expansion of the same themes as the program progresses.

Succeeding "parts" (4 and 5) cover the areas of Requirements Evolution and Affordability, Acquisition Planning, and Risk Management. All the information is important, and all of it is detailed in clear, understandable language.

A2.5   Part 6:   Engineering and Manufacturing

This "part," like others in the DODI, is divided into "sections" or subsets of material which relate to the whole subject. Section A is "Systems Engineering" of the DODI. The instruction outlines, throughout the technical discussions, the "what to do" approach, but it is not a cookbook "how to." This section, and all others, is followed by a matrix which lists points of contact within the office of the Secretary of Defense and the individual services for additional information. Within the Engineering and Manufacturing portion of the directive are the technical disciplines, Reliability and Maintainability (R&M), Electromagnetic Compatibility, and radio frequency, as well as the menu of other considerations to be included in designing the optimal system. *Systems Engineering Management,* by Benjamin Blanchard, is a valuable source document for an understanding of the systems engineering discipline.

Part 6, in addition to technical disciplines, includes "ilities" (e.g., transportability, survivability). These sections direct attention to considerations of Nondevelopmental Items (NDI), Standardization, use of the metric system, and the DOD Parts Control Program.

The thrust of the instruction is to collect all the various topics to be considered in design under "Engineering and Manufacturing." There is no intent to alter the technical procedures themselves. Maintainability/Durability is still MIL-STD-785, 1543, 1796, etc.; System Safety is still MIL-STD-882. The approach, to repeat the words in the instruction, is to construct a management framework within which the technical disciplines will interact. The Systems Engineering section sets the tone.

The established goal that has been traditional is repeated: integrating "the technical efforts of the entire design team to meet program cost, schedule, and performance objectives with an optimal design solution." Integration of the existing disciplines is one primary consideration that is reinforced again in part 6, section A, "Policies." Under 2.a(4)c, "The systems engineering process shall place equal emphasis on system capability, manufacturing processes, test procedures, and support processes."

The instruction does not say that all programs are to be structured in a monolithic fashion, but it does require that each Program Manager consider all influences in determining the technical support needed to achieve the technical objectives.

The Systems Engineering Management Plan, or SEMP, is not a new requirement, thus illustrating the evolutionary nature of the instruction. In the DOD Directive 5000.39, "Acquisition and Management of Integrated Logistics Support for Systems and Equipment," now canceled, the expression of policy was similar. Originally expressed as "policy to ensure that resources to achieve readiness receive the same emphasis as those required to achieve schedule and performance objectives," there is no change in this policy, but the methods to achieve it are streamlined and include the lessons learned in the interim. The language of the procedures section echoes the request stated so often by R&M—to translate the objectives into quantifiable terms.

Both originators and receivers of a Request for Proposal (RFP) are involved in the caution which directs that "contractual requirements will be traceable to operational requirements." The last caution under R&M objectives states, "Predictions will not be used as evidence that the contractual reliability requirements have been met." This should provide the opportunity for innovative solutions to the need for demonstrations. To restate the obvious: A measure of achievement is required to open the gate to the next phase.

Maintainability analyses, required for a complete supportability evaluation (in addition to Life-Cycle Cost impact), should be an iterative procedure, and the first iteration should be completed before detailed design. Further, the analysis is to be conducted under the LSA process. The instruction stipulates that milestone 1 will be the point at which program objectives for R&M "will be established."

The essence of the evaluations is to establish what will be the optimal mix required for the objective. Obviously, to achieve this optimum, all factors must be included. As the instruction makes clear, these evaluations will be done using the LSA.

A2.6   Part 7:   Integrated Logistics Support

The policies and procedures in section A, "Integrated Logistics Support," establish the basis for ensuring "(1) Support considerations are effectively integrated into the system design; and (2) Required support structure elements are acquired concurrently with the system so that the system will be supportable and supported when fielded." (See Tables A2.1 and A2.2.)

The other sections are section B, "Human Systems Integration," and section C, "Infrastructure." Section A replaces DOD 5000.39. In addition, "Instruction for Post-Production Support" (DOD Instruction 4000.26) and "Spares Acquisition Integrated with Production" (SAIP; DOD Instruction 4245.12) have been canceled as stand-alone documents and are now included in the ILS section.

The Integrated Logistics Support Plan (ILSP), which describes the what and when of the ILS procedures under the acquisition and the subjects to be considered are covered extensively under "Procedures" in Section A. The elements of ILS, which are listed in the new instruction, are essentially the same as those in the canceled 5000.39. SAIP and postproduction support are to be included in the ILSP.

The directive that support considerations will be included is amplified by the specific injunction that "a tailored Logistics Support Analysis...will be used iteratively throughout the acquisition program as an integral part of the systems engineering process." In addition, the Logistics Support Analysis Record (LSAR) will be the source for supportability and support data. This establishes the connection between the analysis method, LSA, and the recording of these data, which is designated as supportability related in the LSAR. The inference is clear: Analysis as it matures produces quantifiable results. These results are captured in the LSAR. The relationship between LSA tasks and LSAR data fields is in some cases obvious (e.g., Task and Skill Analysis). The use of standard forms to record analyses when applied throughout a program and at each level ensures consistency.

Attachment 2 of the ILS section details typical issues at each milestone. In "Activities Accomplished by Milestone 2" as an example, the instruction directs that a logistics support analysis program has been initiated. A recurring theme is the identification of cost drivers, a task clearly associated with analyses.

"Human Systems Integration," section B of part 7, can be paraphrased as human considerations as they affect systems and the system design as it affects human considerations. Within this section are included discussions of Manpower, Personnel, and Training concerns. Human Factors Engineering will be established for each system acquisition. An indication of lessons learned is the requirement that a total system training plan shall be developed by milestone 2, as well as a schedule.

Section C, "Infrastructure Support," is directed toward the need to ensure the compatibility of new systems with the infrastructure which will support them and/or identification of unique requirements. The tailoring of the MIL-STD-188 series, "Military Telecommunications Standards," is required for all inter- and intra-DOD component systems and equipment to ensure interoperability and compatibility.

A2.7   Part 8:   Test and Evaluation

"Test and Evaluation," part 8, emphasizes the importance in this process of establishing a method of verifying technical performance specifications and objective achievements. It also states that a verification be defined to ensure the operational effectivity and suitability for intended use. The section is of specific application in the acquisition process because of the repeated emphasis on achieving objectives as a precondition to progressing through the milestone sequence.

The relationship between analyses and test and evaluation is implicit: If the analyses have directed attention to the appropriate measurements, this is evident in the results of these tasks.

A2.8   Part 9:   Configuration Management and Technical Data Management

Part 9 of the instruction addresses (A) "Configuration Management" and (B) "Technical Data Management." Within configuration management concerns, the definition of one category of configuration item is worth noting: Any item required for logistics support and designated for separate procurement is also a configuration item. The increasing incidence of Non-Developmental Items (NDIs) procured as parts of systems emphasizes the need to apply some level of identification to the procured item. In the words of the instruction, "Material requirements shall be satisfied to the maximum practicable extent through the use of nondevelopmental items when such products will meet the user's needs and are cost-effective over the entire life cycle." Some considerations: Again, the instruction states, "Significant consideration must be given to Logistics support when acquiring nondevelopmental items."

Within the configuration control system there is a need to understand the commercial practice. Experience has shown that although manufacturers have a satisfactory configuration control and management system for the commercial application, it is not necessarily effective for support purposes in DOD. Management attention at the government and manufacturer levels, early in the program, will minimize later support headaches. In the commercial product arena, customer support is achieved by a variety of techniques. Some suppliers designate the part number in combination with model number/serial number as a method of control.

Retention of the model number is a recognition of the need for spares, or repair parts. This is a form of configuration management. The basics are in place in most businesses; the next step is to adapt this system to a wider customer need. The payoff is substantial. There is a further benefit to the producer: It makes both repairs and inventory less expensive. Certainly it puts a premium on Concurrent Engineering in that the model sold is reliable, producible, and subject to a minimum of manufacturing redesign. These attributes are good business and for a potential buyer provide a plus.

"Technical Data Management," addressed in section B of part 9, replaces the canceled DODI 5010.12, "Technical Data Management Program," and DODI 4151.9, "DOD Technical Manual Program Management." The thrust of simplification is again emphasized. The use of tailoring requirements to match the needs of the program is specifically encouraged.

The DODI recognizes that an evaluation of contractor in-house manuals may suffice until the design is stable. The transition to a specific format will be as dictated by program needs. It is of interest that the DODI specifically prohibits the acquisition of technical manuals for weapons systems, weapons system components, or support equipments by Data Item Descriptions (DIDs). These will be acquired by line item and have an exhibit attached to the acquisition document. Other manuals and/or management data may still be covered by DIDs.

A2.9   Part 10:   Cost Estimating, Selection of Contractual Sources, and Acquisition Streamlining; Part 11:   Program Control and Review

The subjects covered under part 10 are "Cost Estimating, Selection of Contractual Sources, and Acquisition Streamlining." In the words of the DODI, "these policies...must be judiciously applied. They are not a substitute for good judgment and common sense, nor are they intended to stifle innovation." Part 11, "Program Control and Review," includes the management guidance for programs, contract Performance Measurement, Milestone Reviews, Periodic Reports and Certifications, and Program Plans. Repeated again is the inclusion of supportability as a concern: "Performance parameters shall include supportability."

Part 12 is "Defense Enterprise Programs and Milestone Authorization." Part 13 is "Defense Acquisition Board Process." Both parts are essential background for a well-rounded logistician, but they should be addressed in more depth than is contained within the scope of this document. The designation of enterprise programs, the funding profiles in Chap. 12, and the Cost Analysis Improvement Group (CAIG) Procedures in part 13 are recommended for further review by the reader interested in a broader understanding of the acquisition process.

The revisions do supply what the document describes as a "framework." The inclusion of the variety of disciplines involved in a successful acquisition under one "cover" is a distinct improvement. The overview of a process as the initiating and controlling document establishes a unified approach. There are specific areas which are amplified in application by unique documents, but these are secondary to fitting within the big picture.

A2.10 Summary

There is a significant change in the acquisition process in both 5000.1 and 5000.2. The objectives are sound, and the presentations are clear. The charts and illustrations are helpful guides. As with all documents, their worth in the long run will be demonstrated in the implementation. LSA is clearly an integral part of the procedure. The emphasis on logical, demonstrated achievements as the program progresses from phase to phase, through the milestones, will heighten the recognition of the value of LSA. The community which deals with LSA/ILS is, as of 1992, predominately influenced by U.S. DOD policies.

3

HARDMAN/MANPRINT
Methodology

A3.1 The Purpose of HARDMAN/MANPRINT

The purpose of this appendix is to provide the ILS/LSA manager with an overview of the HARDMAN/MANPRINT methodology (herein referred to simply as HARDMAN) to assist in integrating LSA-HARDMAN activities. Most of the information is taken from Ref. 16. The HARDMAN process has been adopted by the Army and Navy as a standard procedure for performing comparative analyses on manpower, personnel, and training. HARDMAN is an acronym standing for HARDware versus MANpower. It was originally a front-end analysis (prior to milestone 1) performed to predict the manpower, personnel, and training requirements for weapon systems beginning in the preconcept phase. Updates in the following phases may occur. The HARDMAN methodology enables different system alternatives to be compared based on MP&T requirements. By predicting such requirements for each system, HARDMAN identifies the most desirable alternative for development.

This appendix specifically discusses

The HARDMAN approach   (A3.2)

HARDMAN assumptions   (A3.3)

Major steps in the methodology   (A3.4)

HARDMAN Application Plan content   (A3.5)

Summary   (A3.6)

A3.2 Approach

The HARDMAN methodology uses a comparative analysis as a basis for performing tradeoffs. It creates a Baseline Comparison System (BCS) and predicts its MP&T requirements. The BCS is then used to predict the MP&T requirements of the proposed system. The results indicate the most advantageous configuration using MP&T as the criteria for selection.

A3.3 Assumptions

The HARDMAN methodology makes three assumptions in developing its prediction:

1. Changes in weapon technology and the impacts of these changes are always incremental and small in nature.
2. New equipment systems are primarily refinements and reconfigurations of existing technology.
3. The whole system is equal to the sum of its parts.

A3.4 Major Steps in the Methodology

The HARDMAN methodology is composed of five major steps, as shown in Table A3.1. A general description of each step follows:

Step 1: Collect preliminary data/conduct system analysis. Step 1 consists of three major activities. The substeps in each activity are summarized in Table A3.2. In the first activity, data on equipments/systems/subsystems (E/S/S) requirements, concepts, performance goals, and standards are collected and reviewed. The second activity is aimed at creating a BCS. The BCS comprises a comparable system, often a predecessor or a composite of several subsystems, that best matches the requirements, concepts, functions, and performance standards of the new E/S/S. The final activity is the collection of MP&T data on the BCS to be used in step 2. Step 1 is generic to most specialty analyses, including LSA tasks 101, 201, and 203. Updates could interface with LSA task 402.

Step 2: Conduct Comparability Analysis. This step comprises a series of procedures aimed at assessing changes in resource requirements by

TABLE A3.1   HARDMAN Methodology Steps

Step	Title
1	Collect preliminary data and conduct system analysis
2	Conduct comparative analysis
3	Develop MP&T concept
4	MP&T resource requirements
5	Develop program documentation input

TABLE A3.2   Step 1 Substeps

First-level substeps	Second-level substeps
1.1 Collect data for system analysis	1.1.1 Collect/review initial program information: identify program requirements
	1.1.2 Prepare concept summaries of new E/S/S
	1.1.3 Identify new E/S/S performance goals and standards
	1.1.4 List new E/S/S equipment
1.2 Conduct system analysis	1.2.1 Compare new E/S/S and existing system
	1.2.2 Determine baseline comparison system
1.3 Collect data for task and comparability analysis	1.3.1 Identify current manpower requirements for BCS
	1.3.2 Identify current training requirements for BCS

comparing the known parameters of the new E/S/S with the BCS developed in step 1. The substeps are summarized in Table A3.3. The BCS has mature data concerning parameters of the comparable system(s) and the resources required by the existing system(s). Comparing these data with available data concerning the new E/S/S allows the analyst to calculate, in general terms, the differences between the BCS and the new E/S/S for use in step 3. This step can interface with LSA tasks 205 and 301.

Step 3:   Develop an MP&T Concept. Step 3 has four major activities. The subtasks are shown in Table A3.4. First, the installation schedule for the new E/S/S is developed. This installation schedule provides information on the numbers and types of systems in which the new E/S/S will be installed and the number of new E/S/Ss to be installed by fiscal year for each year of installation. Second, the quantitative and qualitative manpower requirements for each new E/S/S configuration are determined and grouped into three categories: organizational, maintenance support, and other manpower. Third, the training requirements for each new E/S/S configuration are determined. These could include team training, initial factory training, skill progression training, training course integration, the applicability of existing training, possible interservice education, and training material re-

TABLE A3.3   Step 2 Substeps

First-level substeps	Second-level substeps
2.1 Determine BCS tasks/skills	2.1.1 List BCS maintenance tasks
	2.1.2 List BCS operator tasks
2.2 Perform task/comparability analysis	2.2.1 Determine new E/S/S maintenance tasks
	2.2.2 Determine new E/S/S operator tasks

TABLE A3.4   **Step 3 Substeps**

First-level substeps	Second-level substeps
3.1 Determine E/S/S configuration and installation schedule	3.1.1 Describe E/S/S configurations and record new E/S/S installation schedule
3.2 Determine unit configuration manpower requirements	3.2.1 Develop operator manpower requirements
	3.2.2 Develop maintenance manpower requirements
	3.2.3 Develop other manpower requirements
3.3 Determine training concepts	3.3.1 Develop training objectives
	3.3.2 Develop training concepts
	3.3.3 Develop training data
	3.3.4 Illustrate training path concept
3.4 Produce MP&T concept document	3.4.1 Prepare MP&T concept document

quirements. Fourth, an MP&T concept document is prepared and coordinated. These data provide the basis for the next step. This step could interface with LSA tasks 302, 303, and 401.

Step 4: Develop MP&T Resource Requirements.   Step 4 is summarized in Table A3.5. MP&T requirements for the new E/S/S are displayed by location and fiscal year. Costs associated with initial training hardware and software and with initial training facilities development are determined by fiscal year. The data developed are summarized in an MP&T resource requirement document, which is updated periodically to reflect changes in the program and more detailed data on the new E/S/S as they become available. This step could interface with LSA tasks 401 and 402.

TABLE A3.5   **Step 4 Substeps**

First-level substeps	Second-level substeps
4.1 Develop organizational/nontraining manpower requirements	4.1.1 Display operator manpower requirements by fiscal year
	4.1.2 Display maintenance manpower requirements by fiscal year
	4.1.3 Display other nontraining manpower requirements by fiscal year
4.2 Develop training resources and costs	4.2.1 Develop annual training input requirements
	4.2.2 Display training planning factors by fiscal year and cost
	4.2.3 Other training material by fiscal year and cost
4.3 Develop training-associated manpower	4.3.1 Display training billets
4.4 Develop training facilities requirements	
4.5 Prepare MP&T resource requirements	4.5.1 Produce MP&T resource requirement document

TABLE A3.6   Step 5 Substeps

First-level substeps	Second-level substeps
5.1 Support acquisition process documents	Navy cites five program-type documents, one of which is the ILS plan
5.2 Support LSA	5.2.1 Support LSA tasks 5.2.2 Support LSAR
5.3 Support manpower documentation	Navy cites ship manpower, squadron manpower, and shore manpower documentation
5.4 Support training planning	Navy cites the Navy training plan and the Naval training acquisition process
5.5 Support fiscal planning	5.5.1 Support sponsor program proposal 5.5.2 Support logistics requirements funding plan 5.5.3 Support HARDMAN information system and Navy manpower planning system

Step 5:   Develop Program Documentation Input. This step is summarized in Table A3.6. It uses the results of steps 3 and 4 to support MP&T input to higher-level program documentation.

A3.5   The HARDMAN Application Plan Content

A typical HARDMAN Application Plan consists of 10 sections:

1. New system identification and description
2. Predecessor system identification and description
3. Anticipated level of detail
4. Program office personnel available for MP&T analysis
5. Funds available for MP&T analysis
6. Anticipated use of other service agencies for MP&T analysis
7. Anticipated use of independent contractor for MP&T analysis
8. Anticipated use of system prime contractor for MP&T analysis
9. Estimated level of effort for MP&T analysis
10. Projected plan of action and milestones for MP&T analysis

A3.6   Summary

This appendix has provided an overview of the HARDMAN process and has listed possible LSA interfaces to assist the ILS/LSA manager in integrating general ILS/LSA/MP&T activities.

LSA Management Team Charter

A4.0 Background

The LSA management team functions under the direction and control of the government logistics organization supporting the program manager (PM). The organization and representation of members on the LSA team will be at the discretion of the individual LMs. It is essential that all ILS elements be sufficiently represented to address all LSA functions.

A4.1 Category Team Review Methods

The current emphasis on digital data transfer and the availability of ADPE in the workplace should be exploited in the LSAR data review and approval process. The LM will maximize the use of CALS.

A4.2 ILS/LSA Management Team Charter

A4.2.1 Purpose

The purpose of the ILS/LSA management team is to assist the acquisition manager in terms of the LSA process and the LSAR data base in tailoring requirements, developing government task inputs, and reviewing and using the various LSA products (task reports and LSAR reports).

A4.2.2 Team membership

All the ILS elements must be represented on the ILS/LSA management team, as well as appropriate System Engineering Specialists (R&M, Human Engineering, Safety). In addition to the LEMs, membership may include representatives from the following areas:

1. PM's ILS office or the SYSCOM ILS office
2. R&M engineering
3. Contracts
4. Configuration Management
5. Test and evaluation
6. LSAR Automatic Data Processing (ADP) analysts

7. Design Engineering

8. Human Engineering

9. Safety Engineering

10. Standardization Engineering

A4.3 Team Organization

The team should consist of a chairman, cochairman, and team members as indicated in team membership. The chairman should be the LM. The cochairman should be from the LM's office.

The responsibilities of the ILS/LSA management team chairman include, but are not limited to, the following:

1. Determines initial team membership and augments or reduces team membership as the process warrants.

2. Schedules and conducts a government-only LSA/LSAR guidance meeting for the purpose of establishing a singular government position prior to the government/contractor LSA/LSAR Guidance Conference.

3. Schedules team meetings as required to develop strategy, prepare contractual documentation, etc.

4. Schedules team meetings as required to establish and/or revise team procedures.

5. Schedules LSA reviews through coordination with the contractor, and notifies team members of the time and place of the review.

6. Ensures that the contractor has available all material needed for the review.

7. Ensures that review team members receive appropriate LSA Data Item Descriptions (DIDs), Statements of Work, LSA functional documentation, and the LSA Plan in order to conduct the review.

8. Reviews and approves consolidated comments and provides technical direction to the contractor through appropriate channels.

9. Ensures, through the Procuring Contracting Officer (PCO), that all agreed-to changes are incorporated into the LSA/LSAR by the contractor.

10. Ensures that LSA/LSAR hard-copy and summary information are available at key ILS events [i.e., physical teardown/logistics demonstration, Draft Technical Manual (DTM) reviews, etc.] for hardware validation.

11. Formulates and secures approval of a charter for the ILS/LSA management team.

12. Ensures that the minutes of all ILS/LSA management team meetings are taken, finalized, agreed to, and distributed to review team members.

13. Ensures that all team comments are consolidated into a single list of comments (these may form an enclosure to the minutes of the team meeting).

Responsibilities of the LSA review team cochairman include, but are not limited to, assuming all responsibilities of the ILS/LSA management team chairman in the chairman's absence.

Responsibilities of the organization or disciplines providing membership to the team include, but are not limited to, the following:

1. Ensures that technically qualified personnel are available to participate in the LSA management process.

2. Ensures that continuity of team membership is maintained to the maximum extent possible.

3. Bears the cost of participating in ILS/LSA management team meetings, including travel, unless the program manager absorbs the cost.

A4.4   Team Schedule

The schedule for ILS/LSA management team meetings is dependent on the complexity of the hardware, progression of the design, and degree of tailoring. Therefore, the schedule must be flexible enough to accommodate the preceding variables rather than being fixed.

A4.5   Planning Actions

The information/data to be provided by the contractor are specified on the SOW and in accompanying CDRL and DIDs. Then the contractor is required to prepare an LSAP (DI-S-7017) in response to the SOW; it should include the following:

1. Description of the contractor's logistics organization, including internal authority and responsibility for conducting the LSA program.

2. Description of how LSA data will flow between the contractor's logistics organizational elements and subcontractors/vendors.

3. LSA program schedule with interfacing milestones.

4. Identification of the specific LSA tasks and the level (depth) of LSA effort to be conducted.

5. Candidate list of the items on which LSA is to be performed.

6. Description of the LCN system to be used in constructing the LSAR file.

The ILS/LSA management team uses the preceding information as the basis for management control between the contractor and the team. The team must pay particular attention to the LSA candidate list (the components/assemblies upon which LSA is to be performed), the level of the analysis to be conducted, and the LCN system to be employed early on. As design progresses, this list is updated to include a more specific identification of the components/assemblies and additions/deletions resulting from design changes. With the updated list, the team will be able to track the contractor's LSA effort. However, adequate review of the LSAR data by the team can only be accomplished if at least the design drawings, and preferably mock-ups or actual hardware, are available during schedule reviews. Access to these aids allows the review team members to simulate the contractor's maintenance and operator task analyses and thus verify the data in the LSAR file. The importance of being able to perform a hands-on demonstration to verify LSAR data cannot be overemphasized.

Performing Organizations

This appendix provides possible performing organizations by subtask, with the rationale and/or comments concerning the recommendations.

Subtask Performing Organization Classifications

Task/subtask	Subtask description	Program, general or peculiar*	Skills*	Performing organization*	Rationale/comments
101	Development of Early LSA Strategy				
101.2.1	LSA strategy	P	E, A	G	The government is responsible for initiating management planning in LSA. An OEM may not exist when this task is first accomplished.
101.2.2	Cost of LSA Program	P	E, A	G	The government is responsible for cost estimate.
101.2.3	Updates	P	E, A	E	Updates may be performed by either government or OEM depending on how early in the acquisition phase the OEM gets involved.
102	Logistics Support Analysis Plan				
102.2.1	LSA Plan	P	E, A	OEM	The OEM must develop the LSA plan as it related to equipment design supportability factors and OEM organizational structure. The OEM has information on the system already at hand. Exceptions are 102.2.1j, which is done by government (it states the government is to furnish data to contractor) and 102.2.1.I, which may be done by either organization, independent of the other, since it addresses government and vendor furnished equipment/material.
102.2.2	Updates	P	E, A	OEM	Performed by OEM since it developed the original and is most familiar with the information contained therein.
103	Program and Design Reviews				
103.2.1	Establish Design Review Procedures	P	E, A	E	Either organization may establish and document review procedures.

374

Subtask Performing Organization Classifications (Continued)

Task/ subtask	Subtask description	Program, general or peculiar*	Skills*	Performing organization*	Rationale/comments
103.2.2	Design Reviews	P	E, A	J	Subtasks 103.2.2, 102.2.3, and 103.2.4 are obviously a joint effort between government and the OEM. Exceptions may be 103.2.2c(5), which deals with design/readiness actions proposed or taken, and 103.2.2d, e, f, and g, which deal with supportability requirements and goals, LSA documentation, and support programs.
103.2.3	Program Reviews	P	E, A	J	
103.2.4	LSA Review	P	E, A	J	
201	Use Study				
201.2.1	Supportability Factors	P	E, A	G	
201.2.2	Quantitative Factors	P	A, I	G	A possible exception here is 201.2.2d, which requires documentation of allowable maintenance periods. This must be determined by the OEM unless directed by the government as a design objective.
201.2.3	Field Visits	P	E, A	E	Either the government or the vendor may conduct field visits depending on whether the OEM is in existence yet. Must be done by government if done early when OEM does not exist.
201.2.4	Use Study Report and Updates	P	A	E	Depends on updates; could be performed by OEM.
202	Mission HW, SW, and Support System Standardization				
202.2.1	Supportability Constraints	P	E, A	OEM	Both parties contribute. However, government contribution is normally part of task inputs. The remainder of the activity, if any, would be OEM.
202.2.2	Supportability Characteristics	P	E, A, I	OEM	
202.2.3	Recommended Approaches	P	E, A	OEM	
202.2.4	Risks	P	E, A	OEM	

375

Subtask Performing Organization Classifications (Continued)

Task/ subtask	Subtask description	Program, general or peculiar*	Skills*	Performing organization*	Rationale/comments
203	Comparative Analysis				
203.2.1	Identify Comparative Systems	P	E, A	E	Depending on information available and internal organization's knowledge, either government or OEM could perform this subtask. Must be government if done early when OEM does not exist.
203.2.2	Baseline Comparison System	P	E, A	E	Rationale is same as preceding subtask.
203.2.3	Comparative System Characteristics	P	E, A	E	Government is preferred, since use of government data bases is generally necessary.
203.2.4	Quantitative Supportability Problems	P	E, A	E	Generally, the government would have the depth of information required for performance of this subtask. However, it could be done by either.
203.2.5	Supportability, Cost, and Readiness Drivers	P	E, A	E	This subtask draws on same data bases as 2.4; government preferred.
203.2.6	Unique System Drivers	P	E, A	J	May need inputs from both OEM and government.
203.2.7	Updates	P	A	E	This is generally done by whoever the previous performing activity was on the other subtasks.
203.2.8	Risks and Assumptions	P	A	E	Rationale is same as preceding subtask.
204	Technological Opportunities				
204.2.1	Recommended Design Objectives	P	E, A	J	For simple equipments, government could perform this subtask; for complex equipments, OEM is generally best. Government could give mission equipment aspects (204.2.1a) to OEM but do logistic element aspect (204.2.1c) internally.
204.2.2	Updates	P	A	J	
204.2.3	Risks	P	A	J	

Subtask Performing Organization Classifications (Continued)

Task/subtask	Subtask description	Program, general or peculiar*	Skills*	Performing organization*	Rationale/comments
205	Supportability and Supportability-Related Design Factors				
205.2.1	Supportability Characteristics	P	E, A	E	There are really three subtasks here. The third one deals with data rights of HW and SW. That portion of the subtask would be performed by the OEM.
205.2.2	Sensitivity Analysis	P	A	E	
205.2.3	ID equip./SW w/o Full design rights	P	E, A	J	For simple equipment, the government can perform this subtask; for complex systems, a joint effort is required.
205.2.4	Supportability Costs and Readiness objectives	P	A	G	
205.2.5	Supportability Design Constraints	P	A	E	Subtasks 2.1 (less data rights), 2.2, and 2.5 should be done by the same party.
205.2.6	Constraints for NATO Adoption	P	A	G	
205.2.7	Updates				
301	Functional Requirements Identification				
301.2.1	Functional Requirements	P	E, A	OEM	Could be joint, but generally OEM.
301.2.2	Unique Functional requirements	P	E, A	OEM	
301.2.3	Risks	P	E, A	OEM	
301.2.4	OEM Tasks	P	E, A	OEM	
301.2.4.1	Results of FMECA	P	E, A	OEM	
301.2.4.2	RCM Analysis	P	E, A	OEM	
301.2.4.3	Other Tasks	P	E, A	OEM	
301.2.5	Design Alternatives	P	E, A	OEM	
301.2.6	Updates	P	A	OEM	

Subtask Performing Organization Classifications (Continued)

Task/ subtask	Subtask description	Program, general or peculiar*	Skills*	Performing organization*	Rationale/comments
302	Support System Alternatives				
302.2.1	Alternative Support Concepts	P	E, A	E	
302.2.2	Support Concept Updates	P	E, A	E	
302.2.3	Alternative Support Plans	P	A	E	
302.2.4	Support Plan Updates	P	A	E	
302.2.5	Risks	P	A	E	
303	Evaluation of Alternatives and Tradeoff Analyses				
303.2.1	Tradeoff Criteria	P	A	E	Except for those subtasks specifically discussed below, the OEM would normally perform the subtasks unless strong reasons exist to do otherwise.
303.2.2	Support System Tradeoffs	P	A	E	Relates to 302; could be performed by government.
303.2.3	System Tradeoffs	P	E, A	OEM	Normally part of OEM system engineering trades.
303.2.4	Readiness Sensitivities	P	E, A	E	Frequently done by government.
303.2.5	Manpower and Personnel Tradeoffs	P	A	E	
303.2.6	Training Tradeoffs	P	E, A	E	
303.2.7	Repair-Level Analysis	P	E, A	E	Data input from OEM, analysis by government.
303.2.8	Diagnostic Tradeoffs	P	E, A	OEM	
303.2.9	Comparative Evaluations	P	A	E	Relates to 203.
303.2.10	Energy Tradeoffs	P	E, A	E	
303.2.11	Survivability Tradeoffs	P	A	E	
303.2.12	Transportability Tradeoffs	P	A	E	
303.2.13	Facilities Tradeoffs	P	A	E	

Subtask Performing Organization Classifications (Continued)

Task/subtask	Subtask description	Program, general or peculiar*	Skills*	Performing organization*	Rationale/comments
401	Task Analysis				
401.2.1	Task Analysis	P	A	OEM	The OEM normally does all or most of 401.
401.2.2	Analysis Documentation	P	A	OEM	
401.2.3	New/Critical Support Resources	P	E, A	OEM	
401.2.4	Training Requirements and Recommendations	P	A	E	There are some differences here; however, in cases of special secure communications devices, equipments, and nuclear systems, the subtask will be performed only by the government. This task could normally be performed by the group doing the training plan.
401.2.5	Design Improvements	P	A	E	
401.2.6	Management Plans	P	A	E	
401.2.7	Transportability Analysis	P	A	J	
401.2.8	Provisioning Requirements	P	A	E	
401.2.9	Validation	P	A	J	Both the OEM and the government validate the LSAR.
401.2.10	ILS Output Products	P	A	E	Subtask 401.2.10, LSAR summary output reports, could be accomplished by the government through a "read-only" capability of the OEM's file or by requiring a periodic update to files resident on its own computers. This has the following advantages: (1) needed reports would not have to be specified years ahead in the contract, thus avoiding the tendency of over specification; (2) rapid response to unanticipated summary requirements.
401.2.11	LSAR Updates	P	A	OEM	

Subtask Performing Organization Classifications (Continued)

Task/subtask	Subtask description	Program, general or peculiar*	Skills*	Performing organization*	Rationale/comments
402	Early Fielding Analysis				
402.2.1	New System Impact	P	A	G	Government has most of the data.
402.2.2	Sources of Manpower and Personnel Skills	P	A	G	Rationale same as preceding subtask.
402.2.3	Impact of Resource Shortfalls	P	A	E	Same organization doing readiness analysis.
402.2.4	Combat Resource Requirements	P	A	G	Could be OEM.
402.2.5	Plans for Problem Resolution	P	A	G	
403	Postproduction Support Analysis				
403.2	Postproduction Support Plan	P	E, A	J	OEM has more detailed knowledge of vendor plans.
501	Support, Test, and Evaluation Verification				
501.2.1	Test and Evaluation Strategy	P	A	E	Subtasks 2.1 and 2.2 tie to overall test and evaluation planning. If done by government, it should also do 2.1, 2.2, and probably 2.3.
501.2.2	Support System Package	P	A	E	
501.2.3	Objectives and Criteria	P	E, A	E	
501.2.4	Updates and Corrective Action	P	A	G	
501.2.5	Supportability Assessment Plan	P	A	G	Government has the information in its data bases.
501.2.6	Supportability Assessment	P	A	G	

*Code definitions: *Program:* general = G; Peculiar = P; *Skill:* Engineering = E; Analytical = A; Information = I; *Performing organization:* Government = G; Either = E; Original equipment manufacturer = OEM; either organization is capable of performing but information from the OEM is required of the government = J.

Readiness-Related Design Definition Sample RFP

A6.1 Readiness-Related Design Definition

A6.1.1 Purpose

The purpose of this requirement is to identify the principal design options and related maintenance concept alternatives that will meet the X wartime availability and deployment requirements for all projected missions.

A6.1.2 Scope

This effort shall include studies that will identify wartime sortie/mission-capable levels achievable for various levels of R&M and logistics (spares, test equipment, etc.) along with projected peacetime operating costs and required maintenance manpower skills.

A6.1.3 Contractor tasks

Requirements. The contractor shall perform readiness-related design studies using, as the baseline, the proposed aircraft configuration and recommended maintenance concepts. Each study shall be performed by varying parameters from the baseline design and/or altering elements of the maintenance concept to determine the effects on wartime sortie/mission capability, deployment requirements, peacetime operating costs, or maintenance skill requirements. At least one design iteration shall result from the BCS developed. Subject to written government approval, design parameters may be the subject of a design redefinition where significant improvement can be gained in wartime availability, reduced deployability requirements, or decreased maintenance manpower numbers or skills.

The contractor shall ensure that the following minimum requirements are met, regardless of mission application:

1. Aircraft must be sustainable in the field under austere basing and maintenance support conditions and must be capable of being routinely operated for an extended period (60 to 90 days) from unimproved sites.

2. Aircraft shall be self-supporting in the field to the maximum practical extent.

3. Under normal tactical operating and support conditions, the aircraft must be designed to sustain a 0.9 operation availability rate.

4. The skill level of personnel necessary to operate and maintain the aircraft and equipment shall not exceed those required for weapon systems that the X replace.

5. To the maximum extent possible, maintenance concepts and logistics (e.g., test equipment, training) shall be uniform among the joint users.

Specific design requirements critical to meeting minimum requirements shall be identified for special emphasis in the System Development (SD) phase. In structuring the design studies, the following elements shall be included:

1. *Current technology* X *configuration:*   Contractors shall utilize a broad spectrum of "real world" operational test and evaluation and sample data (i.e., Army RAM/log test data collection system from appropriate DOD aircraft to establish a quantitative/qualitative R&M description of current technology X configuration, as well as a description of maintenance manpower associated with maintaining these aircraft). The candidate systems for the BCS shall include Army, Air Force, Navy aircraft, and Marine aircraft.

2. *Tradeoff analysis determination:*   The contractor shall apply existing cost-estimating relationships and established mathematical techniques to assess the tradeoff between the wartime sortie/availability equipments, R&M, and operating cost. In general, it is desired that the relationship be kept to the level of depth and detail that is being used in formulating the X design (e.g., two-digit WUC level).

Specifics.   The contractor shall perform a series of studies for each of the capabilities listed below to determine the incremental design changes needed to achieve the capability, the effect of the incremental design change on System Development (SD) cost and schedules, unit production cost and LCC, and the incremental changes in specified R&M and maintenance concepts. (*Note:* Blanks below indicate information to be determined by contractor.)

1. Determine design requirements and R&M goals to meet peacetime 0.90 operational availability rate at minimum support costs.

2. Determine incremental design requirements, R&M goals, and resultant logistics to deploy and operate at austere bases for
   a. Five days at sortie rate _____ .
   b. 30 days at sortie rate _____ .
   c. 60 days at sortie rate _____ .
   d. 90 days at sortie rate _____ .
3. Determine design requirements to permit aircraft to self-deploy and operate at sortie rate _____ with
   a. Self-contained support.
   b. Minimum support limited to no more than _____ .
   number of additional accompanying aircraft (type and number)
4. Determine the capability to operate with periodic inspections limited to
   a. Once every 500 flight hours.
   b. Once every 800 flight hours.

The contractor shall synthesize the results of the preceding studies to define design requirements that best maximize achievement of the desired capabilities. The associated SD cost and schedule, R&M requirements, proposed maintenance concept, and estimates of logistic requirements in terms of weight and cube of spares and test equipment at organizational and intermediate maintenance levels also shall be provided for these design recommendations.

A6.1.4 Documentation

Studies shall be submitted in accordance with DI-A-5029. Individual study reports for each capability shall be submitted no later than 12 months after contract award. The report containing the requirements shall be submitted no later than 20 months after contract award.

A6.2 Support, R&M, and Technology Demonstration

A6.2.1 Purpose

To obtain R&M data on equipment or components critical to achieving the readiness and supportability goals of the *X*. The data gathered from this effort shall be used in determining the *X* system and equipment R&M goals and structuring the program at low risk during SD.

A6.2.2 Scope

Engineering design and test efforts shall be conducted for new equipment or components that could be used to significantly increase *X* R&M, eliminate applicable field R&M problems, reduce scheduled maintenance requirements, or minimize logistic requirements.

A6.2.3 Contractor tasks

General requirements. The contractor shall perform design studies and tests to address all critical elements of R&M design and support capability. At a minimum, design analysis shall be conducted in the areas listed below and be supported by soft mock-ups, laboratory experiments, and analytical assessments as required to provide a low-risk design approach for entering SD. Within 6 months, the contractor shall submit a list of additional design analyses and tests to be performed to determine additional design opportunities that could affect R&M or supportability (paragraph 3.13 of reference document).

Rotor and pylon system layout. Assessment of interrelationships of engine, transmission, rotor controls (rotating and nonrotating), rotor hub, blades, and blade-folding mechanisms shall be conducted to establish a component arrangement that optimizes maintenance support function commensurate with acceptable design functional characteristics (dynamic stability, etc.).

Specific attention shall be directed toward defining onboard and ground-based equipment requirements essential for fault isolation and component replacements. Effort in this area shall be coordinated with the assessments identified in the next paragraph.

Maintenance capabilities with aircraft in a folded condition. A detailed assessment of design features necessary to accomplish maintenance with the aircraft in a folded condition shall be conducted with special attention directed toward the following items at a minimum:

1. Flight control, checkout, maintenance, and rigging (fixed and rotary wing control elements)

2. Major component replacements and/or repair

3. Empennage maintenance

4. Subsystem component placements

Ground-based support equipment versus built-in support equipment. An assessment of built-in maintenance features (platforms, component handling equipment, etc.) shall be conducted to establish the minimum requirements for ground-based support equipment. Specific attention shall be directed toward investigation of alternative approaches for providing access to the rotor/pylon and empennage areas.

Placement of flight control. A detailed investigation and tradeoff assessment of alternative design arrangements for placement of the computer, sensors, actuator, and other elements of the flight control system shall be conducted to establish preferred design arrangements with respect to maintenance and logistics support.

Repair-level assessment. An assessment of the potential for utilizing a two-level maintenance support scheme for avionics, flight control, and engine electronic elements and other mission electronic system components shall be conducted. The assessment shall include estimates of the weight and cube of the logistics needed to implement alternative maintenance schemes.

Standardization of hydraulic and electrical system elements. An assessment of the benefits and costs associated with standardizing line/wire lengths and fittings/connections shall be conducted with emphasis placed on decreasing the number of different parts/pieces making up these systems.

Combat maintenance and damage repair. An assessment of design ring devices (chip detectors, oil pressure transducers, rotor system tachometers, etc.).

Chip detectors. Perform tolerance studies and laboratory tests for chip detector design to establish the achievable capability to minimize the number of false failure indications and sensitivity to normal wear debris.

Auxiliary power unit. Investigate existing APUs to determine if the APU can reliably meet the requirements for powering all onboard systems for maintenance when the aircraft is in either spread or folded configuration, as well as provide sufficient power for engine starting without power interruption. The contractor shall demonstrate that the APU reliability levels are achievable and will support 0.9 operational availability at high sortie rates over extended periods (possibly as high as 60 to 90 days) while operating from austere basing and maintenance support.

Documentation. The results of the design analyses and testing shall be submitted in accordance with DI-A-5029. An initial report shall be submitted 6 months after award of contract identifying additional equipment or components that will be analyzed or tested during this phase and identification of areas that should be investigated in the SD phase. The final results of the design analyses and tests shall be submitted no later than 16 months after contract award. Final results shall document the alternatives considered, including recommendations for changes to the I system specifications and rationale; recommendations for test hardware that will have to be dedicated for reliability, maintainability, and Built-In-Test (BIT) design and test efforts during the SD phase; proposed system and subsystem reliability growth profiles during SD phase and initial production; and BIT maturation program including management controls, proposed demonstration points, and BIT software development schedule.

LSA Tailoring Worksheets

This appendix contains worksheets for recording tailoring selections for major systems, complex equipment, and simple equipment. LSA tailoring decisions are frequently subjective, and reasons for selection may be forgotten unless they are recorded. See discussion in Sec. 4.1.

Worksheet 1: Major System Tailoring

Task/subtask number and title	Application matrix by phase*					Subtask selection	Preforming organization selection	Rationale
	Preconcept	Concept	DVAL	EMD	Prod.			
101 Development of an Early Logistic Support Analysis Strategy								
101.2.1 LSA Strategy	G	G(3)	G(3)	S(3)	NA	NA		
101.2.2 Cost Estimates	NA	G(3)	G(3)	G(3)	S(3)			
101.2.3 Updates	NA	G(3)	G(3)	S(3)	NA			
102 Logistic Support Analysis Plan								
102.2.1 LSA Plan	NA	G	G	G	G			
102.2.2 Updates	G	G	G	G				
103 Program and Design Reviews								
103.2.1 Establish Review Procedures	NA	G(2)	G(2)	G(2)	G(2)			
103.2.2 Design Reviews	NA	G(2)	G(2)	G(2)	G(2)			
103.2.3 Program Reviews	NA	G(2)	G(2)	G(2)	G(2)			
103.2.4 LSA Review	NA	G	G(2)	G(2)	G(2)			
201 Use Study								
201.2.1 Supportability Factors	G	G	G	G	NA			
201.2.2 Quantitative Factors	G	G	G	G	NA			
201.2.3 Field Visits	G	G	G	G	NA			
(a) Use Study Reports	G	G	G	G	NA			
(b) Updates								
202 Mission Hardware, Software, and Support System Standardization	G	G	G	G	NA			
202.2.1 Supportability Constraints	NA	G(2)	G(2)	G(2)	C(2,4)			
202.2.2 Supportability Characteristics	NA	G(2)	G(2)	G(2)	C(2,4)			
202.2.3 Recommended approaches	NA	G(2)	G(2)	G(2)	C(2,4)			
202.2.4 Risks	NA	G(2)	G(2)	G(2)	C(2,4)			
203 Comparative analysis								
203.2.1 Identify comparative systems	G	G	G	NA	NA			
203.2.2 Baseline comparison system	G	G	G	G	NA			
203.2.3 Comparative system characteristics	G	G	G	NA	NA			
203.2.4 Qualitative supportability problems	G	G	G	G	NA			
203.2.5 Supportability, cost, and readiness drivers	G	G	G	G	NA			
203.2.6 Unique system drivers	G	G	G	NA	NA			
203.2.7 Updates	NA	G	G	G	NA			
203.2.8 Risks and assumptions	G	G	G	G	NA			
204 Technological opportunities								
204.2.1 Recommend design objectives	NA	G	G	s	NA			
204.2.2 Updates	NA	G	G	s	NA			
204.2.3 Risks	NA	G	G	s	NA			

205	Supportability and supportability-related design factors					
205.2.1	Supportability characteristics	NA	G	G	NA	NA
205.2.2	Sensitivity analyses	NA	G	G	NA	NA
205.2.3	Data rights	NA	S	S	S	S
205.2.4	Supportability objectives and associated risks	NA	G	G	NA	NA
205.2.5	Specification requirements	NA	G	G	NA	NA
205.2.6	NATO constraints	NA	G	G	NA	NA
205.2.7	Supportability goals and thresholds	NA	NA	G	NA	NA
301	Functional requirements identification					
301.2.1	Functional requirements	NA	G	G	S(1)	C(1)
301.2.2	Unique functional requirements	NA	G	G	S(1)	C(1)
301.2.3	Risks	NA	G	G	S(1)	C(1)
301.2.4	Operations and maintenance tasks					
	(a) FMECA	NA	S	G	G	C
	(b) RCM	NA	S	G	G	C
	(c) Other functional requirement analyses	NA	S	G	G	C
301.2.5	Design alternatives	NA	G	G	G	C
301.2.6	Updates	NA	G	G	G	C
302	Support system alternatives					
302.2.1	(a) Alternative concepts	NA	G	G	NA	NA
	(b) Contractor support	NA	S	S	S	1
302.2.2	Support concept updates	NA	G	G	S	NA
302.2.3	Alternative support plans	NA	G	G	G	C(1)
302.2.4	Support plan updates	NA	S	S	G	C(1)
302.2.5	Risks	NA	G	G	G	C(1)
303	Evaluations of alternatives and tradeoff analysis					
303.2.1	Tradeoff criteria	NA	G	G	G	C
303.2.2	Support systems tradeoff	NA	G	G	G	C
303.2.3	System tradeoffs	NA	G	G	G	C
303.2.4	Readiness sensitivities	NA	G	G	G	NA
303.2.5	Manpower and personnel tradeoffs	NA	G	G	S	NA
303.2.6	Training tradeoffs	NA	G	G	G	C
303.2.7	Repair level analyses	NA	S(1)	G	G	C
303.2.8	Diagnostic tradeoffs	NA	G	G	S(1)	NA
303.2.9	Comparative evaluations	NA	G	G	S(1)	C(4)
303.2.10	Energy tradeoffs	NA	G	G	S	C(4)
303.2.11	Survivability tradeoffs	NA	G	G	G	C(4)
303.2.12	Transportability tradeoffs	NA	G	G	NA	NA

Worksheet 1: Major System Tailoring (Continued)

Task/subtask number and title	Application matrix by phase*					Subtask selection	Preforming organization selection	Rationale
	Preconcept	Concept	DVAL	EMD	Prod.			
401 Task analysis								
401.2.1 Task Analysis	NA	NA	S	G	C			
401.2.2 Analysis Documentation	NA	NA	S	G	C			
401.2.3 New/critical Support Resources	NA	NA	S	G	C			
401.2.4 Training Requirements and Recommendations	NA	NA	S	G	C			
401.2.5 Design Improvements	NA	NA	S	G	C			
401.2.6 Management Plans	NA	NA	S	G	C			
401.2.7 Transportability Analysis	NA	NA	G	S(1)	C(1)			
401.2.8 Provisioning Requirements	NA	NA	S	G	C			
401.2.9 Validation	NA	NA	S	G	C			
401.2.10 ILS Output Products	NA	NA	S	G	C			
401.2.11 LSAR Updates	NA	NA	S	G	C			
402 Early fielding analysis								
402.2.1 New System Impact	NA	NA	NA	G	C			
402.2.2 Sources of Manpower and Personnel Skills	NA	NA	NA	G	C			
402.2.3 Impact of Resource Shortfalls	NA	NA	NA	G	C			
402.2.4 Combat Resource Requirements	NA	NA	NA	G	C			
402.2.5 Plans for Problem Resolution	NA	NA	NA	G	C			
403 Postproduction Analysis								
403.2.1 Postproduction Support plan	NA	NA	NA	NA	G			
501 Supportability test, evaluation, and verification								
501.2.1 Test and Evaluation Strategy	NA	G	G	S	NA			
501.2.2 Support System Package	NA	NA	G	G	S			
501.2.3 Objectives and Criteria	NA	NA	G	G	S			
501.2.4 Updates and Corrective Actions	NA	NA	NA	G	S			
501.2.5 Supportability Assessment Plan	NA	NA	NA	G	S			
501.2.6 Supportability Assessment (Post Deployment)	NA	NA	NA	NA	G			

*G = general use; C = generally applicable for design changes only; S = selective use; (1) = requires interpretation of intent to be cost-effective; (2) = MIL-STD-1388-2A is not the primary implementation document; (3) = done just prior to initiation of the phase.

Worksheet 2: Complex Equipment Tailoring

Task/subtask number and title	Application matrix by phase*					Subtask selection	Performing organization selection	Rationale
	Preconcept	Concept	DVAL	FSD	Prod.			
101 Development of an early logistic support analysis strategy								
101.2.1 LSA Strategy	NA	G(3)	S	NA	NA			
101.2.2 Cost Estimates	NA	G(3)	S	S(3)	NA			
101.2.3 Updates	NA	NA	S	S(3)	NA			
102 Logistic support analysis plan								
102.2.1 LSA Plan	NA	S	S	S	S			
102.2.2 Updates	NA	NA	S	S	S			
103 Program and design reviews								
103.2.1 Establish Review Procedures	NA	S(2)	S(2)	S(2)	S(2)			
103.2.2 Design Reviews								
103.2.3 Program Reviews								
103.2.4 LSA Review								
201 Use study	NA	S	S	S	NA			
201.2.1 Supportability Factors								
201.2.2 Quantitative Factors								
201.2.3 Field visits								
(a) Use Study Reports								
(b) Updates								
202 Mission hardware, software, and support system standardization	NA	S(2)	S(2)	S(2)	S(2)			
202.2.1 Supportability Constraints								
202.2.2 Supportability Characteristics								
202.2.3 Recommended Approaches								
202.2.4 Risks								
203 Comparative Analysis								
203.2.1 Identify Comparative Systems	G	G	G	NA	NA			
203.2.2 Baseline Comparison system	NA	NA	S	S	NA			
203.2.3 Comparative System Characteristics	NA	NA	S	NA	NA			
203.2.4 Qualitative Supportability Problems	NA	NA	S	S	NA			
203.2.5 Supportability, Cost, and Readiness Drivers	NA	NA	S	S	NA			
203.2.6 Unique System Drivers	G	G	G	NA	NA			
203.2.7 Updates	NA	NA	S	S	NA			
203.2.8 Risks and Assumptions	G	G	G	G	NA			
204 Technological Opportunities	NA	G	G	S	NA			
204.2.1 Recommended Design Objectives								
204.2.2 Updates								
204.2.3 Risks								

Worksheet 2: Complex Equipment Tailoring (Continued)

Task/subtask number and title	Application matrix by phase*					Subtask selection	Preforming organization selection	Rationale
	Preconcept	Concept	DVAL	EMD	Prod.			
205 Supportability and supportability-related design factors								
205.2.1 Supportability Characteristics	NA	G	G	NA	NA			
205.2.2 Sensitivity Analyses	NA	G	G	NA	NA			
205.2.3 Data Rights	NA	S	S	S	S			
205.2.4 Supportability Objectives and Associated Risks	NA	G	G	NA	NA			
205.2.5 Specification Requirements	NA	G	G	G	C			
205.2.6 NATO Constraints	NA	S	S	NA	NA			
205.2.7 Supportability Goals and Thresholds	NA	NA	G	NA	NA			
301 Functional Requirements Identification								
301.2.1 Functional Requirements	NA	G	G	S(1)	C(1)			
301.2.2 Unique Functional Requirements	NA	G	G	S(1)	C(1)			
301.2.3 Risks	NA	G	G	S(1)	C(1)			
301.2.4 Operations and Maintenance Tasks Analyses								
(a) FMECA	NA	S	G	G	C			
(b) RCM	NA	S	G	G	C			
(c) Other Functional Requirement Analyses	NA	S	G	G	C			
301.2.5 Design Alternatives	NA	G	G	G	C			
301.2.6 Updates	NA	G	G	G	C			
302 Support System Alternatives								
302.2.1 (a) Alternative Concepts	NA	S	S	NA	NA			
(b) Contractor support	NA	S	S	S	NA			
302.2.2 Support Concept Updates	NA	G	G	G	C(1)			
302.2.3 Alternative Support Plans	NA	S	S	G	C(1)			
302.2.4 Support Plan Updates	NA	S	S	G	C(1)			
302.2.5 Risks	NA	G	G	G	C(1)			
303 Evaluations of Alternatives and Tradeoff Analysis								
303.2.1 Tradeoff Criteria	NA	G	G	G	C			
303.2.2 Support Systems Tradeoff	NA	G	G	G	C			
303.2.3 System Tradeoffs	NA	G	G	G	C			
303.2.4 Readiness Sensitivities	NA	G	G	S	NA			
303.2.5 Manpower and Personnel Tradeoffs	NA	G	G	S	NA			
303.2.6 Training Tradeoffs	NA	G	G	S	C			
303.2.7 Repair Level Analyses	NA	S(1)	G	G	C			
303.2.8 Diagnostic Tradeoffs	NA	G	G	S	NA			
303.2.9 Comparative Evaluations	NA	NA	S	S	C			
303.2.10 Energy Tradeoffs	NA	NA	S	S	C			
303.2.11 Survivability Tradeoffs	NA	NA	S	S	C			
303.2.12 Transportability Tradeoffs	NA	NA	S	NA	NA			

401	Task analysis					
401.2.1	Task Analysis	C	G	S	NA	NA
401.2.2	Analysis Documentation	C	S	S	NA	NA
401.2.3	New/critical Support Resources	C	S	S	NA	NA
401.2.4	Training Requirements and Recommendations	C	S	S	NA	NA
401.2.5	Design Improvements	C	G	S	NA	NA
401.2.6	Management Plans	C	S	S	NA	NA
401.2.7	Transportability Analysis	C(1)	S	S	NA	NA
401.2.8	Provisioning Requirements	C	G	S	NA	NA
401.2.9	Validation	C	G	S	NA	NA
401.2.10	ILS Output Products	C	S	S	NA	NA
401.2.11	LSAR Updates	C	G	S	NA	NA
402	Early fielding analysis					
402.2.1	New System Impact	C	G	NA	NA	NA
402.2.2	Sources of Manpower and Personnel skills	C	G	NA	NA	NA
402.2.3	Impact of Resource Shortfalls	C	G	NA	NA	NA
402.2.4	Combat Resource Requirements	C	G	NA	NA	NA
402.2.5	Plans for Problem Resolution	C	G	NA	NA	NA
403	Postproduction Analysis					
403.2.1	Postproduction Support Plan	C	G	NA	NA	NA
501	Supportability Test, Evaluation, and Verification					
501.2.1	Test and evaluation strategy	NA	S	G	G	NA
501.2.2	Support system package	S	S	S	NA	NA
501.2.3	Objectives and criteria	S	S	S	NA	NA
501.2.4	Updates and corrective actions	S	S	NA	NA	NA
501.2.5	Supportability assessment plan	S	S	NA	NA	NA
501.2.6	Supportability assessment (post deployment)	S	NA	NA	NA	NA

*G = general use; C = generally applicable for design changes only; S = selective use; (1) = requires interpretation of intent to be cost-effective; (2) = MIL-STD-1388-2A is not the primary implementation document; (3) = done just prior to initiation of the phase.

Worksheet 3: Simple Equipment Tailoring

Task/subtask number and title	Preconcept	Concept	DVAL	FSD	Prod.	Subtask selection	Preforming organization selection	Rationale
101 Development of an early logistic support analysis strategy								
101.2.1 LSA Strategy	NA	S(3)	S(3)	S(3)	NA			
101.2.2 Cost Estimates								
101.2.3 Updates								
102 Logistic Support Analysis Plan								
102.2.1 LSA Plan	NA	S	NA	NA	NA			
102.2.2 Updates	NA	NA	S	S	S			
103 Program and design reviews								
103.2.1 Establish Review Procedures	NA	S(2)	S(2)	S(2)	S(2)			
103.2.2 Design Reviews	NA	S(2)	S(2)	S(2)	S(2)			
103.2.3 Program Reviews	NA	S	S	S	S			
103.2.4 LSA Review	NA	NA	NA	NA	NA			
201 Use study								
201.2.1 Supportability Factors	NA	S	NA	NA	NA			
201.2.2 Quantitative Factors	NA	S	NA	NA	NA			
201.2.3 Field Visits	NA	NA	NA	NA	NA			
(a) Use study reports	NA	S	S	S	NA			
(b) Updates	NA	NA	S	S	NA			
202 Mission Hardware, Software, and Support System Standardization								
202.2.1 Supportability Constraints	NA	S(2)	S(2)	S(2)	S(2)			
202.2.2 Supportability Characteristics	NA	NA	NA	NA	NA			
202.2.3 Recommended Approaches	NA	NA	NA	NA	NA			
202.2.4 Risks	NA	NA	NA	NA	NA			
203 Comparative analysis								
203.2.1 Identify Comparative Systems	NA	S	NA	NA	NA			
203.2.2 Baseline Comparison System	NA	NA	NA	NA	NA			
203.2.3 Comparative System Characteristics	NA	S	NA	NA	NA			
203.2.4 Qualitative Supportability Problems	NA	S	NA	NA	NA			
203.2.5 Supportability, Cost, and Readiness Drivers	NA	S	NA	NA	NA			
203.2.6 Unique System Drivers	NA	NA	NA	NA	NA			
203.2.7 Updates	NA	NA	NA	NA	NA			
203.2.8 Risks and Assumptions	NA	NA	NA	NA	NA			
204 Technological opportunities								
204.2.1 Recommend Design Objectives	NA	S	S	NA	NA			
204.2.2 Updates	NA	NA	NA	NA	NA			
204.2.3 Risks	NA	NA	NA	NA	NA			

*Application matrix by phase

205	**Supportability and supportability-related design factors**						
205.2.1	Supportability Characteristics	NA	NA	NA	NA	NA	NA
205.2.2	Sensitivity Analyses	NA	NA	NA	NA	NA	NA
205.2.3	Data Rights	NA	NA	NA	NA	NA	NA
205.2.4	Supportability Objectives and Associated Risks	NA	NA	NA	NA	NA	NA
205.2.5	Specification Requirements	NA	NA	G	G	G	C
205.2.6	NATO Constraints	NA	NA	NA	G	G	NA
205.2.7	Supportability Goals and Thresholds	NA	NA	NA	NA	NA	NA
301	**Functional requirements identification**						
301.2.1	Functional Requirements	NA	NA	NA	G	NA	NA
301.2.2	Unique Functional Requirements	NA	NA	NA	G	NA	NA
301.2.3	Risks	NA	NA	NA	G	NA	NA
301.2.4	Operations and Maintenance Tasks	NA	NA	NA	G	G	C
	(a) FMECA						
	(b) RCM						
	(c) Other Functional Requirement Analyses						
301.2.5	Design Alternatives	NA	NA	NA	S	S	C
301.2.6	Updates	NA	NA	NA	NA	NA	NA
302	**Support system alternatives**						
303.2.1	(a) Alternative concepts	NA	NA	S	S	NA	NA
	(b) Contractor support	NA	NA	S	S	S	NA
302.2.2	Support concept updates	NA	NA	NA	S	S	C(1)
302.2.3	Alternative support plans	NA	NA	NA	NA	NA	NA
302.2.4	Support plan updates	NA	NA	NA	NA	NA	NA
302.2.5	Risks	NA	NA	NA	NA	NA	NA
303	**Evaluations of alternatives and tradeoff analysis**						
303.2.1	Tradeoff Criteria	NA	NA	NA	NA	NA	NA
303.2.2	Support Systems Tradeoff	NA	NA	S†	S†	NA	C
303.2.3	System Tradeoffs	NA	NA	NA	NA	NA	NA
303.2.4	Readiness Sensitivities	NA	NA	S	S	S	NA
303.2.5	Manpower and Personnel Tradeoffs	NA	NA	S	S	S	NA
303.2.6	Training Tradeoffs	NA	NA	S	S	S	C
303.2.7	Repair Level Analyses	NA	NA	NA	S	S	C
303.2.8	Diagnostic Tradeoffs	NA	NA	NA	NA	NA	NA
303.2.9	Comparative Evaluations	NA	NA	NA	NA	NA	NA
303.2.10	Energy Tradeoffs	NA	NA	NA	NA	NA	NA
303.2.11	Survivability Tradeoffs	NA	NA	NA	NA	NA	NA
303.2.12	Transportability Tradeoffs	NA	NA	NA	NA	NA	NA

Worksheet 3: Simple Equipment Tailoring (Continued)

Task/subtask number and title	Preconcept	Concept	DVAL	EMD	Prod.	Subtask selection	Preforming organization selection	Rationale
401 Task Analysis								
401.2.1 Task Analysis	NA	NA	NA	G	C			
401.2.2 Analysis Documentation	NA	NA	NA	G	C			
401.2.3 New Critical Support Resources	NA	NA	NA	NA	NA			
401.2.4 Training Requirements and Recommendations	NA	NA	NA	NA	NA			
401.2.5 Design Improvements	NA	NA	NA	NA	NA			
401.2.6 Management Plans	NA	NA	NA	NA	NA			
401.2.7 Transportability Analysis	NA	NA	NA	NA	NA			
401.2.8 Provisioning Requirements	NA	NA	NA	NA	C			
401.2.9 Validation	NA	NA	NA	S	C			
401.2.10 ILS Output Products	NA	NA	NA	NA	NA			
401.2.11 LSAR Updates	NA	NA	NA	NA	NA			
402 Early Fielding Analysis								
402.2.1 New System Impact	NA	NA	NA	NA	NA			
402.2.2 Sources of Manpower and Personnel Skills	NA	NA	NA	NA	NA			
402.2.3 Impact of Resource Shortfalls	NA	NA	NA	NA	NA			
402.2.4 Combat Resources Requirements	NA	NA	NA	NA	NA			
404.2.5 Plans for Problem Resolution	NA	NA	NA	NA	NA			
403 Postproduction analysis								
403.2.1 postproduction support plan	NA	NA	NA	S	S			
501 Supportability Test, Evaluation, and Verification								
501.2.1 Test and Evaluation Strategy	NA	NA	NA	S	S			
501.2.2 Support System Package	NA	NA	S	S	S			
501.2.3 Objectives and Criteria	NA	NA	S	S	S			
501.2.4 Updates and Corrective Actions	NA	NA	NA	S	S			
501.2.5 Supportability Assessment Plan	NA	NA	NA	NA	NA			
501.2.6 Supportability Assessment (Post-Deployment)	NA	NA	NA	G	S			

*G = general use; C = generally applicable for design changes only; S = selective use; (1) = requires interpretation of intent to be cost-effective; (2) = MIL-STD-1388-2A is not the primary implementation document; (3) = done just prior to initiation of the phase.
†If task 302 is required.

LSA Task Enhancement Sample SOW Language

This appendix discusses a number of LSA-related RFP and contract clauses that have been developed to address frequent problems in the application of MIL-STD-1388-1A. These clauses may be applied to the extent possible consistent with the development effort and cost limitations of each program.

A8.1 ILS Integration Program

Design/supportability integration program. In order to facilitate the consideration of supportability and readiness in design, the contractor will execute a design-supportability integration program. Suggested *contract clause* is as follows:

> *Engineering/LSA focal point.* The contractor's design engineering organization will appoint an LSA focal point who will coordinate the design interfaces with LSA tasks and subtasks which require design engineering input on performance.

Dissemination of all support-related design and support system requirements. As an integrated package for all contractor design personnel, this area raises the stature of design management and designers by treating support-related design as an entity rather than the fragmentation currently in specifications. Figure 4.21 is an illustration of the scope of such requirements and their fragmentation in a typical acquisition program. Suggested *contract clause* is as follows:

> *Design review discussion of support-related design requirements.* Contractor design personnel will address, as an entity, all support-related design requirements at government/contractor design reviews.

Design engineering/LSA seminar. This meeting promotes a greater understanding of LSA between the two contractor communities. Suggested *contract clause* is as follows:

> *Design seminar.* The contractor will conduct a joint design engineering/LSA seminar shortly after contract award. The seminar shall be at least one day's duration. The seminar will emphasize the relationship of LSA

to design activity and describe the LSA tasks and subtasks to be performed, the specific approaches planned for execution, the specific role of different organizations and specialists in LSA task accomplishment, and preparation of LSAR data, the specific data relationship and interfaces between organizations with regard to task accomplishment and LSAR data generation, and the planned internal distribution and use of the LSA task reports and the LSAR.

Requirement for input and consideration of supportability requirements and O&S costs in all contractor tradeoff studies. Many current contracts do not require consideration of O&S costs or support requirements in design tradeoff studies. Suggested *contract clause* is as follows:

> *Supportability/cost tradeoffs.* All contractor design tradeoff studies will explicitly document the impact on supportability-related design requirements, system readiness requirements, and support investment and operating and support cost.

Contractor ILS/engineering coordination and integration. To ensure review of proposed designs for supportability considerations, ILS/LSA need to comment prior to final design decisions and ensure access to the most current data.

> Contractor logistics personnel shall review and sign off all proposed engineering drawings and specifications prior to approval and release.

Colocation of contract support personnel. This clause is intended to promote dialogue and data interchange by causing close location of design, R&M, and LSA personnel:

> *Colocation.* The contractor will, where possible, colocate technical R&M, LSA, and ILS element personnel to foster and expedite design-support dialogue and information exchange.

Contractor engineering personnel attendance at LSA/LSAR reviews. *This clause is intended to ensure that contractor design personnel are available for discussion purposes at LSA/LSAR reviews:*

> *Contractor engineering personnel attendance at LSA/LSAR reviews.* Appropriate prime contractor engineering/other personnel shall be available for government LSA/LSAR reviews for in-depth discussion of support-related design issues with government personnel.

A8.2   LSA Plan-Related Clauses

Contractor use of LSA/LSAR products called out in the LSAP. Many of the LSA/LSAR products should be disseminated to contractor design and ILS personnel. Examples are the use study and identification of support system alternatives. However, the LSA plan requirements in MIL-STD-1388-1A do not address this subject. The contractor should be required to treat the LSA products as an internal management tool as well as a deliverable. Suggested *contract clause* is as follows:

> *Contractor use of LSA products.*   The contractor's LSA plan will include a detailed description of proposed internal usage of any required MIL-STD-1388-1A task reports, LSA records, and LSA output reports. Usage descriptions shall include both the design and ILS communities.

Listing of all supportability design requirements in the LSAP.   This forces contractor ILS personnel to understand the full depth of support-related design requirements. Suggested *contract clause* is as follows:

> *Support-related design requirements in the LSA plan.*   The contractor's LSA plan will summarize in one location all support-related design requirements and all support-system objectives. A simple but all-inclusive list of such design requirements is as follows: availability, reliability, maintainability, diagnostics, accessibility, repair times, transportability, redundancy, interchangeability, parts, logistics, test equipment, safety, and human engineering.

Contractor LSA performing organization.   This change is designed to identify in the LSAP the specific contractor organization that will perform the 1A standard tasks and subtasks and prepare the LSAR records. Suggested *contract clause* is as follows:

> *LSA performing organization.*   The contractor's LSA plan will document the specific role of the organizations and specialists in performing the required LSA tasks and subtasks and preparing the LSAR.

Required format for contractor LSA task reports.   Such format is currently at the contractor's discretion. This permits use of formats and outlines useful to obscure the extent to which the report satisfies contract requirements. The standard format suggested in the following contract clause improves ease of government task review and makes it more difficult to hide report inadequacies.

> *Required format for MIL-STD-1388-1A tasks.*   The contractor will use the following format for all required LSA tasks:
> A. Introduction
>   1. General
>   2. SOW requirements
>   3. Task input
>   4. Task outputs
>   5. General methodology

B. Task title
   1. Subtask A, subtask B, etc.
   2. Requirement
   3. Discussion
C. Conclusions
D. Recommendations
E. Appendices

A8.3   LSA-Related Clauses

Engineering-FMECA-LSAR-LSA elements data scheduling.   Incompatibilities and inconsistencies between these schedules can result in lower-quality data, schedule slippage, and/or increased costs. While the LM is responsible for compatibility in the RFP and contract, such situations also can occur in contract execution. The intent of the following contract clause is to require the contractor to bring such circumstances promptly to the attention of the LM to minimize the impact.

Data stream compatibility.   The contractor shall analyze the schedules for engineering data (drawings, etc.), FMECA, RCM data, LSAR data, and ILS element data development in terms of the consistent, sequential development necessary to ensure accuracy of data on which each data stream is based. The resulting generation stream will be compared with delivery dates specified in the RFP or contract. The results of this analysis, and the likely results of any schedule incompatibilities, will be identified to the LM in the proposal, LSA plan, the ILS/LSA guidance conference, or program, design, and LSA reviews as appropriate.

FMECA-RCM-LSAR compatibility.   Structural differences (functional versus top-down drawing) can negate a one-to-one correspondence between the LSAR and R&M data and make review and accountability of LSA data more difficult. It may be desirable to perform both a system functional FMECA for fault isolation and a top-down drawing FMECA for LSAR use. The following contract clause is intended to require contractors to identify incompatibilities or inconsistencies.

FMECA, RCM, and LSAR compatibility.   The contractor will review the requirements for FMECA, RCM, and LSAR to determine any structural gaps (functional or top-down drawing) between the three data sets. Any inconsistencies or gaps and the likely consequences in terms of data inaccuracies and their impact on LSAR data will be reported to the LM in the proposal, LSA plan, the ILS/LSA guidance conference, and program, design, and LSA reviews as appropriate.

Location of LSAR read-only terminal in the program office.   This clause requires the prime contractor to locate a terminal with read-only capability provided by supplied software in the program office to review and monitor progress of the LSAR. Use of such terminal permits greater in-process review and may help reduce on-site travel and review. (Also, see Appendix 12.)

Read-only LSAR capability. The contractor will provide the project office with a read-only computer terminal to assist in the review and monitoring of the contractor's development of the LSAR data base.

A8.4   Low-Reliability Improvement Opportunities

Low-reliability opportunities. The Comparative Analysis Task (203) requires contractors to use existing data bases, i.e., 3-M and VAMOSC, to identify readiness and support costs on comparable existing systems/equipments to assist in preventing repetition of current problems. An alternate or additional approach, whether a data base exists or not, is to use R&M analyses of the proposed system/equipment to identify readiness and O&S cost drivers through study of low-reliability, high-unit-cost items (O&S cost improvement) and low-reliability, low-unit-cost items (readiness improvements). Suggested *contract clause* is as follows:

Review of low-reliability hardware. The contractor shall identify and analyze low hardware reliability areas for improvement opportunities. Low-reliability items of high cost will be identified as a target for support cost reduction. Low-reliability items of low cost will be targets for reliability and system readiness/operational availability improvement.

CALS-Related LSA/LSAR Requirements

This appendix contains sample SOW and CDRL language for LSA/ LSAR and associated data requirements for R&M, RCM, and LORA using the automation, integration, on-line access and digital delivery approaches embodied in the CALS program. Samples are in the following order:

1. Automation
2. Integration
3. On-line access
4. LSA/LORA task reports/plans
5. CDRL examples
6. Summary

A9.1 Automation

A9.1.1 Sample SOW language: LSAR automation

The contractor shall establish and maintain a validated LSAR automated data processing system capable of input, storage, and retrieval of LSAR data in accordance with MIL-STD-1388-2B as part of the CITIS. The LSAR shall be utilized as a source for ILS element information. The contractor may use an internally developed and validated LSAR automated data processing system or an independently developed and validated LSAR automated data processing system. The validated LSAR automated data processing system shall comply with paragraph 4.2.2 of MIL-STD-1388-2B and shall be used for the preparation of LSAR output reports as specified in the CDRL.*

Comment. Implementation of CALS could increase the use of the LSAR in concept formulation and demonstration-validation. While this is true conceptually, how to do this in a cost-effective manner is open to question. Many of the LSAR output reports are used primarily to identify detailed support resources. With the exception of long-lead-time factors or resources needed to support testing, such identification

---

*Source: Modification of language in MIL-HDBK-59A, 20 December 1988.

may not be necessary prior to EMD. Some LSAR output reports are potentially useful for design feedback. See the discussion of LSAR in DEMVAL in Chap. 4. Table 6.1 identifies each of the current LSAR output reports, their major use (e.g., design influence, QA, etc.), and their chief users.

Several approaches can be used to specify LSAR requirements prior to SD:

1. Don't, unless good reasons exist to do so.
2. Confine the requirements to the exception data records, i.e., personnel, top-level support and test equipment, top-level training equipment, facilities, and transportation.
3. Confine the requirement to output summaries useful for design influence (note that CALS and concurrent engineering may accomplish this purpose in other ways).
4. Ask the contractor to include his or her approach to use of the LSAR prior to SD in the LSA plan or proposal.

A9.1.2   Sample SOW language: Reliability-centered maintenance (RCM) automation

An RCM automated worksheet process following the specified requirements shall be used by the contractor to store and process RCM data/worksheets. A complete RCM audit trail shall be maintained throughout the equipment's life cycle. This audit trail will permit traceability of preventive maintenance tasks back to specific engineering failure modes listed in the LSAR. The automated RCM analysis process must have the capability of delivering results digitally on magnetic tape or computer disk and through an interactive electronic interface with the LSAR data base. The contractor shall utilize the RCM automated worksheet process on a contractor-developed, DOD-validated RCM automated process with the capability of automatically transferring compatible data to the government RCM software. The automated RCM analysis data/worksheets must be delivered to the government in accordance with MIL-STD-1840A and the CDRL.

A9.1.3   Sample SOW language: Level of repair analysis (LORA) automation

The contractor shall employ an automated system to perform LORA in accordance with specified requirements. The economic analysis will be performed utilizing service-provided or validated LORA model software. The contractor shall establish a data base as a repository for LORA input data and LORA output report files generated by execution of the LORA model software. The LORA data base will be integrated with the automated LSAR data base to maintain traceability of LSA data used as input to the LORA and the output results of the LORA documented in the LSA for development of the LSA-024 maintenance plan report. Approved LORA input data files and LORA output reports shall be delivered to the government via nine-track magnetic tape or PC compatible format floppy disk as specified in the contract.

Comment. Language may need tailoring, since services vary in LORA approaches.

A9.2   Integration

A9.2.1   Sample SOW language: R&M-LSAR integration

The contractor shall document in the LSA plan the automated linkage procedures to be used to ensure LSAR documentation requirements for initial and updated reliability and maintainability (R&M) information through use of appropriate R&M data sources. Algorithms or transformations that must be applied to source data elements to conform to LSAR documentation requirements shall be documented in the procedures. Traceability between the LSAR and individual R&M data sources and the preservation of appropriate data flows while maintaining established LSAR data element relationships and interdependencies shall be clearly demonstrated in the procedures.*

Comment.   Consider "ask" or mixture of "ask and tell" RFP strategy (see Chap. 3).

A9.2.2   Sample SOW language: RCM-LSAR integration

The contractor shall establish automated links between the LSAR and Reliability-Centered Maintenance (RCM) analysis if RCM is specified elsewhere in this (RFP, SOW).

A9.2.3   Sample SOW language: LORA-LSAR integration

The contractor shall establish automated links between the LSAR and Level of Repair Analysis (LORA) as specified elsewhere in this (RFP, SOW).

---

*Source: Modified language from MIL-HDBK-59A.

A9.3   On-Line Access

A9.3.1   Sample SOW language: On-line LSAR access

Sample SOW language: Read-only capability. The contractor shall provide the government with both on-site and remote on-line access to the LSAR data base. This access will be a read-only capability and will allow the government entry into both the contractor's submitted and approved LSAR files.

The contractor software shall provide the requiring activity the capacity to display on a cathode-ray tube (CRT) the specified reports and input screens in 80- and/or 132-character format.

Sample SOW language: Predefined query capability. The contractor shall establish and maintain automated sets of LSAR data for the management and control of the LSAR. As a minimum, the contractor shall maintain a set of LSAR working data for in-process review and a set of government-approved LSAR data. The LSAR data contained in the working (in-process review) set shall be LSAR that has been subjected to internal contractor review procedures and frozen pending government review and approval. The LSAR working data shall be updated in accordance with the schedule in the LSA plan regardless of the approval status of their content since the last update. Upon government approval, LSAR data contained in the working set shall be transferred to the government-approved LSAR data set. All government-directed changes resulting from the LSAR review process shall be incorporated prior to relocation of the data. The government-approved LSAR data shall be cumulative of all government-approved LSAR data.

The contractor shall provide the government with interactive access to both the working and approved LSAR data sets. The contractor shall provide the means for controlling access capability. The interactive access capability shall include the ability to interrogate, retrieve, review, and print the following:

1. Predefined standard LSAR summaries using established standard LSAR report-selection procedures contained in the applicable Data Item Descriptions.

2. Any of the following government-specified reports: (*Specify content, format, and sequence of each report.*)

The software will provide the capacity for terminal display of the specified queries or data files in 80- and/or 132-character format and will include the capability to print the results of the queries on a local printer at designated locations. The user shall have the capability to specify queries by data set. User options shall include interactive queries from the working data, the approved data, or a combination of both.

The contractor shall provide government with the interactive access capability (Specify periods of required access, i.e., 0800–1600 Eastern Standard Time daily, 24-hour continuous, etc.). Government use of the access capability shall be limited to (*Specify access usage requirements,* e.g., in CPU minutes/month, total connect time, etc.). Access shall be limited to the following locations: (*Specify locations.*)

The contractor shall establish telecommunications capability using one or more of the following methods and shall establish a means for ensuring completeness and accuracy of data transmissions:

1. Point-to-point dedicated lines
2. A mutually acceptable commercial time-sharing or packet-switching network
3. Telecommunications equipment and networks compatible with OSI using FIPS 146
4. The Defense Data Network (DDN)
5. Another mutually acceptable method as defined in the contractor's proposal

In addition, the contractor shall provide

1. The hardware for each of the designated locations (if required)
2. Maintenance for contractor-furnished equipment and software (if required)
3. Training for (*Specify number*) operators at each designated location
4. (*Specify number*) set(s) of automated data processing system operator manuals and user documentation per location.

Sample SOW language: Ad hoc query capability. Add the following to paragraph A9.3.1:

> The contractor shall provide the government with the capability for on-line ad hoc query (report generation). Ad hoc reporting capabilities shall be defined by the contractor's LSAR automated data processing system software and presented in the LSA portion of the contractor proposal. As a minimum, the ad hoc report generation shall be capable of keying on and displaying the following LSAR data elements: LSA Control Number (LCN), Alternate LSA Control Number Code (ALCN), Ultimate Operating Capability (UOC), Part Number, Item Name, Task Frequency, Federal Supply Code for Manufacturers (FSCM), Quantity per Assembly,

and Unit of Measure Price (*Add additional data elements as required to this list*).*

Sample SOW language: Training. The contractor shall provide training to designated government personnel on operation of contractor LSA software and shall provide operator manuals and user documentation.

Comment. The government should specify the number of people and sites of training as GFI to enable contractor to bid cost-effectively. This paragraph could be consolidated with paragraphs 1.3.2.2 and 1.3.3.2.

A9.3.2 RCM

Sample SOW language: On-line RCM access. The contractor will provide the government with both on-site and on-line access to the RCM automated process (and AE data base). This access will be a read-only capability and will allow the government access to both the contractor's working files and approved worksheet/data files. The contractor will provide the government with automated data processing security, data retrieval, review of data on screen, remote site printing of standard reports and screen images, and on-line comment capabilities.

Comment. The language covers both R&M and Age Exploration (AE). Tailor if only one is required.

Sample SOW language: Training. The contractor shall provide training to designated government personnel on use of developed automated RCM process and AE data base software.

Comment. The government should specify the number of people and sites of training as GFI to enable contractors to bid cost-effectively. The on-line access and training paragraphs for LSA, RCM, and LORA could be integrated if two or more are used.

A9.3.3 LORA

Sample SOW language: On-line LORA access. The contractor shall provide on-line access to the LORA data base for review of LORA input data and LORA output reports. The contractor ADP system shall include the capability of transfer of government comments/approval of the LORA output reports. An archive capability to maintain traceabil-

---

Source: MIL-HDBK-59A.

ity of iterations of the LORA process throughout the equipment life cycle shall be developed by the contractor. On-line capability for transfer of LORA input data and output reports to government hardware shall be part of the contractor ADP system. The government shall have the capability of printing any LORA data or output reports at a remote government printer. The contractor shall maintain appropriate security of the LORA data base and ADP system to safeguard access to archived and approved data. Approved LORA input data files and LORA output reports shall be delivered to the requirer as specified in the CDRL.

Sample SOW language: Training.   The contractor shall provide training to requirer personnel in the operation of the remote terminal and necessary tasks to accomplish the on-line review, comment, and approval of the LORA data.

Comment. The requirer should specify the number of people and sites of training as GFI to enable contractors to bid cost-effectively.

A9.4   Sample SOW Language: LSA and LORA Task Reports and Plans

The contractor shall provide access to or digitally deliver in a manner mutually agreeable to the requirer and the contractor such LSA and LORA plans and task reports as designated in the CDRLs.

Comment.   There are a wide variety of such reports (including studies, technical reports, specifications, etc.) which will be covered by CALS standards in the nineties time frame. However, "local" arrangements are now possible if the requirer decides that this is cost-effective for certain reports (see paragraph 1.5.4). The requirer could identify the reports desired on-line or delivered digitally and/or could "ask" the contractor in section L of the RFP to propose specific reports for such access/delivery. Specific requirements should be entered in block 16 of the CDRL for each report involved. Most of the LSA task reports not recorded in the LSAR and the LSA and LORA plans are good candidates for this requirement.

A9.5   Sample CDRLs

A9.5.1   LSA and LORA task reports and plans

For reports/plans as a document image file deliverable, enter the following in block 16 of DD form 1423:

LSA and LORA plans and task reports (other than the LSAR) shall be submitted as document image files in accordance with MIL-R-28002, "Raster Graphics Representation in Binary Format, Requirements For," dated December 20, 1987, type I data, and MIL-STD-1840, "Automated Interchange of Technical Information," dated December 22, 1987. There shall be one raster image per file. The delivery mode shall be magnetic tape (level 3) (level 4), CPI (1600) (6250) in accordance with MIL-STD-1840A.

For reports/plans as a processable data file, enter the following in block 16 of DD form 1423:

LSA and LORA task reports and plans shall be submitted using software (for example, WORDSTAR, WORD PERFECT, ASCII, other) and delivery mode (magnetic disk, magnetic tape) mutually agreeable to the contractor and DOD program office.

It is possible to use IGES group I and SGML in the second example; however, delayed order/delivery or interim contractor support would have to be employed until requirer destination systems are in place.

A9.5.2   LSAR

A sample CDRL requirement for on-line LSAR from one DOD acquisition program follows:

LSAR data shall be submitted on-line beginning 6 MAC. The contractor shall submit a magnetic tape, 9 track, ODD parity, 1600 BPI, EBCDIC code, containing all LSAR master files every 180 days.

In this case, the requirer specified on-line real-time and contractor-furnished equipment and dedicated modems and lines. This should be placed in the SOW. The magnetic tape technical details cited above may vary in specific acquisitions.

A9.6   Summary

This appendix has provided examples of SOW and CDRL language and requirements for LSA and associated requirements which address the use of the CALS concepts of automation, integration, on-line access, and digital delivery.

Sample Common Core LSA SOW

This appendix contains a sample LSA SOW which, when tailored LSA tasks and subtasks are added, will suffice for most acquisitions of LSA. Besides the addition of tasks and subtasks, a few other items may need to be customized or recovered according to the needs of the particular acquisition; these items are enclosed within brackets.

A10.1 Logistics Support Analysis (LSA)

A10.1.1 General requirements

The contractor shall establish and maintain an LSA program in accordance with MIL-STD-1388-1A and MIL-STD-1388-2A/2B as defined herein. The purpose of this LSA program is defined as follows:

1. [Cause supportability requirements to be an integral part of design.]
2. [Define support requirements that are optimally related to each other.]
3. [Define the support required during the operational phase.]
4. [Prepare attendant data products.]

The LSA program shall be the single, analytical effort interfacing the system engineering process to identify logistic support requirements. [It shall include design analysis/feedback, support resource analysis/feedback, and shall encompass all aspects of supportability including reliability, maintainability and safety and human factors engineering to ensure that design features enhance cost-effective operation and support throughout the life cycle of the hardware.]

A10.1.2 LSA program planning

The contractor shall submit an LSAP, either separately or as part of the Integrated Support Plan (ISP), in accordance with DI-L-7017A. The LSAP is to be based on this SOW and the CDRL (DD form 1423), attachment 1, and must satisfy task 102, MIL-STD-1388-1A. The LSAP must be responsive to a high level of government involvement and must show how the LSA will be controlled to:

1. Provide schedule milestones to properly time phase the accomplishments of the requirements contained herein.

2. [Influence hardware design for enhancement of logistic support features.]

3. Establish an LSA feedback loop to correct deficiencies.

4. Establish data flow and responsibilities for the execution of the LSA/LSAR requirements herein.

5. Preclude duplication of effort.

6. Establish a government-approved LSA Control Numbering (LCN) System.

A10.1.3   LSA guidance conference

The contractor shall host/participate in a joint government/contractor guidance conference within [30 to 60 days] after contract award to establish government/contractor interfaces and to finalize the tailoring of LSA tasking and data requirements. Agreements reached on substantive changes to the form and substance of the LSA program will require amendment of the LSAP and possibly making modifications to the contract. Any subsequent updating or changes to the LSAP or identified tasks will be submitted to the government for approval prior to implementation.

A10.1.4   LSA documentation

General. LSA documentation shall be generated from the analysis tasks and studies and will, when applicable, be stored in an automated form based on the standard DOD LSAR data tables. The LSAR will utilize an LCN system to establish a systematic top-down hardware breakdown of the equipment into its subsystems, components/assemblies, and parts. The contractor's proposed LCN system, with an explanation of comprising elements, shall be contained in the LSAP and will require government approval. Supplemental data, such as functional block diagrams, decision trees, troubleshooting charts, sketches, etc. shall be incorporated into LSA hard-copy storage by entering the key fields (LCN, alternate LCN code, reference number, usable on code, and task code when applicable) on the supplemental data. To ensure a basis for logistics resource planning and tradeoff decisions prior to freezing hardware design, the contractor shall initiate the LSAR and process LSAR data tables concurrently with equipment design activities. The contractor shall update the LSAR on a continuing basis throughout the contract period to reflect changes in support requirements resulting from

1. Changes to equipment design or mission/operational requirements.

2. Logistics support improvements or the correction of deficiencies discovered through analysis of data from field and test data collection systems, test reports, and formal design reviews and by the contractor's validation of LSAR documentation.

3. Approved, recommended changes that result from government data review.

The LSAR will constitute the central file for validated, integrated, and design-related logistics data pertaining to the acquisition program. At the completion of the acquisition program, the LSAR master files will be delivered to the acquisition program manager. [*Comment:* Options could be either the hard-copy, computer print-out, and/or magnetic tape.]

Data storage and retrieval.   The contractor shall establish and maintain an automated data processing system to record, store, and process LSAR data. The government may make available, as Government-Furnished Equipment (GFE), the 1388-2A data system, or the contractor may elect to use a contractor-developed system. Most contracts will now require MIL-STD-1388-2B. Only systems with letters of validation on file with the DOD LSA support activity will be authorized.

[*Comment:* DOD policy encourages use of contractor-developed, automated LSAR systems when such systems offer cost and other advantages. It may be wise to require identification of the LSAR system to be used and the basis for its selection, particularly in terms of cost.]

Review and approval of LSA documentation.   The contractor shall host LSA/LSAR review sessions at the contractor's plant. During these sessions, the government will review the contractor's progress in accomplishing the LSA tasks, evaluate how results have affected the design process, and approve the data in the LSAR. The LSA analyses, LSAR data, LOR studies, and other analyses shall be made available for government review at least 30 days prior to the scheduled reviews.

Data and documentation delivery.   The contractor shall submit LSA documentation to the government in accordance with the requirements of the CDRL (DD form 1423). The contractor shall use the LSA data base as source data for the development of deliverable CDRL requirements as specified in the CDRL and any other deliverable that can be generated completely or partially from this data.

If the contractor considers that the LSAR will not contain sufficient information to permit development of the documents listed in the CDRL, the procuring activity should be advised, the data deficiencies that exist should be described, and a means of generating the necessary data proposed.

A10.1.5 LSA program requirements

The following MIL STD-1388-1A LSA tasks or subtasks shall be performed by the contractor to support this acquisition program:

LSA Tasks (A list of the LSA Tasks determined to be applicable by the LM)

LSA Subtasks (A list of the subtasks selected from the tasks that are selected by the LM)

Accomplishments of the preceding tasks or subtasks will be limited to

1. Task (-) narrative:
2. Task (-) narrative:
3. Task (-) narrative:

(To be added by the LM as necessary)

A10.2   LSAR

10.2.1 General

The LSAR shall quantify and compile the logistics support requirements for preoperational and operational support of the system/equipment. The LSAR shall define the logistics support requirements to the degree possible in terms of

1. Maintenance plan
2. Manpower and personnel
3. Supply support
4. Support and test equipment
5. Training and training devices
6. Technical data
7. Computer resources support
8. Packaging, handling, storage, and transportation
9. Facilities

A10.2.2 LSA candidate list

The LSAP shall identify an LSA candidate list. LSA candidates include items that impose a logistics burden and items for which the government lacks a maintenance capability and logistics support determination. The contractor shall perform the LSA on the government-approved LSA candidate list to the lowest level of repair. Systems, subsystems, end items, components, assemblies, subassemblies, support and test equipment, and training equipment that require documentation of operational and logistics support parameters and requirements and all items for which the government does not have an existing maintenance capability shall be candidates for LSA. Maintenance capability as used in this context includes, but is not limited to, trained personnel, transportation and handling, logistics technical data, support and test equipment, supply support, and facilities. The selection of LSA candidate items shall be governed by the following procedures: The contractor shall prepare an initial list of LSA candidates in consonance with the criteria below. The list shall include LCN, NSN and/or manufacturer's part number, and item name as available. The initial list of candidate items shall be included in the initial LSAP submission and shall be revised by the contractor as design engineering progresses and the results of the RCM, FMECA, and LOR analyses are approved. The following material shall be candidates for LSA:

1. Contractor-furnished installed equipment items that can or will be inspected, tested, repaired, maintained, or overhauled as part ("on-equipment" maintenance) of the system/equipment.

2. Contractor-furnished installed equipment items that can or will be tested, repaired, maintained, or overhauled separately ("off-equipment" maintenance) from the systems, end item, component, assembly, or subassembly with which they are functionally associated.

3. Contractor-furnished noninstalled equipment to include support and test equipment and training equipment.

4. Installed and noninstalled GFE items when such analyses are required to interface GFE with contractor-furnished equipment or when usage/environment will be different, and/or to determine total support requirements of the end item(s).

5. Installed and noninstalled GFE items for which government-furnished data are inadequate or incompatible and where such data are necessary to fulfill contract requirements.

6. Connecting and installation hardware, bracketry, standard hardware items, bulk material, and simple parts, which are not econom-

ically repairable, shall be included in the LSA documentation for the next higher assembly.

[*Comment:*   Any components to be used in this design that are currently in the DOD inventory or are commercial off-the-shelf items for which suitable manuals and parts lists exist should be excluded from the list as candidates for study. If needed, the time, skills, etc. required to perform these tasks can be accounted for by merely filling out a 2A C06 card and a D04 card on 2B C tables for each task.]

A10.2.3   Data validation

The contractor shall establish (and describe in the LSAP) internal procedures for validating the adequacy and technical accuracy of LSA documentation. The contractor shall conduct LSA data validation continually to correct and amplify the LSAR and shall ensure data feedback from OT&E, R&M demonstrations, support and test equipment compatibility tests, and technical publications validation actions that update and provide corrections to the affected LSA data.

Guidance and General Language for Sections L and M and Source Selection Planning

This appendix contains suggested language for section L, "Instructions to Offerors (ITO)," and section M, "Evaluation Criteria," and related source selection planning. This material should be tailored/ augmented to develop specific program supportability/ILS/LSA RFP requirements.

A11.1 General Coverage

The material is organized in the following sequence:

Guidance and sample language, section L   (A11.2)

Guidance and sample language, source selection planning   (A11.3)

Guidance and sample language, section M   (A11.4)

Summary   (A11.5)

A11.2 Section L

A11.2.1 General

Section L of the RFP provides overall guidance to bidders with regard to proposal submissions. Table A11.1 shows the overall outline of a recent RFP.

TABLE A11.1   **Example Content of Section L**

1.0   Introduction and background of RFP
2.0   Scope of effort required by RFP
3.0   Type of contract
4.0   Physical format (number of pages, single or double space, etc.)
5.0   Proposal classification and distribution
6.0   Content requirements: General
7.0   Volume I: Technical/management proposal: Specific requirements
8.0   Volume II: Price proposal: Specific requirements

In some major procurements, it is not unusual to see Volume I in this example become more than one volume (e.g., separate volumes for technical, ILS, and management). The instructions for those volumes are further defined in section L. Table A11.2 shows the required content of the technical approach for the management section in a recent RFP.

A11.2.2 Integrating supportability/ILS/LSA into the ITO

An individual preparing or assisting in preparing RFP materials must fit the material into the framework of the overall approach used on the specific RFP. Possible candidate general sections include system engineering, ILS, and management sections. It may be desirable to place general ITO material in one section and detailed ITO material in another section. For example, the general material could go in the management section, R&M/supportability design requirements in a systems engineering section, and LSA matter in an ILS section. Regardless of the area of placement of ITO guidance, it obviously must be related to the supportability/ILS/LSA requirements in section C, "Description/Specification/Statement of Work." The amount and length of the ITO material can vary widely. One can use a minimal SOW requirement while simply stating in a short paragraph in section L that the bidders will submit their proposals for implementation of supportability/ILS/LSA. The other extreme is to provide ITO guidance on every supportability/ILS/LSA SOW requirement. A logical approach in practice is to focus ITO guidance on the areas of major concern or interest. A detailed discussion of possible areas is contained in Sec. 114.4.

TABLE A11.2 Example Content of Management and Technical Approaches

Technical

1.0 Summary of technical approach
2.0 Technical design
3.0 Intra/interface design
4.0 Test program

Management

1.0 Summary of management approach
2.0 Corporate and program organization
3.0 Method and control
4.0 Quality assurance
5.0 Previous experience
6.0 Personnel and resources
7.0 Facilities
8.0 Implementation plan

The supportability/ILS/LSA material developed for section L, like the SOW requirement, may vary by type of program and program phase. During the concept phase of a new development, the program office may wish to focus efforts on design-LSA functional integration efforts and the contractor's ability to provide on-line access to a wide variety of documentation. In the case of an NDI program, emphasis may be on digital delivery. A program currently in FSD may be interested in on-line access and digital delivery. A key factor in preparing ITO material is the RFP strategy the government wishes to pursue. This involves the decision to tell the bidders what the government must have or to ask the bidders what they propose to do. In some current procurements, the government is stating what it would like and asking the contractor to discuss the viability and effectiveness of those desires and to suggest alternate approaches to meet the government's desires. The following requirements tend to fall into the "ask" or a mix of the "ask and tell" approaches:

- Degree of design-R&M-RCM-LORA-LSA integration
- Ad hoc LSAR system capability
- Degree of interactive access to contractor data bases and analysis tools
- Use of the LSAR (prior to EMD)
- Digital delivery of the LSA task reports
- Other focusing/enhancements

As a practical matter, the DOD program office should state any mandatory requirements in these areas in the SOW. Section L should instruct the bidders to state their proposed approach in those SOW areas not considered mandatory.

A11.2.3   Suggested departure-point language for section L

LSA proposal/plan. Paragraphs _____ of section C contain the desired supportability/LSA requirements for this solicitation. Paragraphs _____ are mandatory requirements. The remaining paragraphs _____ are considered subject to change based on the contractor's evaluation of their cost-effectiveness and contractor-proposed alternate requirements. The contractor's proposal should provide comparisons of costs, cost avoidances, and benefits for alternate approaches. The proposal should identify significant costs and benefits and associated risks over the life of the program. The contractor shall submit a proposed supportability/LSA plan as part of the proposal. The content of this plan shall be in accordance with the SOW, CDRL, and the contents of

this section. The contractor's proposal/plan shall address paragraphs
_____ in section C (*Identify any major areas cited of interest here*).

A11.3   Source Selection Planning

A11.3.1   General

Source selection planning is a major step in the procurement process.
It involves a number of actions which affect supportability/LSA RFP
requirements preparation. Included in this section are discussions on:

Source selection criteria, factors, and subfactors

Criteria and factor weighting

Development of standards

Rating adjectives

Rating and scoring methods

A11.3.2   Source selection criteria, factors, and subfactors

Background.   The development and definition of source selection criteria are difficult but key parts of the source selection process. Such
criteria include evaluation categories and factors and their relative
importance. Some acquisition centers use alternate terminology for
this breakdown. Table A11.3 illustrates this process. These criteria
must be set forth in the Source Selection Plan (SSP), identified in the
RFP, and evaluated in the source selection evaluation. The criteria
are the basis for preparing section M, "Evaluation Criteria."

There are no restrictions on the kinds of evaluation criteria that
may be used, as long as they are disclosed in the RFP and are related
to the purposes of the acquisition program. The specific criteria used
will depend on the particular circumstances. They generally fall into
the following functional disciplines or categories: technical (design
and production capability), management and business, schedule, and
cost. The choice of categories most critical to the program and their
relative importance represents the first classification of criteria. The
categories selected as essential to the selection process can be broad in

TABLE A11.3   Hierarchy of Categories, Factors, and Subfactors

Categories (areas)
   Factors (items)
      Subfactors (factors)

scope; however, they should be selected so that the evaluation may be limited to aspects necessary to the success of the program. The broad categories selected may then be broken down into factors and subfactors.

Criteria.   The criteria selected may vary with the program phase. The categories chosen for a concept phase might be

1. Military worth
2. System life-cycle cost
3. Development risk

When the same program is about to enter the demonstration/validation phase, the categories might change to

1. Technical
2. Management
3. Cost

From a supportability/LSA perspective, the top-level categories might be

1. Management
2. Technical
3. Benefits
    a. Costs/cost avoidances
    b. Readiness

Examples of how to translate these categories into factors and subfactors are given later in this section.

Factors and subfactors.   Categories such as the preceding are broken down into factors and subfactors in some cases. Table A11.4 illustrates

TABLE A11.4   Typical Contractor Support Service Source Selection Criteria Structure

Proposal evaluation criteria		
Technical category 1	Management category 2	Cost category 3
Factors	Factors	Factors
Understanding of program	Organization	Cost government
	Project management	Cost realism
Technical capability	Cost management	
Key positions and staffing	Corporate experience	
Special requirements	Past performance	
Facilities	Security	

this step for an RFP for contractor support services. With some modification this structure might be applied to the LSA portion of an RFP for a hardware system. The use of too many factors should be avoided, since it leads to dilution of the ability to differentiate between factors. Each evaluation factor should be clearly defined. Once the "evaluation factors for award" are disclosed in the RFP in section M, they must be used in the evaluation and cannot be changed without modifying the RFP. Subfactors may be disclosed in the RFP, depending on what subfactors, if any, are chosen.

From a supportability/LSA perspective, a departure point for a list of technical factors and subfactors might be

1. Clarity and specificity of approaches to LSA task and LSAR execution
2. Appropriateness of models/algorithms to be used
3. Newly automated areas (e.g., RCM, LORA, AGE, etc.)
4. Functional integration (e.g., RCM, FMECA, LORA, etc.)
5. On-line access to the LSA task reports and the LSAR
6. Digital delivery
7. Other: any points of special interest

A departure point for a list of management factors and subfactors could include

1. Organization
2. Supportability/LSA plan
3. Use of team approach to LSA/LSAR execution
4. Corporate experience
5. Key positions and staffing (résumés)
6. Clarity and completeness of subcontractor LSA/LSAR requirements
7. Other

The breakdown of the supportability/LSA benefit category is challenging from a "bottom line" evaluation viewpoint. Beginning with EMD, DOD generally expects LSA to reduce both acquisition and/or operating and support (O&S) costs and/or to improve readiness through more timely and accurate data. LSA, in company with the concurrent engineering, total quality management, and CALS initiatives, is also supposed to save time in development, production, and reprocurement, as well as in support activities. Translating the preceding into specifics, however, may be difficult.

With this in mind, benefits must still be addressed. Costs may be broken into acquisition costs and O&S costs. Appendix 27 contains an example breakout of such costs. Acquisition costs may be broken further into contractor and government costs. In terms of contractor acquisition costs, there are two options. The first is to consider contractor acquisition costs and cost avoidance as an internal input into the contractor overall cost bid. In this view, further cost details are unnecessary for an evaluation.

Another option in terms of contractor costs/savings is to examine any detail which may be visible in this regard, such as the contractor bid on specific LSA/LSAR contract line items. However, some LSA costs/benefits may be incorporated into functional contract line items such as technical manuals, etc. and thus not be visible.

A third approach to evaluation of the contractor's LSA effort is the *surrogate approach*. The use of R&M as surrogates for O&S costs is an example of this approach. For a surrogate approach, one would select the most appropriate surrogates from the technical factors cited earlier in this section.

Government acquisition savings due to the contractor's LSA approach could be evaluated. The government may change its budget for travel, etc. based on the contractor proposal for LSA/LSAR interactive access and digital delivery.

The evaluation of O&S costs as a factor has similar options. It may be possible to directly compare contractor O&S cost estimates or to use contractor inputs in an O&S cost model run by the requirer. The surrogate approach is also possible in the O&S cost area.

Variations of factors by phase may be desirable from an LSA perspective. Table A11.5 illustrates how three possible LSA technical categories might vary by phase. Other categories are obviously possible.

TABLE A11.5 LSA Factor Variation by Phase

LSA factor	Phase			
	Concept	DEMVAL	EMD	Production
Design-LSA	x	x		
Support system design	x	x	x	
Detail support resource identification		x	x	x
CALS-related LSA				
Automation		x	x	
Integration	x	x	x	
On-line access	x	x	x	
Digital delivery			x	x
Other categories are obviously possible.				

A11.3.3 Weighting

Since some of the award factors will normally have more impact on the selection decision than others, it may be necessary to weight award factors so that their relative importance is reflected in the final evaluation. Weights are normally developed and assigned by the Source Selection Advisory Council (SSAC) in conjunction with their participation in the preparation and review of the Source Selection Plan (SSP). Weights will be included in a separate document from the SSP and not divulged to the Source Selection Evaluation Board (SSEB) or the potential offerors. Weighting should be established and approved prior to issuance of the RFP.

As an illustration of weighting, an alternative to the example in Table A11.5 might be to use all three categories in each phase but change their relative importance to fit the phase. Table A11.6 illustrates variations of weighting by phase for a new program.

This example assumes that points are assigned to each category. Forty points are assigned to the LSA area in this example (some acquisition centers do not believe such detailed numerical scoring is useful, since it may lend false precision to subjective judgments). In the example, dividing the number of points for the factor by the total available points for the area factors gives the relative weight. In the example above for the concept phase, integrated design-LSA teams is 20/70. Support system design is 10/70. Thus integrated design-LSA teams is twice as important as support system design. One approach for stating relative weights in section M is discussed in paragraph A11.4.3. The factors and weights above are only examples. More factors and different weights may be appropriate in a given application. The weighting of factors obviously can be continued at the subfactor level.

TABLE A11.6 Weighting Variations

LSA factor	Phase			
	Concept	DEMVAL	EMD	Production
Plan/management	10	10	10	10
Design-LSA integrated teams	20	16	10	5
Support system design	10	11	10	15
Detailed logistic support identification	0	3	10	10
CALS-related LSA				
Automation		10	7	5
Integration	18	10	8	5
On-line access	10	7	8	10
Digital delivery	2	3	7	10

A11.3.4   Development of standards

General.   Standards, which are sometimes called *evaluation considerations* or *evaluation elements,* should be developed by the SSEB in conjunction with the evaluation categories and factors. Standards provide a means of disseminating uniform guidance to evaluators on how to rate an offeror's response with respect to a given factor. They focus the evaluation on each individual factor and assist in achieving consistent and impartial evaluations. A standard defines, describes, or otherwise provides a basis for considering a particular aspect of an evaluation factor. Several standards may thus be developed to evaluate one factor.

Quantitative standards.   In some areas standards can be defined by a readily measurable form or in the degree or percentage of attainment of a required threshold or stated goal. For example, system performance might be broken down into the following subfactors:

- Speed
- Range
- Reliability
- Maintainability

The RFP requirement for the "speed" might read

> Within 3.5 seconds after launch, the weapon must reach a sustained speed of 980 knots. Speeds in excess of 980 knots are believed to be attainable and will be given additional award preference.

The standard against which proposals would be evaluated for the subfactor "speed" could thus be written: "Standard—sustained speed must equal or exceed 980 knots." The guidance to the evaluator for subfactor "speed" might indicate that an acceptable rating should be given those proposals offering a sustained speed of 980 knots. Additional rating points could be assigned to those proposals offering a sustained speed in excess of 980 knots, either on a percentage basis or in a series of speed ranges above the minimum acceptable level. Standards and evaluator guidance for range, reliability, maintainability, and other performance factors could be similarly addressed.

There are several quantitative standards that are theoretically possible in potential LSA requirements. Examples are R&M, O&S cost, LCC, and readiness. Such factors could be further subdivided; for example, reliability could be separated into mission reliability and logistics reliability (mean time between maintenance) using areas specified in the SOW and ranked according to importance. R&M, LORA, and RCM automated linkages with LSAR could be possible subfactors.

Qualitative standards. Quantitative standards are not possible in many technical and management areas. In these cases, the standards may take the form of attributes framed as questions or narrative statements for consideration by the evaluator.

An example of the use of the statement approach in lieu of questions could be

> The proposed LSA program should emphasize the "team" approach with a small LSA staff that works with LSA focal points in the various functional organizations rather than a larger office performing LSA activities off-line from the functional organizations.

The rating instructions for the subfactor "LSA program" could require the evaluators to provide nominal ratings for those offerors' proposals which are responsive to the RFP and which have no major deficiencies. Lower ratings could then be assigned to those proposals considered less adequate and higher ratings to those proposals which are more than adequate. Standards and rating instructions for other imprecisely measurable factors could be similarly addressed.

An example of the use of questions as standards is illustrated by the following questions:

> How does the contractor propose to integrate LSA with design and R&M?
> What are the ad hoc report capabilities of the contractor's automated LSAR system?

A11.3.5   Rating adjectives

Use of adjectives.   For many source selections, a rating scheme is used to translate narrative evaluations (which highlight strengths, weaknesses, deficiencies, and risks with respect to each award factor) into some sort of rating or scoring. Methods for rating and scoring are discussed in paragraph 11.3.6. Rating or scoring lends itself to final modification by the relative weights and aids the final selection process. A set of adjective definitions suitable for technical and management factors will not necessarily be suitable for the cost category. Accordingly, a given source selection process may use two sets of rating adjective definitions, one for technical and management factors and another for cost factors.

Examples of rating adjectives.   The following is an example of rating adjectives that could be used for selected technical and management factors and categories:

- *Exceptional*—Exceeds specified performance or capability in a beneficial way; high probability of success; no significant weakness.

- *Acceptable*—Meets standards; good probability of success; weaknesses can be readily corrected.

- *Marginal*—Fails to meet standards; low probability of success; significant deficiencies but correctable.

- *Unacceptable*—Fails to meet a minimum requirement; needs a major revision to the proposal to make it correct.

A11.3.6   Rating and scoring methods

General.   There are many rating and scoring methods. These include quantitative (numerical), semiquantitative (check-plus-minus criteria, green-red-yellow criteria, pass-fail criteria), qualitative (narrative), or a combination of any of the preceding.

Numerical methods. The numerical method involves a preestablished numerical scale for each specific factor susceptible to direct numerical measurement. An example of this method was given previously in Sec. 1.3.

Narrative methods.   A narrative rating method involves the use of rating adjectives such as *outstanding, good,* etc. for each factor to provide a means of comparing a proposal to the established standard. It is not sufficient that the narrative states that something is good or inferior. The evaluator must first indicate in the narrative what is being offered, how it meets the standard, what its strengths, weaknesses, and deficiencies are, what, in the evaluator's opinion, must be done to remedy the deficiencies, what impact the deficiencies have on the offeror's proposal, and what impact their correction will have on the proposal.

Combination methods.   The numerical approach may be combined with the narrative approach. A common combination method requires the evaluator to first provide a narrative description of each proposal, factor by factor. This can be accomplished by the use of worksheets on which the evaluator describes the attributes and deficiencies of the proposal. Evaluators complete the narrative description prior to assigning (by themselves or by others) any rating to the factors. The rating methodology may make use of the intermediate step of assigning rating adjectives. The final scoring step may be to then assign a score, based on the rating adjective and a predetermined scoring method established for the factor. An example of assignment of numerical scores to rating adjectives would be to assign 5 to an outstanding proposal factor, 4 to an excellent or good proposal, 3 to a good proposal, 2 to a fair or marginal proposal, and 1 to an unsatisfactory proposal. Color coding is an alternative to assigning numbers to rating adjectives. An example of practice at one acquisition center is to code exceptional blue, acceptable green, marginal yellow, and unacceptable as red.

A11.4   Guidance and Sample Language for Section M

A11.4.1   General

The material to appear in section M, "Evaluation Criteria," is based on the criteria developed for the Source Selection Plan. There are two general approaches: The first is to duplicate the source selection plan criteria and weighting. The argument for this approach is that not only are offerors entitled to know the basis for award and the order of importance of criteria to the DOD, but it is to the government's advantage to make these points clear to properly orient contractor proposals. The second approach is to summarize the criteria and weights in some form. The argument for this view is that it discourages "gaming" by the bidders.

While the precise outline of section M will vary between U.S. DOD acquisition centers, a common outline is as follows:

- Introduction
- Basis for award
- Scope of award

A11.4.2   Sample language for the introduction and basis for award

The following illustrates the types of language that can be found in section M:

Introduction. Evaluation of the proposals submitted for (*Specify acquisition phase*) of the _____ program will be accomplished using the criteria set forth below.

Basis for award. Selection of the contractor(s) will be made on the basis of an assessment of each offeror's ability to satisfy the requirements of the solicitation by the Source Selection Authority (SSA). In essence, the assessment will involve a determination by the government of the overall value of each proposal judged in terms of capability in concert with cost to the government, recognizing that subjective judgment on the part of the government evaluators is implicit in the entire source selection process. This assessment will include evaluation of general considerations as well as the results of the evaluation of proposals against criteria. Examples of general considerations include past performance, proposed contractual terms, and conditions and results of Preaward Surveys (PAS) or Manufacturing Management/Production Capability Reviews (MM/PCRs). Throughout the evaluation, the government may consider "correction potential" when a deficiency is discovered. Proposals that are found to be unrealistic in terms of program commitments, or unrealistically low in cost or price, will be deemed reflective of an inherent lack of competence or of failure to comprehend the complexity and risks of the proposed contractual requirements, and this may be grounds for proposal rejection. The government reserves the right to award to other than the lowest proposed acquisition price (or life cycle cost). The SSA will make a determination of the overall merit of each proposal in terms of its potential to satisfy the needs of the DOD. The following considerations (or areas) are listed in descending order of importance and will be observed in making the integrated assessment:

Technical

Logistics

Manufacturing

Management

Cost*

If the preceding areas were more extensive, CALS could be listed as a major approach, as shown below:

1. Program cost
2. Operation capability
3. Prototype/weapon system transition
4. Adequacy of the program
5. Supportability
6. Contractor capability
7. Technical approach or technical excellence
8. Production/manufacturing feasibility
9. Maintainability in an austere operation environment
10. Availability and technical competence of the design team
11. Any other specific considerations, such as supportability/LSA, CALS, and Concurrent Engineering.

---

*Note: The Cost area is not rated but must be ranked in order of importance.

A11.4.3 Sample language for the scope of evaluation

The following illustrates the type of language that can be placed in the scope of evaluation portion of section M:

A detailed evaluation will be made of the offeror's proposal, and an on-site inspection or preaward survey may be conducted to provide information to the SSA for use in determining the source(s) in accordance with the above. The following areas corresponding to the order of importance shown in the basis of award above are further broken out in respective factors and subfactors.

Sample language for the LSA factors to be evaluated might be

1. Plan/management
2. Approach and personnel assigned to LSA subtasks
3. Models/algorithms to be used
4. Automated systems (LSA, R&M, RCM, LORA)
5. Functional integration (LSA, R&M, RCM, LORA)
6. Interactive access (LSA, R&M, RCM, LORA)
7. Digital delivery

These factors are of equal importance (or, factors 1, 2, and 3 above are worth twice factors 4 and 5). The LSA factor plan/management is further broken into the following subfactors:

1. Organization
2. Management
3. Subcontractor approach
4. Others

A11.5 Summary

This appendix has presented guidance and example language for section L, source selection planning, and section M. The actual language for these materials should relate to the requirements expressed in section C of the RFP. The language will vary by the type and phase of the specific acquisition. Different strategies for supportability/LSA may be used for section L and section M material. The development of supportability/LSA source selection plan criteria and weights is a key step in the source selection process. The material for sections L and M must be consistent with the overall approach to be used for these sections on the specific procurement.

Supportability/LSA Contract Incentives

This appendix focuses on incentives available today for possible use in conjunction with supportability/ILS/RFP requirements. Specifically, this section discusses

Contractor motivation   (A12.1)

Specific contract incentives available   (A12.2)

Award fee incentives   (A12.3)

IMIP incentives   (A12.4)

VE incentives   (A12.5)

Design-to-cost incentives   (A12.6)

Performance (R&M, RIW, etc.) incentives   (A12.7)

Relating incentives to other considerations   (A12.8)

DOD program management commitment   (A12.9)

Summary   (A12.10)

A12.1   Contractor Motivation

There are a number of potential noncontractual factors that could motivate contractors to implement an aggressive supportability/ LSA effort. Examples of factors that are beyond contract incentives include

1. A contractor's desire to win the contract (or if competition exists)
2. A contractor's desire to modernize and thus enhance competitive position
3. A contractor's desire to "save" the program (if the program is in trouble or is threatened by a competing program which has a similar mission objective)

While some aspects of using the LSA standards may be cost-effective in initial contract application, other aspects will not. Investing in LSA is not always attractive to a contractor for a number of reasons, such as

Inertia

Lack of perceived customer interest

Poor apparent return on invested capital

Reduction in future profits and value of future sales due to reduction of future costs

Program budget/schedule instability due to more general turbulence in DOD budgeting

In some cases, the DOD needs to employ incentives to help overcome such inhibitions. Various contract incentives exist today which may be useful for this purpose.

A12.2   Contract Incentives

The major types of existing contract incentives that could enhance contractor motivation to aggressively implement supportability/ILS/LSA are shown in Table A12.1. The classifications in the table represent normal practice. Many of these incentives could be combined or modified in such a way that the table classifications would change. While each of them is discussed in more detail later in this section, several factors are evident from the table. First, most contract incentives apply to within-scope contract actions. No contract change is necessary because they involve matters entirely within the scope of the contract. Normally, only Value Engineering (VE) incentives require a change to the contract.

Second, most of the incentives are program unique (i.e., they must be specifically tailored to the program on which they are applied). Third, most incentives are subject to statutory profit limits. Only the Industrial Modernization Incentive Program (IMIP) and the VE incentive consider incentive sharing to be independent of statutory profit limits, an attractive feature to contractors. Fourth, only the VE and IMIP incentives allow sharing of savings on future contracts for the

TABLE A12.1   Possible LSA Contract Incentives

Type	Contract change required	Program unique	Profit limitation	Sharing confined to current contract
Award fee	No	Yes; considerable administration	Yes	Yes
Performance incentives	No	Yes	Yes	Yes
IMIP	No	Yes	No	No
VE	Yes	No, standard clauses	No	No
Design-to-cost/LCC	*	Yes	Yes	Yes†

*Can address contract change if clause so states.
†Some design-to-cost incentives have been tailored to affect future profits.

TABLE A12.2   **Possible LSA Incentive Clause Selection by Phase**

Type of incentive	Concept	DEMVAL	FSD	Production
Award fee	X	X		
Performance			X	X
IMIP			X	X
VE			X	X
Design-to-cost/LCC			X	X

same action. This is attractive to contractors because it ensures them a fair share of such savings when the savings on the current contract is insufficient to motivate a contractor to implement the change.

Another key factor in choosing a particular incentive is the acquisition phase. Table A12.2 summarizes this aspect. The table again is intended to show major thrusts rather than absolute limitations, since unique tailoring could change the classifications. The major reasons for the classifications in the table are

1. A major use of award fee is for areas not easily measured quantitatively.

2. Performance specification and/or measurement may be premature or difficult prior to EMD, since R&M, RIW, DTC/LCC, IMIP and VE, for example, depend on quantitative measurement.

To summarize the two tables, award fee appears to be a leading candidate prior to EMD, R&M, RIW, DTC/LCC, VE, and IMIP appear attractive in EMD and production.

A12.3   Award Fee Incentives

A *cost-plus-award-fee contract* is a contract that provides for a fee consisting of (1) a base amount fixed at inception of the contract and (2) an award amount that the contractor may earn in whole or in part during performance and that is sufficient to provide motivation for excellence in such areas as quality, timeliness, technical ingenuity, and cost-effective management. The amount of the award fee to be paid is determined by the government's judgmental evaluation of the contractor's performance in terms of the criteria stated in the contract. This determination is made unilaterally by the government and is not subject to the disputes clause.

The cost-plus-award-fee contract is suitable for use when

1. The work to be performed is such that it is neither feasible nor effective to devise predetermined, objective, incentive targets applicable to cost, technical performance, or schedule.

2. The likelihood of meeting acquisition objectives will be enhanced by using a contract that effectively motivates the contractor toward exceptional performance and provides the government with the flexibility to evaluate both actual performance and the conditions under which it was achieved.

3. The additional administrative effort and cost required to monitor and evaluate performance are justified by the expected benefits.

The criteria and rating plan should motivate the contractor to improve performance in the areas rated, but not at the expense of at least minimum acceptable performance in all other areas.

Award-fee contracts provide for evaluation at stated intervals during performance so that the contractor will periodically be informed of the quality of its performance and the areas in which improvement is expected. Partial payment of the fee corresponds to the evaluation periods. This makes the incentive which the award fee can create effective by inducing the contractor to improve poor performance or to continue good performance. See FAR 16.3013-3 for limitations and FAR 1604.2 for general information.

Examples of management factors that might be used in the award fee evaluation are

1. Quality of LSA plan

2. Program management support

3. Subcontractor management

4. Use of integrated teams

Examples of technical factors that might be used in the award fee evaluation are

1. Degree of design-LSA integration

2. Use of the LSA input to design

3. Use of the LSAR (prior to EMD)

4. Degree of interactive access to LSA products

5. Degree of digital delivery of LSA task reports

To summarize, award fee incentives are frequently used when quantitative measurement is difficult or impractical. This aspect is potentially useful for supportability/LSA incentives prior to EMD. There is generally considerable administrative effort associated with award fee incentives. The reader is cautioned that the criteria selected for this sample may not be appropriate for his or her particular program or program phase.

A12.4   IMIP Incentives

IMIP incentives are described in DOD 5000.44-G, "Industrial Modernization Incentives Program (IMIP)," dated August 1986. The Industrial Modernization Incentives Program's primary purpose is to motivate the private sector to invest in modernization projects to improve productivity and enhance the reliability of products and services purchased by the Department of Defense.

One of the tools available to motivate defense contractor investment is a financial incentive intended to offset the disincentives of cost-based pricing and the profit policy. IMIP incentives apply to those investments a contractor would not make without such an incentive or those which can be accelerated through use of an IMIP incentive. IMIP incentives may be organization- or enterprise-wide rather than limited to a specific program. This is probably good, since some contractor ILS/LSA applications appear to be enterprise-wide. Two other features of IMIP incentives are that savings sharing can apply beyond the current contract and are not subject to normal profit limitations. Contractor investment protection against program instability can be provided through use of a contractor investment clause in conjunction with the IMIP provision. One potential drawback of IMIP is that considerable study and negotiation are required, since the incentive must be tailored to a specific situation. A second potential drawback is difficulty in measuring benefits. This difficulty is due in part to the current structure of accounting systems, which are growing less useful as we shift from a product, labor-intensive mode to a process, capital-intensive mode. This difficulty, however, may not be cause to reject IMIP possibilities. Measurement is sometimes a problem in the VE area, yet DOD is currently approving some 800 VE change proposals worth almost $500 million annually. Thus it appears desirable to work with what can be done now rather than to await better, but long-term solutions to the measurement problem. IMIP incentives have many desirable characteristics. They may be used prior to EMD in selected areas where reasonable measurement is possible. There is significant flexibility in their applications. Since IMIPs are plant-wide, one may already be in place. This could significantly reduce the time required to process a specific proposal.

A12.5   VE Incentives

VE incentives are described in detail in paragraphs 48 and 52-248-1, -2, and -3 of Federal Acquisition Circular 84-42, dated March 2, 1989. There are two kinds of VE clauses: the incentive clause and the program requirement clause. The VE incentive clause is voluntary and is not funded by the government. It applies only to contract changes. VE clause sharing can apply to O&S savings (called *collateral savings*) as well as acquisition savings. Since the VE program requirement clause is funded by the government, it applies to both changes requiring a contract modification and those which do not. Sharing rates for savings on contracts with a program clause are less than the sharing rates on contracts with an incentive clause. The fact that (1) savings can be shared on future contracts as well as the present contract and (2) shared savings are not under the profit limitations makes VE clauses potentially very useful as a vehicle to motivate support-related contract changes. The clauses are easy to apply, since they can be incorporated by reference. Used with IMIP incentives, they can provide contractors with a very attractive incentive package for both in-scope and out-of-scope change.

A12.6   Design-to-Cost/LCC Incentives

Design-to-cost incentives are program unique and are primarily suited to motivating improvement in the manufacturing area, but they may include Life-Cycle Cost (LCC) elements. These incentives can be used to motivate contractors to achieve or better a specified production unit cost. Most adjust the EMD contract profit according to the contractor production unit cost achieved as compared with the production unit cost goal established in the EMD contract. Design-to-cost incentives generally apply only for in-scope changes, but they can be modified to include VE-type changes. Design-to-cost incentives may address O&S costs in some fashion. This could include a surrogate (e.g., R&M) approach or a direct estimate of O&S costs, including field measurement of O&S costs, thus in effect creating an LCC incentive.

A12.7   Performance (R&M, RIW) Incentives

Technical performance incentives are briefly described in FAR 16.402-3 and FAR 16.402-2. They must be tailored to a specific program situation. An example of a technical performance incentive is the achievement of some desired level of R&M. RIWs also could be considered. Program-unique incentives should be well thought out to avoid overlap or other potential problems.

A12.8   Relating Incentives to Other Considerations

A last consideration is the obvious need to consider contract incentives for supportability/LLS/LSA in the context of the other aspects of the program. Other contract factors, such as plans for multiyear procurement, can affect incentives.

A12.9 DOD Management Commitment

A key point in making contract incentives effective is for the requiring program manager to make clear to the contractor his or her commitment for supportability. History has shown that the attitude of requiring program office personnel toward contract incentives can "make or break" their effectiveness, regardless of the quality of the incentive. Without management commitment, incentives are not likely to heavily affect the program.

A12.10 Summary

A number of contract incentives are available today to improve the contractor motivation to emphasize supportability/ILS/LSA requirements. Such contract incentives, in most cases, need to be carefully tailored to the specific program. Top program management commitment is needed for such incentives to be effective.

Suggested LSA Guidance Conference Agenda

A13.1 Conference Preparation

Prior to the date of the guidance conference, the LM has a number of responsibilities that he or she must carry out. Chief among these are the establishment of the government's ILS management team and co-ordination of an agenda with the contractor's LSA manager.

A13.1.1 Establish ILS management team

The LSA process involves so many disciplines that few individuals can be expected to have the required detail knowledge of all its facets. Therefore, the most feasible method for managing an LSA program is by a team effort. This team should consist of ILS element managers, representatives from the various engineering disciplines (R&M, design, configuration management, QA, etc.), representatives from the end-user community, and their contractor counterparts. This team is known as the *ILS management team* (ILSMT) and is headed by the LM. The LSA data that will be developed should be used by, and useful to, all ILS element managers. The involvement of these various disciplines will vary based on the program and its life-cycle phase. The ILSMT should be augmented or reduced accordingly.

Establish review procedures. One of the functions of the government members of the ILSMT is to review the contractor's submittals of LSA and ILS element deliverables. It is each logistics element manager's (LEM) responsibility initially to be familiar with the contract requirements and ensure that necessary ILS deliverables are on contract. For the guidance conference, the LM and his or her LEMs should review the LSA plan if it is available. This may be the case if one was requested in the RFP or the project is a continuing program. In addition, in the case of a continuing program, the ILSMT members should review any previously submitted ILS deliverables. The LM should ensure that future ILS products are distributed in a timely manner to the applicable team members for review and coordinate any government comments to these deliverables. Ideally, there will be a government-only meeting prior to the actual review conferences with the contractor. This allows the government to iron out any questions/ disputes between the various ILS-element managers prior to meeting with the contractor and present a unified consensus to the contractor at the guidance conference.

Review the LSA plan. The LSA plan identifies the contractor's approach to implementing an LSA program. This plan should comply with MIL-STD-1388-1A and DI-ILSS-80531. In particular, the following items should be addressed: (1) the contractor's ILS organization, (2) the interfaces among the various contractor ILS and system engineering disciplines, (3) LSA candidates, and (4) the approach to accomplishing the LSA tasks called out in the contract. If an LSA plan is not available prior to the LSA guidance conference, the contractor should address these elements at the conference.

A13.1.2 Coordinate agenda

The LM should coordinate with the contractor's LSA manager to establish agenda items and responsibilities for the guidance conference. In addition, the LM also should request agenda items from his or her LEMs.

A13.2 Conference Agenda

The LM cochairs the guidance conference with the contractor's LSA manager. Some of the agenda items discussed below will be the responsibility of the government and some belong to the contractor. The responsibilities assigned below are only suggestions. Factors such as life-cycle phase of the program, and contractor/government knowledge of the program and/or ILS/LSA should play an important role in assigning responsibilities for each topic and/or tailoring this list to the specific project.

A13.2.1 Government responsibility

The topics in this section have been given government responsibility because they concern program objectives and goals, as well as standardization of LSA documentation. The information given under each topic heading is provided to assist the LM in identifying the type of data that should be included with each agenda item. These are not meant to be verbatim sample presentations, since each program will be unique.

A13.2.2 LSA program objectives

Both ILS and LSA have the objectives of improving supportability, reducing costs, and increasing system effectiveness. In its simplest form, the LSA process is a set of related tasks and subtasks performed to achieve the LSA objectives. These tasks are designed to help the logistics analyst accomplish these goals by

1. Establishing supportability design requirements
2. Developing viable support concepts and support system alternatives
3. Evaluating design/support/operational concept alternatives
4. Identifying detailed logistics resource requirements that satisfy readiness requirements
5. Verifying achievement of supportability requirements

LSA is involved in each phase of the acquisition cycle. The ILS development goals for each phase are more detailed and dependent on the results of the previous analyses. Typical ILS/LSA products for each phase are

1. *Concept exploration:* This includes readiness and cost improvement targets, support concept alternatives, and supportability-related design and support-system objectives.
2. *Demonstration and validation:* This includes a firm support concept, firm supportability-related design goals and thresholds, and readiness and support-system parameter objectives.
3. *Engineering and manufacturing development:* This phase is heavily influenced by maintenance planning to identify detailed logistics support resources and to correct deficiencies discovered during testing. Identification of detailed corrective and preventive maintenance and other tasks is essential. Logistics element analyses are conducted, and firm maintenance plans are developed. Development of provisioning lists is done so that support items may be ordered with production.
4. *Production and deployment:* This involves continued assessment of supportability and readiness, improvement of problem areas, and planning for postproduction support.

In actuality, each acquisition program is unique, and its ILS/LSA program requirements may deviate from the preceding examples. This is particularly true in the case of accelerated acquisitions or programs which are initiated in other than the concept exploration phase; most support equipment acquisition is a good example of these types of acquisitions. The LM must be aware of these differences and explain the goals for the specific program at the guidance conference so that all members of the ILSMT (government and contractor) review/perform the LSA tasks with the same perspective.

14

LSA Task Input Organizations

This appendix lists suggested organizations to provide the task inputs (e.g., GFI) listed in the LSA standard. See paragraph 3.4.3 of the standard.

ILS Input Organizations

Task input no.	Brief description	Design interface	Maintenance	Supply	Manpower, personnel, training	Support equipment	PHS&T	Tech. data	Computer resources	Facilities
101	Development of an Early LSA Strategy									
101.3.2	Support Constraints	X	X	X	X	X	X		X	X
101.3.3	Data Base Availability	X	X	X	X	X	X	X	X	X
102	LSAP									
103	Program and Design Reviews									
201	Use Study									
201.3.4	Available Source Documentation	X	X	X	X	X	X	X	X	X
202	Mission Hardware/Software and Support System Standardization									
202.3.1	Supportability Constraints	X		X		X	X		X	
202.3.2	Planned Logistic Support	X	X	X	X	X	X		X	
202.3.3	Standardization Requirements			X		X	X			X
203	Comparative Analysis									
204	Technological Opportunities									
204.3.2	Technology Evaluations and Improvement: Information	X	X			X			X	
205	Supportability and Supportability-Related Design Factors									
205.3.1	Program Documentation	X	X	X	X	X	X		X	
301	Functional Requirements Ident.	X								
301.3.2	Detailed RCM Procedures		X	X	X	X				
301.3.3	System/equipment: Indenture Levels		X		X	X			X	
301.3.4	Levels of Maintenance		X		X	X			X	
301.3.5	Additional Documentation		X		X				X	
302	Support System Alternatives									
303	Evaluation of Alternatives and Tradeoff Analysis									
303.3.5	Constraints: Personnel				X					
303.3.6	Manpower/Personnel: Factors				X					
303.3.11	Personnel Classification: Task inventory		X							

Code	Task									
401	Task Analysis						X	X	X	X
401.3.1	System/Equipment: Identification						X	X	X	X
401.3.2	Indenture Levels: Identification		X		X	X	X			
401.3.3	Levels of Maintenance: Identification					X	X			
401.3.4	Logistics Support Resources: Shortages	X	X	X	X	X	X			
401.3.6	Supplementation Documentation	X	X	X	X	X	X			
401.3.8A	Personnel Training: Existing				X					
401.3.8B	Standard Support and Test Equipment Lists					X				
401.3.8C	Facilities: Available	X								
401.3.8D	Training Devices: Available						X			
401.3.8E	Transportation: Existing			X						
401.3.9	Personnel Capabilities: Required					X	X			
401.3.10	Personnel: Limits					X	X			
401.3.11	Task Frequencies: Annual		X		X	X				
402	Early Fielding Analysis				X					
402.3.2A	Manpower/Personnel Skills: Sources					X				
403	Postproduction Support Analysis						X			
403.3.1A	Sources of Supply					X				
403.3.1B	Lifetime of System/Equipment	X	X		X	X	X			
403.3.1C	R&M Data	X	X		X	X	X	X		
403.3.1D	Repair Alternatives: Costs	X	X		X	X	X	X	X	X
403.3.3	Supply Consumption	X			X					
501	Supportability Test Evaluation and Verification									
501.3.2.	Standard Reporting System	X	X	X	X	X	X	X	X	X
501.3.8	Supportability Plan	X	X	X	X	X	X	X	X	X

445

LSA Task/Subtask Checklist

This appendix provides checklists for the LSA tasks and subtasks.

The checklist may be used for two purposes by the APML and/or the ILS/LSA management team. The first purpose is a review to aid in the RFP tailoring process. Examine each subtask definition and determine (1) if each is appropriate for the acquisition, (2) if so, where in the CDRL is the requirement contractually imposed and satisfied, and (3) if required but satisfied from a source other than the LSA program, what is the source of the satisfaction?

The second purpose of the checklist is to aid in reviewing LSA products as the LSA program progresses. Some LSA tasks and subtasks may be satisfied from sources other than the instant program.

Use of MIL-STD-1388-1A, "LSA," in contracts is emphasizing the need for review of these contractor products in associated disciplines. A review methodology was developed. The basic review approach developed consists of a review of the SOWs, DIDs, CDRLs, and other appropriate governing documents using the attached LSA task review checklists.

The questions on the checklists are derived from the subtasks in MIL-STD-1388-1A. Use of the checklist simply facilitates review and aids in thoroughness.

In practice, it has been found useful to (1) modify the checklist based on the SOW and other governing documents, (2) review the LSA tasks, making notes as the reviewer proceeds, (3) fill out the checklist to ensure thoroughness and accuracy (write in the paragraph number of the document satisfying the requirements), (4) summarize findings, classifying omissions as major or minor, and (5) prepare appropriate action documentation.

As a final reminder, the purpose of the task should be kept in mind, and the task should be reviewed from the perspective of achieving that general goal as well as satisfying the individual detailed requirements of each subtask (e.g., judgment should be used in assessing the impact of deficiencies).

Project:
Task 101:   Development of an early LSA strategy: Checklist (MIL-STD-1388-1A)
Purpose: Develop a proposed LSA program strategy for use early in an acquisition program; identify LSA tasks and subtasks which provide the best return on investment.

Date:

Task/subtask	SOW/CDRL impact	Satisfied by	Comments
101.2.1 LSA strategy			
(a) Potential supportability objectives identified?			
(b) Cost of performing tasks and subtasks estimated?			
(c) Probable design, maintenance concept, and operational approaches identified?			
(d) Availability of data to LSA subtasks discussed?			
(e) Potential impact of performing LSA tasks discussed?			
(f) Cost-effectiveness of each subtask estimated?			
(g) Availability of data to LSA subtasks discussed?			
(h) Potential impact of performing LSA tasks discussed?			
(i) Cost effectiveness of each subtask estimated?			
(j) Are task inputs identified?			
101.2.2 Cost evaluation			
Estimate cost-effectiveness of performing each task/subtask under 101.2.1.			
101.2.3 Updates as required.			
101.3 Are inputs identified?			

Project:
Date:

Task 102: LSA plan: Checklist (MIL–STD–1388–1A)

Purpose: Develop an LSAP to identify and integrate all LSA tasks, identify management responsibilities, and outline the approach toward accomplishing analysis tasks.

Task/subtask	SOW/CDRL impact	Satisfied by	Comments
102.2.1 LSA plan			
(a) Description of how the program meets system/logistics requirements?			
(b) Management structure; organization interrelationships?			
Line			
Staff			
Policy			
(c) Identification of 1A subtasks to be performed; *how* each will be performed; major tradeoffs under 303.2.3, if applicable?			
(d) Estimated start and completion date of each task (subtask?) and LSA program activity?			
(e) LSA tasks and data interfaces, as applicable:			
(1) System/equipment design program			
(2) System/equipment reliability program			
(3) System/equipment maintainability program			
(4) System/equipment HE program			
(5) System/equipment standardization program			
(6) System/equipment parts control program			
(7) System/equipment safety program			
(8) System/equipment PHS&T program			
(9) System/equipment initial provisioning			
(10) System/equipment testability program			
(11) System/equipment survivability program			
(12) System equipment technical publication program			

449

Project:

Task 102: LSA plan: Checklist (MIL-STD-1388-1A)

Purpose: Develop an LSAP to identify and integrate all LSA tasks, identify management responsibilities, and outline the approach toward accomplishing analysis tasks.

Date:

Task/subtask	SOW/CDRL impact	Satisfied by	Comments
(13) System/equipment training and equipment program			
(14) System/equipment facilities program			
(15) System/equipment support equipment program			
(16) System/equipment T&E program			
(f) Identification of WBS items to have LSAR performed?			
(g) LSACN structure explanation?			
(h) Method for dissemination of supportability-related requirements to designers and other personnel?			
(i) Methods for dissemination of supportability-related requirements and controls on subcontractors?			
(j) Government data to be furnished to contractor.			
(k) Procedures for configuration control and updating and validating LSAR data?			
(l) LSA requirements on GFE/GFM and subcontractors/vendors?			
(m) Procedures to evaluate status and control of each task (subtask?) and organization responsible for each task?			
(n) Procedures, methods, and controls for identifying/recording design problems affecting supportability, corrective actions required, and status of action to resolve problem?			
(o) Description of data collecting system used to document, disseminate, and control LSA and related design data?			
(p) Description of LSAR ADP system to be used and validation status.			
102.2.2 LSA plan updates?			
102.3 Are all these inputs identified?			

Project:

Task 103: Program and design reviews: Checklist (MIL-STD-1388-1A)

Purpose: Establish a requirement to provide for official review of released design information with LSA program participation; ensure that LSA program is proceeding in accordance with contractual milestones.

Date:

Task/subtask	SOW/CDRL impact	Satisfied by	Comments
103.2.1 Review procedures identified			
103.2.2 Is status of LSA on all design review agendas?			
(a) LSA by task and WBS?			
(b) Assessment of drivers?			
(c) Corrective action considered, proposed, or taken?			
(d) Review of requirements?			
(e) Progress toward goals?			
(f) LSA documentation?			
(g) Problems impacting support?			
(h) Identify supportability design recommendations and rationale for selection.			
(i) Other topics as applicable.			
103.2.3 Is LSA status on program review agendas?			
103.2.4 Are results of LSA reviews documented?			
103.3 Are all inputs identified?			

Project:
Task 201: Use study: Checklist (MIL-STD-1388-1A)
Purpose: Identify and document the pertinent supportability factors related to the intended use of the new system/equipment.

Date:

Task/subtask	SOW/CDRL impact	Satisfied by	Comments
201.2.1 Supportability factors identified and documented (such as):			
(a) Mobility requirements			
(b) Deployment scenarios			
(c) Mission frequency and duration			
(d) Basing concepts			
(e) Service life			
(f) Interactions with other systems			
(g) Human capabilities/limitations			
201.2.2 Document quantitative data (such as):			
(a) Operating requirements			
(b) Missions/time period			
(c) Mission duration			
(d) Number of operating days, _____ , etc.			
(e) Number of units supported			
(f) Transportation factors			
(g) Allowable maintenance periods			
(h) Environmental requirements, hazardous materials/waste			
201.2.3 Field visits			
201.2.4 Use study documentation updates			
201.3 Identify inputs			
(a) Mission and use information			
(b) Locations			
(c) Type units			
(d) Depot locations			
(e) Etc.			

Project:

Task 202: Mission hardware, software, and support system standardization: Checklist (MIL-STD-1388-1A)

Purpose: Define supportability and supportability-related design constraints based on existing and planned logistics support resources; provide supportability input into standardization efforts.

Date:

	Task/subtask	SOW/CDRL impact	Satisfied by	Comments
202.2.1	Quantitative supportability-related design constraints based on support standardization considerations?			
202.2.2	Supportability, cost, and readiness characteristics of mission hardware and software standardization approaches?			
202.2.3	Standardization approaches which have utility due to cost, readiness, or supportability?			
202.2.4	Risk associated with each constraint?			
202.2.5	Are all inputs identified?			

453

Project:

Task 203: Comparative analysis review: Checklist (MIL-STD-1388-1A)

Purpose: Select or develop a BCS representing the new system/equipment for (1) projecting/making judgments concerning the feasibility of supportability parameters and identifying targets for improvement and (2) determining supportability, cost, and readiness drivers.

Date:

	Task/subtask	SOW/CDRL impact	Satisfied by	Comments
203.2.1	Are the existing system/subsystems identified appropriate for comparison to new system alternatives?			
203.2.2	Is there a BCS for each significantly different new system/equipment alternative?			
203.2.3	(a) Are the values established at system and subsystem level (if appropriate)?			
	(b) Are O&S costs, ILS resource requirements, R&M, and readiness values addressed?			
	(c) Are adjustments made for differences between the old and new system?			
203.2.4	Have qualitative support problems on comparable systems/equipment been identified?			
203.2.5	Have the supportability, cost, and readiness drivers of comparative systems or BCS been identified?			
203.2.6	Have "drivers" for new system/equipment features with no comparable basis in the comparative systems been identified?			
203.2.7	Have all germane subtasks affected by change necessitating update been addressed?			
203.2.8	Have risks (uncertainty) or assumptions for 203.2.4, 203.2.5, and 203.2.6 been identified?			
203.3	Are all inputs identified?			

Project:
Date:
Task 204: Technological opportunities: Checklist MIL-STD-1388-1A)
Purpose: To identify and evaluate design opportunities for improvement of supportability/characteristics and requirements in the new system/equipment.

Task/subtask	SOW/CDRL impact	Satisfied by	Comments
204.2.1 Establish design technology approaches as follows:			
(a) Identify potential design improvements			
(b) Estimate impacts of each			
(c) Identify design to be applied			
204.2.2 Update design objectives			
204.2.3 Identify risks			
204.3 Are all inputs identified?			

Project:

Task 205: Supportability and supportability-related design factors: Checklist (MIL-STD-1388-1A)

Purpose: Establish (1) quantitative supportability characteristics from alternative design and operation concepts and (2) supportability and supportability-related design parameters for inclusion in appropriate program documents.

Date:

Task/subtask	SOW/CDRL impact	Satisfied by	Comments
205.2.1 Supportability characteristics			
(a) Are quantitative supportability characteristics identified?			
(b) Are they expressed in terms of feasible support concepts, R&M parameters, system readiness, O&S cost, and logistics support resource requirements?			
(c) Peacetime and wartime?			
205.2.2 Sensitivity analysis			
(a) Sensitivity analysis on drivers performed?			
205.2.3 Identify proprietary data			
(a) Data rights and impact thereof identified? Include alternatives.			
205.2.4 Supportability objectives			
(a) Objectives established?			
(b) Risks of achievement addressed?			
(c) Supportability risks with new technology identified?			
205.2.5 Qualitative and quantitative design constraints identified? Recommendations for appropriate documents for inclusion? Are they documented in LSAR?			
205.2.6 NATO constraints identified?			
205.2.7 Objectives updated? Goals established? Where placed?			
205.3 Are all inputs identified?			

Project:
Date:

Task 301: Functional requirements identification: Checklist (MIL-STD-1388-1A)

Purpose: Identify the operations and support functions to be performed for each system/equipment alternative under consideration; then identify tasks to be performed in order to operate and maintain the new system/equipment in its intended environment.

	Task/subtask	SOW/CDRL impact	Satisfied by	Comments
301.2.1	Are operational and maintenance functions identified and documented? Peacetime and wartime?			
301.2.2	Are unique or driver functions identified?			
301.2.3	Risk (uncertainty) assessment?			
301.2.4	Corrective maintenance tasks identified? Preventive maintenance tasks identified? Non-PM, CM tasks identified? Are they documented in LSAR?			
301.2.5	Based on other subtasks, are design deficiencies identified? Alternatives proposed?			
301.2.6	Updates. All impacts included?			
301.3	Are all inputs identified?			

457

Project:

Date:

Task 302: Support system alternatives: Checklist (MIL-STD-1388-1A)

Purpose: Establish support-system alternatives for evaluation, tradeoff analysis, and determination of best system for development.

Task/subtask	SOW/CDRL impact	Satisfied by	Comments
302.2.1 System alternatives			
(a) Are viable alternative system-level support concepts identified?			
(b) Are all ILS elements addressed?			
(c) Are innovative (nonstandard) approaches identified?			
(d) Is contractor support (interim, part, or total) considered?			
302.2.2 Updates			
(a) To subsystem level?			
(b) Are drivers and unique functional requirements identified?			
302.2.3 Alternative support plans			
(a) Design and document viable support plans			
(b) Are support plans commensurate with level of detail?			
302.2.4 Updates. Do they address all pertinent impacts?			
302.2.5 Are risks of such alternatives addressed?			
302.3 Are all inputs identified?			

Project:

Task 303: Evaluation of alternatives and tradeoff analysis: Checklist (MIL-STD-1388-1A)

Purpose: Determine the preferred support alternative(s) for each system/equipment alternative; participate in alternative system tradeoffs to determine the best approach which satisfied the need with the best balance among drivers.

Date:

Task/subtask	SOW/CDRL impact	Satisfied by	Comments
303.2.1 For each tradeoff:			
(a) Criteria identified?			
(b) Models, etc. identified?			
(c) Boundary conditions identified?			
(d) Weighting factors identified?			
(e) Sensitivity analysis performed?			
(f) Risks and assumptions identified and documented?			
(g) Peacetime and wartime?			
(h) Impact on existing systems addressed? (See task 402)			
(i) Life-cycle support including postproduction support considerations addressed? (See task 403)			
303.2.2 (a) Support alternative tradeoffs and evaluations conducted? (See task 302)			
(b) New or critical support resources identified (including restructured personnel job classes) and documented?			
303.2.3 Design, operational, and support tradeoffs			
303.2.4 Readiness sensitivity conducted?			

459

Project:

Task 303: Evaluation of alternatives and tradeoff analysis: Checklist (MIL-STD-1388-1A)

Purpose: Determine the preferred support alternative(s) for each system/equipment alternative; participate in alternative system tradeoffs to determine the best approach which satisfied the need with the best balance among drivers.

Date:

Task/subtask	SOW/CDRL impact	Satisfied by	Comments
303.2.5 Manpower implications of alternative mission-system concepts evaluated.			
(a) Numbers			
(b) Job classifications			
(c) Skill levels			
(d) Experience requirements			
(e) Overhold			
(f) Error rates			
(g) Training requirements			
303.2.6 Training, tradeoff conducted?			
(a) Design, operations, training, job design			
(b) Job shifting			
(c) Different tech. pub. concepts			
(d) Mixes of formal training, OJT, unit training, simulators			

460

Project:

Task 401: Task analysis: Checklist (MIL-STD-1388-1A)

Purpose: Analyze required O&M tasks to identify (1) logistics support resource requirements, (2) new or critical logistics support requirements, (3) transportability requirements, and (4) support requirements that exceed established parameters, and to provide (1) data to support participating in development of design alternatives and (2) source data for preoperation of ILS documents.

Date:

Task/subtask	SOW/CDRL impact	Satisfied by	Comments
401.2.1 Task analysis conducted for each operation and maintenance task requirement?			
401.2.2 Entered in LSAR?			
401.2.3 New or critical resources identified? (Note: Also in Task 303)			
401.2.4 Training requirements and best training mode identified and documented?			
401.2.5 Design improvements			
(a) Have total logistics resource requirements for each task been identified and those failing to met goals flagged?			
(b) Have tasks which might be improved been identified?			
(c) Are design improvements to (b) under way/identified?			
401.2.6 What management actions have been identified to minimize risk on new or critical logistic support resources?			
401.2.7 Transportability analysis			
(a) Has a transportability analysis been conducted?			
(b) If MIL-STD-1366 limits are exceeded, has an LSAR record been prepared?			
(c) If (b), are design alternatives under consideration?			
401.2.8 Initial provisioning documented in LSAR?			
401.2.9 Has key information in LSAR been validated by performance of O&M tasks on prototypes? Was the information in 401.2.1 used in these evaluations?			
401.2.10 Have required LSAR output reports been prepared?			
401.2.11 Has the LSAR been updated?			
401.3 Are all inputs identified?			

Project:

Task 402: Early fielding analysis: Checklist (MIL-STD-1388-1A)

Purpose: Assess the impact of new system/equipment on existing systems; identify source of M&P; determine impact of failure to obtain the necessary logistic support resources; determine essential resource requirements for combat environment.

Task/subtask	SOW/CDRL impact	Satisfied by	Comments
402.2.1 (See 301.2.1) Has impact on existing mission and support systems been addressed?			
(a) Depot workload and schedule			
(b) Provisioning and inventory			
(c) ATE availability and capability			
(d) Manpower and personnel factors			
(e) Training programs and requirements			
(f) POL requirements			
(g) Transportation systems			
(h) Changes to support existing mission systems			
402.2.2 Have existing M&P sources been analyzed to determine sources of M&P? Will existing systems be impacted?			
402.2.3 Has the impact on system readiness due to failure to obtain required logistics support in quantities specified been assessed (do not duplicate 303)?			
402.2.4 (a) Have survivability analyses to determine combat usage impact on logistics support resources been conducted? Do they include:			
(1) Threat assessment			
(2) Combat scenarios			
(3) System/equipment vulnerability			
(4) Battle damage repair capabilities			
(5) Component essentiality in combat			
(b) Have recommended combat logistics support resources (supply storage lists for combat) and sources been identified and documented?			
402.2.5 Have plans been developed to address problems identified above?			
402.3 Are all inputs identified?			

Project:

Task 403: Postproduction support analysis: Checklist (MIL-STD-1388-1A)

Purpose: Analyze life-cycle support requirements prior to closing of production lines to ensure that adequate logistics support resources will be available during the remaining life.

Date:

Task/subtask	SOW/CDRL impact	Satisfied by	Comments
403.2 (a) Assessment of expected useful life?			
(b) Have support items that present problems due to inadequate source of supply after production shutdown been assessed?			
(c) Have alternatives for solution to problems in (b) been analyzed?			
(d) Plan with funding requirements for remaining program life?			
(e) Does plan address:			
(1) Manufacturing			
(2) Repair centers			
(3) Data modifications			
(4) Supply management			
(5) Configuration management			
403.3 Are all inputs identified?			

Project:
Date:

Task 501: Supportability test, evaluation, and verification: Checklist (MIL-STD-1388-1A)

Purpose: Assess achievement of specified supportability requirements; identify reasons for deviations from projections; identify methods of correcting deficiencies and enhancing system readiness.

Task/subtask	SOW/CDRL impact	Satisfied by	Comments
501.2.1 Test and evaluation strategy			
(a) Has a T&E supportability strategy been developed and included in system T&E plans?			
(b) Is it based on:			
(1) Supportability requirements?			
(2) Supportability drivers?			
(3) Supportability issues with high risk?			
(c) Is test length (quantities) cost and statistical risk assessed?			
(d) Have limitations and their impact on accuracy been identified?			
501.2.2 Develop a system support package to include:			
(a) Test requirements			
(b) Maint. allocation chart			
(c) Tech. pubs.			
(d) Spares			
(e) Training devices			
(f) Special common tools			
(g) TMPE			
(h) Manpower			
(i) Training courses			
(j) Transportation handling equipment			
(k) Calibration procedures/equipment			
(l) Mobile and fixed facilities			
(m) Embedded software			
(n) Other support equipment			

464

501.2.3 Objectives and criteria					
(a) Have T&E results been assessed?					
(b) Has the extent of improvement to meet goals been identified?					
(c) Are corrections to problems identified?					
(d) Has the support plan and LSAR been updated based on results?					
(e) Have the efforts of these updates on supportability parameters been quantified?					
501.2.4 Updates					
(a) Have standard data systems been analyzed?					
(b) Have shortfalls in measuring test objectives been identified?					
(c) Has a viable plan for collecting field supportability data been developed and documented?					
501.2.5 Supportability assessment					
(a) Is supportability data from reporting systems being analyzed?					
(b) Are goals being met?					
(c) Have causes for deviations and corrective actions been identified? Was cost considered in this analysis?					
(d) Are recommended improvements documented?					
501.2.6 Analyze feedback and identify areas for improvements					
501.2.7 System support package component list required?					
501.3 Are all inputs identified?					
Charts:					
(a) ILSDS-LSA relationship					
(b) Six ways to influence design					
(c) Cases considered. Let them prioritize					
(d) Charts:					
(1) Design					
(2) Support system					
(3) Logistic requirements					
(e) Charts: Strip down MIL-STD table on tailoring					
(f) Review checklists					
(g) Pyramid					

16

LSAR Review

The purpose of an LSAR review is twofold. First, the data must be entered correctly and in a timely manner. Second, the reviewer must use these data to identify support problems with the equipment. Following is a questionnaire that may be useful to the reviewer in determining the adequacy of the LSAR. This questionnaire is very generic and will not cover all the nuances of a specific program. Therefore, it should only serve as a guide, and each program should expand on it according to its need. Also, the point of the LSAR review is not just to determine if all the data elements have been filled in but also to determine whether the item can be supported effectively.

A16.1   Sample LSAR Review Checklist

	Yes	No

Cross-Functional (X) Tables

1. Has the government-furnished data from the DD form 1949-3 been entered into the XA, XB, and XC tables? ☐ ☐
2. Is the LCN structure correctly documented in the XB table? ☐ ☐
3. If there is both a functional and a physical LCN, have they been correctly mapped in the XG table? ☐ ☐
4. Has all the information been entered into the XH table for all the CAGE codes in this program? ☐ ☐

Operational and Maintenance Requirements (A) Tables

1. Has the government-furnished data from the DD form 1949-3 been entered into the AA, AB, AC, AD, AE, AG, AH, AI, AJ, and AK tables? ☐ ☐
2. Are the requirements being satisfied? LSA-003, "Maintenance Summary," and LSA-023, "Maintenance Plan Summary," are useful to identify the roll-ups of these data? ☐ ☐
3. Are the data in the XA and AI tables being used in the LORA and other models? ☐ ☐

Item Reliability, Availability, and Maintainability Characteristics; Failure Modes, Effects, and Criticality Analysis; and, Maintainability Analysis (B) Tables

1. Has a FMECA been conducted? If yes, use LSA-056, "FMECA Report," to answer the following: ☐ ☐
   a. Have all failure modes been identified and completely documented (i.e., cause, effect, detection method, predictability)? ☐ ☐
   b. Do failure modes identify the way in which the item under analysis can fail (e.g., a resistor can fail to "open" or it can short out)? ☐ ☐
   c. Do the failure symptoms provide a means for identifying that an item has failed (e.g., when a capacitor fails there is no output signal)? ☐ ☐
   d. Do the failure rates roll up to the next higher assembly? Do the failure mode ratios add to 1.00? ☐ ☐
   e. Has the safety hazard severity code been determined correctly? ☐ ☐
   f. Does the item function information sufficiently define the item's function? ☐ ☐
2. Using LSA-58, "Reliability and Maintainability Analysis," answer the following:
   a. Have all the logistics concerns been considered? If there are any comments in the "RAM Logistics Considerations Narrative," have they been resolved? ☐ ☐
   b. Have maintenance tasks been identified to correct each failure mode? Has an estimated MTTR been entered? ☐ ☐
3. Using LSA-050, "Reliability-Centered Maintenance Summary," answer the following:
   a. Has the RCM logic been identified? ☐ ☐
   b. Has an RCM analysis been conducted for all failure modes with an SHSC of 1 or 2? ☐ ☐
   c. If a preventive maintenance task is called for by the RCM analysis, has it been identified? ☐ ☐
   d. Are the total manhours for preventive maintenance within the constraints of the A table requirements?
4. Have the MTBF and MTTR, as a minimum, been entered for each LCN? ☐ ☐
5. If the MTBF is different from the MTBMA, is a preventive maintenance task documented or is there some other explanation (e.g., redundancy)?

Yes   No

Task Inventory, Task Analysis, Personnel and Support Requirements (C) Tables

1. LSA-019, "Task Analysis Summary," and LSA-023, "Maintenance Plan," may be useful to answering the following questions:

   a. Do the task codes reflect the recommended maintenance concept documented in the BB tables?

   b. Has the task frequency been entered? Has it been computed correctly? Does the measurement base reflect the basis for the task frequency computation?

   c. If the facility recommended requirement code (FTRNRQCA) is Y, have the facility (F) tables been addressed?

   d. Does the tool/support equipment requirement code (TSEREQCA) accurately reflect the types of tools called out in the task narrative?

   e. If the training equipment requirement code (TRNRQCCA) is Y, have corresponding support equipment and training materiel requirements been developed?

   f. Has a hazardous maintenance procedure code (HAZMPCCA) been entered for each task?

   g. Does the sequential task narrative adequately describe the task to be performed?

   h. Has the LSA manager established a "model" narrative that is followed for all corrective maintenance? Does the narrative have a logical beginning and end (e.g., does it start with a fault-locate/isolate procedure and end with a test)?

   i. If other tasks are referenced, as they at the same maintenance level?

   j. Are the times to perform each subtask of the task being accounted for properly? Do the sums of the mean man-minutes and mean elapsed minutes for the subtasks equal the mean manhours and mean elapsed time, respectively, for the task?

   k. If the hazardous maintenance procedure code (HMPC) is other than D, is there an appropriate warning or caution included in the narrative? If there is a warning or caution, does the HMPC reflect it?

   m. Has the correct skill specialty been assigned to perform the task?

   n. Are all the support items required to perform the task listed? Do they have the proper item category codes (ICC) assigned? Are all the support items listed used in the task?

   o. Does the quantity per task correctly reflect the number of support items needed to perform the task?

Support Equipment and Training Materiel Requirements (E) Tables

1. Has each new or unique item of support equipment been documented on LSA-070, "SERD," or LSA-071, "Support Equipment Candidate List"?

2. If there are calibration requirements documented in the task narratives, does LSA-076, "CMRS," adequately identify the UUT parameters that need to be measured?

Facilities Considerations (F) Tables

1. If a task has requirements for a new or modified facility, are those requirements documented on LSA-012, "Facility Requirements"?

Personnel Skill Considerations (G) Tables

1. Have tasks that require new or modified skills been identified? If so, have the requirements for that skill (e.g., mental, physical, and educational qualifications) been documented? LSA-075, "Manpower, Personnel, and Training," is a useful report to use to review these items.

	Yes	No

Packaging and Provisioning Requirements (H) Tables

1. LSA-008, "Support Items Validation Summary," and LSA-009, "Support Items Validation," ☐ ☐
   can be useful in answering the following questions:
   a. Have reference numbers and CAGE codes been entered for each support item? LSA-008 ☐ ☐
      lists the support items from the CG and CI tables, while LSA-009 obtains these data
      from the H tables.
   b. Has a correct item name and ICC code been entered for each item? ☐ ☐
   c. Does the SMR code correspond to the maintenance concept? ☐ ☐
2. Using LSA-027, "Failure/Maintenance Rate Summary," determine whether the MRRI and ☐ ☐
   MRRII have been developed correctly.
3. Do LSA-025, "Packaging Requirements Data," and LSA-026, "Packaging Developmental ☐ ☐
   Data," accurately reflect the packaging requirements?
4. Is LSA-030, "Illustrated Parts Breakdown," adequate to support the technical manual de- ☐ ☐
   velopment?

Transportability Engineering Analysis (J) Tables

1. Does LSA-085, "Transportability," identify all end items and/or major components that ☐ ☐
   have transportability requirements? Does it document those requirements?

Recommendations

After reviewing this LSAR, are there any design or support recommendations that should be investigated further?

_____

_____

_____

_____

_____

_____

_____

_____

_____

_____

_____

_____

_____

_____

Estimating the Costs of the LSA Tasks

The tables which follow are intended to give the ILS manager a general idea of the relative cost of the different LSA tasks. Tables A17.1 through A17.4 portray this data for 25 and 200 LCN candidates for electrical and mechanical systems. Table A17.5 contains the relative costs of the task 401 (Task Analysis) subtasks, since task 401 is usually roughly 50 percent or more of the entire cost of LSA. The data is extracted from MRSAP 700-11, Cost Estimating Methodology for LSA (CELSA). The ILS manager can construct similar tables for different numbers of LCNs from the CELSA if this is desirable.

TABLE A17.1   Relative LSA Task Cost by Phase (Electrical—25 LCN Candidates)*

Task	Preconcept	CD/V	DEMVAL	EMD	Production
101	0.23	0.03	0.01	0.0+	0.02
102		0.04	0.03	0.1	0.02
103		0.22	0.13	0.12	0.52
201	0.24	0.07	0.04	0.0+	
202		0.07	0.03	0.03	0.08
203	0.54	0.10	0.04	0.01	
204		0.04	0.01	0.01	
205		0.06	0.03	NA	
301		0.09	0.05	0.04	
302		0.05	0.02	0.01	
303		0.20	0.10	0.07	
401			0.52	0.61	
402				0.04	
403					0.11
501		0.03	0.03	0.04	0.24
Total hours	221	2053	3562	3889	741

*Numbers by individual tasks are the percent of total hours. Data taken from the average estimates for the various numbers of LCNs in CELSA.

TABLE A17.2   Relative LSA Task Costs by Phase (Mechanical—25 LCN Candidates)*

Task	Preconcept	CD/V	DEMVAL	EMD	Production
101	0.17	0.03	0.0+	0.0+	0.02
102		0.05	0.03	0.0+	0.02
103		0.18	0.13	0.11	0.55
201	0.25	0.07	0.02	0.0+	
202		0.06	0.03	0.03	0.07
203	0.58	0.12	0.04	0.01	
204		0.04	0.01	0.0+	
205		0.07	0.07	NA	
301		0.11	0.05	0.06	
302		0.06	0.02	0.01	
303		0.20	0.10	0.07	
401			0.46	0.63	
402				0.03	
403					0.09
501		0.02	0.02	0.03	0.24
Total hours	322	2200	4218	4853	

*Numbers by individual tasks are the percent of total hours. Data taken from the average estimates for the various numbers of LCNs in CELSA.

TABLE A17.3   Relative LSA Task Cost by Phase (Electrical—200 LCN Candidates)*

Task	Preconcept	CD/V	DEMVAL	EMD	Production
101	0.14	0.03	0.01	0.0+	0.03
102		0.03	0.02	0.0+	0.03
103		0.19	0.11	0.10	0.50
201	0.24	0.07	0.03	0.0+	
202		0.06	0.03	0.03	0.09
203	0.62	0.10	0.03	0.03	
204		0.04	0.01	0.01	
205		0.07	0.03	0.01	
				NA	
301		0.10	0.06	0.06	
302		0.06	0.03	0.03	
303		0.24	0.10	0.10	
401			0.52	0.63	
402				0.02	
403					0.11
501		0.02	0.02	0.02	0.02
Total hours	640	4265	7640	8388	1388

*Numbers by individual tasks are the percent of total hours. Data taken from the average estimates for the various numbers of LCNs in CELSA.

TABLE A17.4   Relative LSA Task Cost by Phase (Mechanical—200 LCN Candidates)*

Task	Preconcept	CD/V	DEMVAL	EMD	Production
101	0.16	0.03	0.04	0.0+	0.02
102		0.04	0.02	0.0+	0.02
103		0.17	0.08	0.08	0.51
201	0.23	0.07	0.02	0.0+	
202		0.06	0.02	0.02	0.09
203	0.61	0.10	0.04	0.01	
204		0.04	0.01	0.01	
205		0.08	0.03	NA	
301		0.05	0.05	0.05	
302		0.07	0.02	0.01	
303		0.20	0.09	0.07	
401			0.60	0.71	
402				0.03	
403					0.12
501		0.02	0.02	0.02	0.23
Total hours	601	3755	9331	10414	1183

*Numbers by individual tasks are the percent of total hours. Data taken from the average estimates for various numbers of LCNs in CELSA.

TABLE A17.5   Relative Task 401 Subtask Cost

Subtask	Electrical, 25 LCNs		Mechanical, 25 LCNs		Electrical, 200 LCNs		Mechanical, 200 LCNs	
	DV	EMD	DV	EMD	DV	EMD	DV	EMD
401.1	0.31	0.42	0.23	0.44	0.32	0.42	0.28	0.50
401.2	0.12	0.13	0.11	0.12	0.12	0.14	0.12	0.12
401.3	0.03	0.02	0.03	0.02	0.05	0.02	0.03	0.02
401.4	0.04	0.03	0.04	0.03	0.05	0.03	0.04	0.03
401.5	0.04	0.03	0.04	0.02	0.04	0.02	0.04	0.03
401.6	0.02	0.01	0.01	0.01	0.02	0.01	0.01	0.01
401.7	0.02	0.00	0.01	0.00	0.0	0.00	0.01	0.00
401.8	0.17	0.12	0.25	0.15	0.17	0.12	0.21	0.11
401.9	0.05	0.03	0.04	0.02	0.05	0.03	0.04	0.03
401.10	0.08	0.08	0.08	0.07	0.08	0.08	0.08	0.07
401.11	0.12	0.12	0.14	0.12	0.12	0.12	0.13	0.08
Total hours	1850	2385	2139	3042	3942	5322	5594	12600

JVX Comparative Analysis Example

This appendix summarizes the comparative analysis performed on the JVX aircraft. It illustrates the application of subtask 203.2.5, Supportability, Cost, and Readiness Drivers. For the JVX, the BCS was developed for

1. The air vehicle L&M parameters
2. Operating and support costs
3. The following ILS elements:
    a. Support equipments
    b. Spares concept
    c. Maintenance concepts
    d. Publications
    e. MP&T

A18.1 BCS Construction

Candidate aircraft were identified for selection of subsystems to be used in developing a composite air vehicle BCS. Aircraft system configurations were obtained from Design Engineering. These configurations were compared with current Army, Navy, and Air Force aircraft configurations at the system two-digit WUC and subsystem three-digit WUC levels and utilized to construct an air vehicle BCS. The BCS consists of a composite of R&M parameters selected from different existing systems of the several candidate aircraft.

The comparative analysis develops a BCS by identifying existing systems and subsystems (hardware and software, operational and support) that are useful for comparative purposes with new system/equipment alternatives. This is a continuous process from Preliminary Design (PD) stage I and PD stage II through FSD, as the initial air vehicle BCS is expanded and updated with further definition in design. Alternative BCSs, developed for the purpose of comparing different support parameters, are evaluated relative to the current stage of design.

A18.2 Air Vehicle R&M Parameters

Table A18.1 illustrates the air vehicle R&M drivers. The different drivers of a system affect cost and readiness. The implied reliability

TABLE A18.1   Prioritized Unscheduled Maintenance Manhours
(High 10 Maintenance Manhours/FH)

WUC	System	MMH/FH
34	Rotary wing	1.821
	Rotor head	1.310
	Swashplate	0.309
	Rotor blades	0.194
11	Airframe	1.232
	Wing sweep system	0.250
	Nacelles	0.226
	Cabin enclosure	0.197
22	Power plant	0.823
13	Landing gear	0.764
14	Flight control	0.636
	Wing sweep system	0.273
	Actuators (Swashplate servo cyl.)	0.172
46	Fuel	0.554
15	Drive	0.527
	Accessory gearbox	0.367
74	Cockpit management	0.294
	Inertial navigational unit	0.122
	Electro-optical display	0.104
42	Electrical power	0.291
44	Lighting	0.235

and availability have significant effects on mission time, time to repair, and overall cost of maintaining the system.

A18.3   Operating and Support Cost

The life-cycle cost (LCC) BCS provides data to estimate operation and support costs in the following areas: (1) maintenance labor, (2) consumables, (3) depot maintenance, and (4) replenishment spares costs. The LCC BCS subdivides into two categories: (1) reliability and maintainability (R&M) BCS and (2) a cost (repair and spare parts) BCS. This dual BCS is necessary, since cost data are not readily available for all the aircraft systems selected in the R&M BCS. An example of this is the A-10 aircraft, which has been used extensively in the R&M BCS. While R&M data for this aircraft are readily available from AF 66-1 data, costs of repair are not in sufficient detail to estimate O&S costs at the two-digit (or greater) WUC level in either Air Force VAMOSC or K 51 cost data systems. Therefore, Navy-Marine aircraft systems have been substituted in the cost BCS (i.e., S-3A, E-2B/C, CH-53D, etc.) to provide the costs of repairs and maintenance actions.

Areas of support cost, other than direct maintenance, have been estimated at the aircraft level. Specific examples of these are support equipment maintenance, energy consumption, and software support.

A18.4 ILS-Element Comparative Analysis

The air vehicle composite BCS is input to the logistics support resource BCSs (i.e., support equipment, spares concept, maintenance concept, publications, manpower, personnel, training, and LCC).

A18.4.1 Support equipment (SE) BCS

The SE BCS identifies and substantiates the specific SE items most appropriate to each maintenance requirement identified during the LSA process. SE is compiled into a single-baseline list organized by subsystem down to the two-digit WUC level. The SE BCS includes common as well as peculiar items for each subsystem.

Related items in the SE BCS are evaluated against each aircraft maintenance/service requirement identified through the LSA process. The evaluation includes all aspects of use, such as size, weight, ease of application, complexity, skill requirements, durability, need, and performance acceptability, as well as initial cost and cost of ownership, within the limits of such data as secured from item suppliers and/or users. The end product is a list of SE, common and peculiar, determined to be most appropriate for cost-effective, efficient, and timely support and rapid servicing and turnaround. The result of this comparative evaluation is the identification of an existing SE item which best meets the unique support and operational requirements or the decision to design and develop new SE for those requirements which cannot be supported effectively with existing SE.

A18.4.2 Spares concept

Spares engineering has redefined the baseline systems to three-, four-, or five-digit WUC indenture levels as required to identify major repairable components. For each component listed, the following list of provisioning elements and quantitative allowance factors was researched for inclusion to the baseline:

1. Source, maintainability, and recoverability codes
2. Maintenance task distribution
3. Production lead time
4. Estimated unit price
5. Repair turnaround time
6. Maximum allowable operating time
7. Initial outfitting list quantity
8. Attrition quantity

9. Pipeline quantity (ashore versus afloat)

10. Unit weight

11. Unit cube

12. Essentiality code

As engineering design decisions are initiated, the baseline data base becomes a valuable tool in improving supply support concepts. The listing of components is updated periodically to remain consistent with the current stage of design.

A18.4.3   Maintenance concept

A BCS was constructed and used as a means of optimizing the development of the aircraft maintenance concept to interface with existing support systems. A continuing effort expands the BCS to lower indenture levels as the development of the new system becomes more defined. The risks and assumptions associated with the BCS are identified and developed to define the risks associated with the maintenance concept. All qualitative supportability problems recognized in the BCS are used as guidance to eliminate identical problems in the aircraft maintenance concept as it interfaces with existing support systems.

A18.4.4   Technical documentation

The impact of introducing the JVX into the U.S. military inventory, relative to technical publications, can be assessed by a baseline comparison of the publications. Properly conceived, this comparison establishes parallels between the JVX and mature weapons systems in the areas of existing and projected resources and assumed repair levels. It identifies duplication of data and influences the aircraft design, when practicable. This lessens the effect of technical manuals as an LCC driver.

Development of the publications baseline comparison is an iterative process which continues through FSD and into pilot production. Aircraft systems are compared on a WUC level against similar mature systems. A listing of all known technical manuals required to support the appropriate mature systems, subsystems, or components is developed, and a comparative assessment against JVX systems is made to determine the following:

1. The applicability of listed general or GFE data to JVX configuration, including the identification of needed changes or expanded coverage.

2. The correlation of the maintenance concept between the selected mature systems in the BCS and the JVX to determine the need for either similar or different type, level, and scope of coverage.

3. The practicality of implementing design or maintenance concept changes in lieu of developing and implementing awkward or difficult maintenance tasks in technical manuals.

4. The presentation methods of listed mature manuals to qualitatively and quantitatively assess their desirability when developing specific presentation approaches or features for JVX technical manuals.

A18.4.5 Manpower, personnel, and training

A BCS is developed which identifies the personnel grades and skills required to operate and support a system or systems similar to the JVX. The requirement for the manpower to support the JVX is developed and compared with the BCS to determine if existing manpower authorizations, skills, and qualifications are adequate to provide support for the JVX or if changes will be required. Also contained in the developed BCS are present training curriculums evaluated to assist in the development of appropriate JVX curriculums and methods and optimize training of support personnel.

Manpower, personnel, and training is integral to determining what it will take in terms of manpower and the skill level of personnel to keep a system operating in accordance with the mission scenario. Training provides the means to increase skill levels to the point required for adequate performance of tasks related to the specified system and equipment. Some of the factors for consideration are

1. Amount of manhours used in keeping the system operational at peak and off-peak levels.

2. Type of personnel and experience needed to operate and maintain the system.

3. Training required to bring personnel to a level of achieving smooth and efficient operation and maintenance of the system.

Draft RCM Tasks

A19.1   Task 100: Design Influence

101   Purpose. To ensure that RCM philosophies are considered during design, since maintenance cannot correct inadequate reliability but can only optimize the inherent design in amount.

102   Task input. Design considerations, logistics concepts, and technological opportunities are inputs to the design process which reduce the need for extensive maintenance actions as well as facilitate cost-effective RCM-based maintenance procedures.

102.1   Design requirements. The following are examples of design requirements that shall be designed into the equipment:

1. Maintenance ratio

2. Mean preventive maintenance time

3. Damage tolerance

4. Equipment life

5. Mean time between preventive maintenance

6. Availability (inherent and achieved)

102.2   Logistics concepts. The following are examples of maintenance concepts which shall be designed into the equipment:

1. Manhours of preventive maintenance available

2. Skill levels of maintenance available

3. Available inspection tools and techniques

102.3   Technological opportunities. Advancements in current technological opportunities shall be designed into the equipment. Examples of these are

1. State-of-the-art materials

2. Built-in-test/built-in-test-equipment

103   Task description. Design considerations that facilitate preventive maintenance shall be established as follows:

1. Age-limiting characteristics which cause hard-time removals should be designed out of equipment and replaced with potential failure characteristics that facilitate detection before a functional failure occurs.

2. The design should incorporate potential failure mechanisms and graceful degradation. Potential failure indicators such as wear detectors and audible alarms shall be incorporated when possible.

3. Hidden functions which affect safety, mission effectiveness, or costly economics should be designed out of the equipment or made evident to the operators by devices such as warning lights, gages, alarms, or condition indicators.

4. Mission or safety items that fail catastrophically should be designed with redundant systems.

5. Items that require maintenance inspections shall be designed to facilitate maintenance techniques requiring the shortest amount of time. Items such as fluid sampling valves and inspection ports will ease the RCM maintenance concept.

6. Design for fault tolerance.

7. The necessity for using safe-life concepts for structures shall be avoided; damage tolerant and durable designs are preferred.

104 Task output. Recommendations for design improvements/changes that facilitate preventive maintenance.

104.1 Summary report of assessment. A summary report shall be provided addressing all areas in an equipments design phase which are considered in the realm of RCM methodologies.

104.2 Design improvement recommendations. A listing of all proposed design recommendations shall be provided, accompanied by associated risks, tradeoffs, and justifications.

104.3 LSA interface. Means for implementing the proposed design changes shall facilitate incorporation into the LSA process.

A19.2 Task 500: Age Exploration

501 Purpose. To continually update and refine preventive maintenance requirements throughout the life of the equipment.

502 Task input

502.1 Reliability data (life and failure data)

502.2   Maintenance data (from existing and planned inspections/removals)

502.3   RCM analysis results from task 300 which required default answers for determining scheduled maintenance tasks

503   Task description

503.1   Develop an age exploration plan to establish how the age exploration program will be accomplished

503.2   Determine AE candidates.   Do the following:

1. Analyze operational failure data for adverse failure trends as applicable.
2. Require that each significant item that used a default value during the RCM analysis is automatically considered to be an AE candidate as applicable.

503.3   Collect AE data.   Do the following:

1. Use existing data collection processes.
2. Develop separate data collection tasks as required for age exploration.

503.4   Document age exploration candidates, data collection methods, and any tasks developed to collect data

503.5   Screen and prioritize AE tasks

504   Task output

504.1   Age exploration plan.   This plan should contain the following types of information:

1. Sample size
2. Study period
3. Sample location(s)
4. Candidates
5. Sample intervals (continuous or periodic)
6. Costs
7. Documentation
8. Data collection method
9. Data collection tasks (elements)

10. Data collection source(s)

11. Other elements as required

504.2   Age exploration candidates

504.3   Updated or new data to be used in the RCM decision logic task 300 to develop refined preventive maintenance requirements. The refined Planned Maintenance (PM) may include any of the following:

1. Redesign application

2. Cancellation of existing PM tasks

3. Modification of existing PM tasks by maintenance level and/or periodicity

Engineering Judgment
Tradeoff Technique

A20.1 Background

The systematic application of the judgment of a group of experienced personnel for making a selection from alternate problem solutions is an elementary use of statistical procedures. In such cases, years of experience data are manipulated mentally, and applicable statistics such as an estimated mean or evaluation rating result. This approach to decision making is rapid and economical. The accuracy of the approach is a function of the experience and objectivity of the participating personnel.

A20.2 General Approach

In this approach, the most significant parameters involved in a decision are selected and weighted in terms of importance to the final solution. The degree to which each parameter satisfies solution requirements is then determined with regard to some established scale, this value is multiplied by the weight factor, and resulting values are summed. The solution with the most favorable sum is selected.

Table A20.1 contains an engineering judgment type of tradeoff. The purpose of the tradeoff was to select the most effective quick-access splice for an Army missile. A baseline configuration had been established during the validation phase, but further study revealed that the baseline did not meet established time requirements for missile mating operations.

The following basic steps were followed in conducting the tradeoff:

1. Identify the alternatives.

2. Determine evaluation parameters.

3. Determine weight factor assignment.

4. Develop a matrix.

5. Establish quantitative and qualitative values.

6. Summarize results.

TABLE A20.1   Engineering Judgment Tradeoff

1.0   Objective

To select a missile splice method.

1.1   Alternatives

a. Spline splice (baseline)
b. Internal teeth ring assembly (alternative 1)
c. External teeth ring assembly (alternative 2)
d. Internal expansion ring (alternative 3)
e. Vee band (alternative 4)

1.2   Evaluation Parameters

Mating time is a prime consideration. The weight of the missile is critical, since it affects range; therefore, splice weight is also critical. Other factors to be considered are cost, design margin, technical risk, ground support equipment, logistics, and human factors.

1.3   Weight Factors

Weight factors are established for the evaluation parameters as shown in the tradeoff matrix (at end of table). These factors are assigned based on engineering judgment and the mutual agreement of tradeoff participants. Participants represent the functions of system engineering, reliability, maintainability, maintenance engineering, support elements, finance, human factors, and manufacturing. The term *logistics* in the tradeoff matrix represents both maintenance engineering and the support elements. A weight of 10 is assigned to the time and weight parameters. Lesser weights, as shown, are assigned to the other parameters.

1.4   Matrix

The tradeoff matrix shows measurement units for evaluation parameters, when applicable. The "Weight Factor" column contains quantitative, qualitative, and weight factor lines. The quantitative lines contain estimated parameter values for each alternative. The qualitative lines contain scoring values from 0 through 10 for each factor based on the quantitative factor value and any other qualitative considerations. The qualitative values are assigned based on the degree to which the parameter satisfies the requirement, with 10 being the highest possible score. A 0 in any column indicates unacceptability of the candidate with regard to the indicated parameter. Therefore, the candidate total is listed as 0, regardless of the values generated by it and the other parameters. The weight factor lines contain the product of the weight factor and the qualitative scoring value.

1.5   Quantitative and Qualitative Values

Quantitative values for time, weight, cost, and design margin were developed as follows:

a. *Time.*   Times were derived from recorded data and analysis. The time for the baseline was obtained from tests that had been conducted on an existing, similar configuration. Time-line analysis was used for the other candidates. Alternative 3 was abandoned at this point in the tradeoff because it was determined that it was unsatisfactory with regard to time. However, prior to this decision, certain other alternative 3 data had been generated, and are included in the matrix.

b. *Weight.*   Weights were calculated with dimensional and material data obtained from conceptual drawing.

c. *Cost.*   Costs were based on estimated manufacturing and material costs.

d. *Design margin.*   Design margin refers to the design safety factor and was derived from an analysis of material and conceptual design data. Qualitative values for each parameter were developed by appropriate functional groups. For example, human factors, maintainability, and maintenance engineering assigned a value for time, maintenance engineering, and the support elements assigned a factor for logistics, etc. Each person participating in the evaluation of a parameter assigned a qualitative value between 0 and 10, and the sum of the values was averaged and entered into the matrix.

1.6   Summary

Weight factors were multiplied by scoring values, and the results were summed as shown in the tradeoff matrix. Alternative 3 had been eliminated previously because of excessive time requirements. The baseline alternative was eliminated for the same reason. Alternative 4 was eliminated because of excessive time requirements and an unacceptable design margin. The study, therefore, results in the selection of alternative 1 on the basis of a total qualitative factor of 210 as compared to 176 for alternative 2.

Evaluation parameters	Weight factor	Tradeoff Matrix — Alternative candidates				
		Baseline	1	2	3	4
Time, min	Quan.	3 to 5	0.5 to 1	0.5 to 1	—	2 to 4
	Qual.	0	5.0	5.0	—	0
	10	0	50	50	—	0
Weight, lb	Quan.	20.3	17.3	45.6	55.0	37.8
	Qual.	4.9	5.8	2.2	1.8	2.6
	10	49	58	22	18	26
Cost, $/splice	Quan.	1170	1799	1889	1323	1626
	Qual.	8.5	5.5	5.2	7.6	6.2
	5	43	28	26	38	31
Design margin	Quan.	0.23	0.39	0.15	—	0
	Qual.	1.23	1.39	1.15	—	0
	4	5	6	5	—	0
Technical risk	Qual.	3.0	1.0	2.0	0	3.0
	5	15	5	10	—	15
GSE considerations	Qual.	2.0	5.0	5.0	3.0	4.0
	3	6	15	15	9	12
Logistics	Qual.	5.0	7.0	8.0	—	9.0
	3	15	21	24	—	27
Human factors	Qual.	4.0	9.0	8.0	—	6.0
	3	12	27	24	—	18
Totals		0	210	176	—	0

The example tradeoff considered cost as an evaluation parameter; it is, in effect, a cost-effectiveness tradeoff. This type of tradeoff is particularly valuable when only conceptual data are available for the candidates and there is an urgent requirement to narrow the field and concentrate on a limited number of candidates.

Readiness Analysis Examples

This example is adapted from an actual case. The item involved is an electronic warfare (EW) aircraft; its mission is to escort fighter aircraft. The program modified an existing aircraft to accommodate a new design EW subsystem. The primary readiness parameters of interest were A_o for peacetime and sortie rates per day (the number of missions an aircraft can fly per day) for wartime surge and sustained conditions.

A21.1   A "Simple" Readiness Simulation

A review group developed a simple modeling approach for determining sensitivity of wartime sortie rates to EW subsystem reliability. Two models were used in tandem. The models and their inputs/outputs are summarized in Fig. A21.1.

The first model in this figure is a sparing to A_o model used for determining peacetime readiness. The second model is a simple simulation model developed for this analysis.

The simulation model is defined as simple because the aircraft was assumed to be one entire Line Replaceable Unit (LRU) comparable with a Navy Weapons Replacement Assembly (WRA). All actual LRUs in the EW subsystem were used, totaling 35. Figure A21.2 shows the results of the analysis.

The analysis considered both optimistic and pessimistic assumptions as well as several false removal rates. The likely range of achieved reliability was selected based on Demonstration Testing/Operational Testing (DT/OT) reliability results for the lower point and DT/OT results plus reliability improvement results for the upper point. Projection of this range upward provides a comparison of likely sortie rates against the goal. In this case, it did not appear that the goal would be achieved unless the identified reliability improvements were made. Accordingly, the program was restricted to limited production until the improvements were incorporated and tested. It is considered unlikely that these improvements would have been made so rapidly without the top management visibility of this analysis. As a final note on this analysis, the subsystem reliability becomes asymptotic past a certain point. At this point, airframe reliability begins to dominate EW subsystem reliability. No further investment in reliabil-

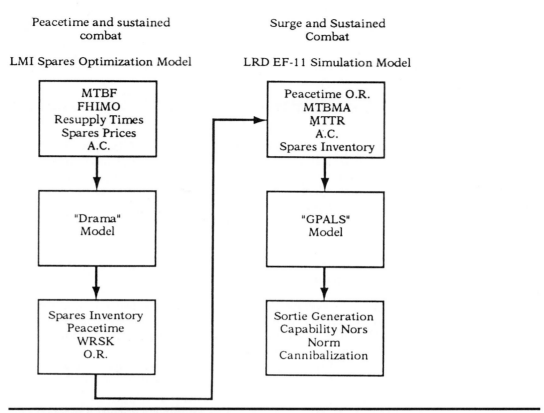

Figure A21.1   Availability analysis process.

ity past this point is warranted. This illustrates the use of system readiness analysis as a guide to the amount of reliability needed.

A21.2   A "Complex" Simulation

The preceding simple simulation was sufficient for top management purposes. More complex simulations can provide additional insights. Figure A21.3 illustrates this. The figure depicts the impact of three variables (spares, reliability, and O-level maintenance) on sortie rates. Note, for example, the sum from $36M to $168M, obtained from increasing spares investment. By adding the cost considerations of various improvement alternatives, the decision maker can determine which improvements are the most cost-effective.

Additional insights are possible depending on the capability of the specific complex simulation model used. In this case, for example, the

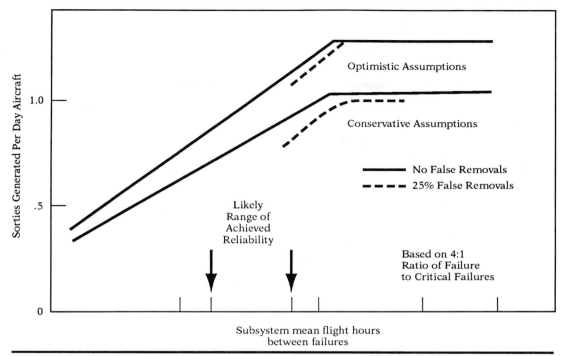

Figure A21.2 Surge sortie generation capability vs. Subsystem reliability.

Organizational Manpower Variation	Spares (Initial + Wrks) Variation			Mfhbf Variation
	$36M	$45M	$168M	
4 Man Maintenance Crew Per Shift	0.50	0.52	0.54	2.7 h
8 Man Maintenance Crew Per Shift	0.60	0.63	0.76	
4 Man Maintenance Crew Per Shift	0.57	0.59	0.61	3.7 h
8 Man Maintenance Crew Per Shift	0.64	0.69	0.79	

Figure A21.3 Sustained sortie/per day generation capability versus manpower and spares.

analysis could predict the change in sortie rate and manhours per month for each deficiency identified. These were analyzed in conjunction with the estimated cost of each fix to prioritize and select the order of implementation.

A21.3   Summary

This brief case has illustrated some of the various possible uses of readiness analysis by a logistics manager. Note that many of the uses involve program management decisions beyond the sole purview of the logistics manager. Use of this type of analysis frequently results in better decisions from a supportability perspective.

Diagnostics SOW Language

A22.1 Background

Historically, fielding systems which have satisfactory diagnostic capabilities have been a problem. In response to this issue, at the request of the Office of the Secretary of Defense, the NSIA's Integrated Diagnostic Group developed this appendix of sample SOW language for the various phases of weapon systems.

The principal attributes added by this new Statement of Work are

- An engineering analysis (including gathering of field data) for a previously fielded weapon system(s) to determine diagnostic capability performance deficiencies experienced

- Identification of specific risk areas which require design attention

- A format for specifying diagnostics requirements to achieve 100 percent unambiguous FD/FI coverage

- A requirement for preparation and implementation of a diagnostics capability maturation plan, including assets required, activities required, and data collection

- Thorough analysis of the design of the embedded diagnostics to be completed by CDR

- Design analysis and specification of the external diagnostics capability, including overlap, by CDR

- A requirement for demonstration of the diagnostics capability, including a thorough, statistically valid sample in selected areas of the system

A22.2 General Guidance

A22.2.1 Provisions for implementing integrated diagnostics are required throughout an RFP

It should be recognized that no one document or provision can effectively implement integrated diagnostics. Provisions are needed in many parts of the RFP to implement the concept. The Statement of Work defines the program tasks which must be accomplished. The detailed diagnostics requirements are contained in the System Specification. The deliverables are controlled by the CDRLs and DIDs. Other provisions in the RFP, which assist in the proper implementation of integrated diagnostics, are the Evaluation Factors for Award and the Instructions to Offerors. Integrated diagnostics provisions which apply to the entire weapon system and each diagnostic element must be interrelated and interconnected to ensure the integration of these diagnostics elements. Even though it is inevitable that diagnostics provisions will appear in many different places in the RFP, the sample Statements of Work tend to consolidate these requirements in one section.

A22.2.2 Present military standards can be used effectively to implement integrated diagnostics

The group recognized that present military standards can implement integrated diagnostics. This is not to say that some improvements in military standards are not required, but implementation does not have to await revision of these standards (e.g., MIL-STD-499, MIL-STD-1388, MIL-STD-785, MIL-STD-470, and MIL-STD-2165). All these standards are programatic in nature and, if invoked, require that a plan (i.e., system engineering, LSA, reliability, maintainability, testability) be prepared. No additional plans are required. Tailoring of appropriate DIDs can be used to emphasize additional diagnostics documentation requirements.

A22.2.3 Diagnostics design is an integral part of the system engineering process

It is recognized that to develop and deploy a weapon system with an adequate diagnostics capability requires that the design of the diagnostics be an integral part of the weapon system design, i.e., part of the system engineering function. Where a system engineering management plan, such as that required by MIL-STD-499, is prepared, the DID describing this plan should be tailored to address integrated diagnostics implementation. Integrated diagnostics provisions incorporated in the Statement of Work or System Specification should not be referenced solely in the supportability section.

A22.2.4 A goal should be established for concurrent development and delivery of the total diagnostics capability with the weapon system itself

Concurrent development and delivery of the total diagnostics capability with the remainder of the weapon system is a goal which should be emphasized in the RFP. This has been accomplished in some instances. However, it is recognized that this goal cannot always be met because the diagnostics capability sometimes cannot be completed prior to the freezing of the weapon system design. This is especially true for off-line diagnostics (e.g., test program sets). On the other hand, it is obvious that it is extremely difficult to demonstrate and evaluate a weapon system's diagnostics capability as separate parts (e.g., embedded versus off-line). Requiring only evaluation of the embedded diagnostics during scheduled weapon system test and evaluation phases will tend to encourage contractors to provide more off-line diagnostics capability—entirely opposite current DOD trends. Concurrent delivery of the entire diagnostics capability encourages contractors to employ CAE/CAD/CALS techniques for design of their diagnostics capability. Incentive provisions in the RFPs are a recommended means for encouraging a contractor to converge on the goal of concurrent delivery. In any case, the contractor should be required to explain why he or she cannot meet this goal.

A22.2.5 Diagnostics for GFE and CFE are part of the entire diagnostics capability and thus are the ultimate responsibility of the prime contractor

To ensure delivery of an adequate diagnostics capability, the prime contractor, ultimately, must be responsible. For CFE, the prime contractor must allocate diagnostics requirements for subsystems to subcontractors and audit their performance to ensure that these requirements are met. For GFE, the government should be responsible for specifying and guaranteeing the diagnostics capability of these equipments. If the prime contractor cannot satisfy the total diagnostics requirements because of inherent GFE diagnostics performance, DOD should be informed of this situation and alternatives proposed.

A22.2.6 Quantitative diagnostics requirements must be specified for the entire diagnostics capability as well as for each diagnostic element

The Statement of Work or System Specification must include quantitative diagnostics requirements for the system and for each diagnostic element (e.g., test times, technical information delivery times, formal training time) of the system. These requirements must be derived from weapon system mission and operational requirements. Preventive maintenance, as well as corrective maintenance, times should be specified and allocations for diagnostics times established. The allocation process should place quantitative requirements for each of the diagnostic elements.

A22.2.7   Maturation should be an integral part of the diagnostics demonstration process

The diagnostics capability of the weapon system as a whole, as well as for each diagnostic element, must be demonstrated at established milestones in the weapon system acquisition process. Diagnostics capability growth requirements must be established for each of these milestones. When a specific element of the diagnostics capability does not yet exist, this capability should be simulated. Maintainability demonstration techniques (e.g,, MIL-STD-471) should be supplemented by a diagnostics maturation process, which extends until the performance of the diagnostics capability meets established contractural diagnostics requirements. A plan for this maturation process should be prepared early in weapon system acquisition. Incentive and warranty provisions can be included in the RFP as a means for encouraging the contractor to meet the milestones of the maturation plan.

A22.2.8   A capability to report, track, and measure diagnostics performance throughout the length of the contract should be established

Program reporting, tracking, and measurement should emphasize diagnostics performance throughout the length of the contract. This is an integral part of the diagnostics maturation process. This data base should be constructed using an agreed-on metric base and should be compatible with the service data base to be employed once the system is transitioned to the government.

A22.2.9   Management responsibility for incorporating integrated diagnostics in weapon system development should be a critical factor in proposal evaluation

Proposal evaluation shall emphasize both management and technical aspects of the implementation of integrated diagnostics. The management structure should indicate how the integration of the diagnostics shall take place throughout the length of the contract.

A22.3   Concept Exploration Phase

A22.3.1   Diagnostics approach

The contractor shall define the diagnostics approach to be provided for maintenance of each system alternative. Methodologies shall be established to translate system mission, performance, and mobility objectives into diagnostics performance requirements. These requirements shall reflect the various mission scenarios, mission-critical functions, and operational and logistics constraints. The goal is to detect and isolate all faults to a replaceable unit with a minimum of false alarms and false removals using a mix of diagnostics capabilities built into the system and external to the system. The mix that makes up the diagnostics capabilities shall be selected from design techniques (e.g., built-in test, fault tolerance, status monitoring, partitioning, test points), external hardware and software (e.g., automatic and manual test equipment), technical information (e.g., technical manuals, information systems, and operator displays), and training (e.g., skill levels, formal schooling, on-the-job training). The contractor shall provide the basis for the proposed diagnostics approaches which, as a minimum, shall include a baseline analysis of fielded systems, identifying current diagnostics performance (e.g., fault detection, TPS run-time, etc.) and functional areas where diagnostics are difficult (e.g., requires high skill levels, high performance test equipment, long diagnostic times). In addition, the contractor shall identify diagnostics design feasibility issues and related designs that will require prototyping and testing during the next phase to reduce risk, prove out feasibility issues and related designs that will require prototyping and testing during the next phase to reduce risk, prove out feasibility, or demonstrate new diagnostics capabilities.

A22.3.2   Recommended list of deliverables

These items are to be identified in the CDRL of the applicable RFP package:

Current diagnostic capability baseline analysis results

Recommended system and subsystem level diagnostics performance and approaches

Diagnostics implementation feasibility/risk-reduction proposals

A22.4   Demonstration/Validation Phase

A22.4.1   Integrated diagnostics specification development

The contractor shall perform detailed comparability and design analysis and risk-reduction efforts necessary to develop a specification provision for the allocation of diagnostics capabilities to be used for fault detection/isolation, repair verification, performance or condition monitoring, and damage assessment and enable the weapon system to meet maintenance and operational goals with a minimum of false removals. Diagnostics capabilities shall be selected from design techniques (including built-in test, fault tolerance, status monitoring, partitioning, test points), external hardware (e.g., automatic and manual test equipment and maintenance aids), technical information (e.g., technical manuals, information systems, and operator displays), and training materials (e.g., formal schooling, on-the-job training). The capabilities selected may be designed into the system as part of the system or may be provided separately to maintenance personnel, as required, to meet mission and level of repair objectives.

Based on the results of the analyses and risk-reduction efforts, the contractor shall specify the diagnostics capabilities to be provided with the system at each level of maintenance and how these will be allocated (see Table A22.1), to include such capabilities as:

Mode of operation (e.g., status monitoring) and areas where there is diagnostics ambiguity or overlap

Operational test strategies, fault tolerance, prognostics, and fault-model assumptions

Performance in terms of mean time to diagnose, fault coverage, false alarms, and false removals

Physical and functional equipment partitioning requirements

Physical (weight, volume) and functional (five memory) limitations

Diagnostics capability interface requirements

Options for augmenting government-furnished equipment (GFE) diagnostics capabilities

Reliability of the BIT and external diagnostics hardware

A22.4.2   Detailed diagnostics comparison analysis

As part of the development of the integrated diagnostics specification provisions, the contractor shall perform a comparison analysis using the baseline fielded system at each level of field maintenance to include analysis of the causes of excessive diagnostics times, undetected faults, false alarms, and false removals. The contractor shall identify, to the greatest extent practicable, the sources of these causes and describe how the new, propose system design and diagnostics capabilities will result in improvements. As a minimum, this analysis should characterize whether the causes of diagnostics problems are inherent to the design (i.e., partitioning, connectors, etc.) or due to maintenance

TABLE A22.1 Diagnostics Performance Specification

Level of maintenance	Diagnostic capability*	Fault detection coverage†§	Fault isolation coverage†§	Mean time to diagnose	False alarm rate‡	False removal rate‡	Other requirements
Organizational	Status monitor	____%	____%	____	____	____	____% of fault coverage by status monitor for mission-critical functions
	Bit	____%	____%	____	____	____	
	Manual test	____%	____%	____	____	____	
	Maint. aids/manual troubleshooting	____%	____%	____	____	____	BIT memory allocation not to exceed X words
	Total	100%	100%	____	____	____	Technical information access time
Intermediate	External ATE/expert system	____%	____%	____	____	____	
	Manual test	____%	____%	____	____	____	ATE limited
	Total	100%	100%	____	____	____	To X lb Y ft^3
Depot	External ATE	____%	____%	____	____	____	
	Manual test	____%	____%	____	____	____	
	Total	100%	100%	____	____	____	

*Listed by way of example.
†Unambiguous percentage of fault coverage for each capability shown. Total ATE (Automatic Test Equipment) each level of maintenance should add to 100 percent of the identified replaceable items for that level.
‡Relate rates to operational usage (e.g., 1 false alarm per monitoring hour).
§For each diagnostic capability listed, indicate whether P = primary mode or S = secondary or augmenting mode.

procedures, lack of "vertical" testability (e.g., cone of tolerance, compatibility between levels of maintenance), or transients.

The contractor will provide quantitative assessments of diagnostics capabilities identifying current capabilities, extrapolations to proposed capabilities, and the engineering analysis that is the basis for the extrapolation. The contractor shall determine where there are overlaps or ambiguities in diagnostics capabilities used for maintenance of fielded systems and how these will be addressed for the propose system. When deficiencies in the GFE preclude meeting the diagnostics requirements, the contractor shall develop alternatives. In addition, the contractor shall identify the weight and volume of the major external test equipment, type and extent of technical information, and maintenance skill levels and training requirements for currently fielded systems and provide an estimate of these quantities for the proposed diagnostics capability and an explanation of the basis for this estimate.

A22.4.3   Diagnostics risk reduction

As part of the design, prototype, and test and demonstration activities proposed (the basis of the proposal shall be risk areas determined in concept exploration), the contractor shall determine the feasibility of achieving diagnostics capability performance improvements.

A22.4.4   Diagnostics maturation plan

This plan shall include the contractor's proposal for next-phase activities in the areas of analysis, growth, and demonstration of the entire diagnostics capability (hardware and software). The plan also shall include the resources required for maturation activities (e.g., prime hardware, laboratory facilities).

The contractor shall design a diagnostics data system which extends from demonstration and validation through full-scale development, production, and deployment. The data system shall be designed so that the performance of the diagnostics capability can be ascertained at any point during the acquisition, production, and deployment of the weapon system. It shall be compatible with the established DOD data system, which will be employed after the maintenance of the weapon system becomes the responsibility of the government.

A22.4.5   Recommended list of deliverables

These items are to be identified in the CDRL of the applicable RFP package:

Proposed diagnostics system and subsystem performance specification provisions and allocations and design requirements (see Table A22.1).

Diagnostics maturation plan to include system testing, design analysis, and data collection

Results of risk-reduction tasks

Results of comparison analysis

A22.5   Full-Scale Development Phase

A22.5.1   Integrated diagnostics design

As part of the system design, the contractor shall incorporate testability features in the system and provide external diagnostics capabilities that will satisfy the diagnostics capability performance requirements contained in the system specification (per notional specification developed in the demonstration and validation phase).

A22.5.2   Diagnostics design analysis

The contractor shall implement a structured design analysis process to assess in detail the ability of the diagnostics capability design to meet the system diagnostics performance specification (e.g., fault coverage, mean time to diagnose, false removal, etc.), analyze the inherent testability of the preliminary design, identify the areas where the primary means of diagnostics may lead to an ambiguous result and ways the ambiguity will be resolved, identify areas where there is a redundant (overlapping) diagnostics capability planned, and verify that the detailed design of diagnostics is in accordance with the functional allocation established during the previous program phase. As a minimum, the analytical task shall be performed and delivered as described below:

Design analysis of diagnostics built into the system.   The contractor shall complete a structured analysis of system design implementation to identify the functional areas in which diagnostics capabilities (allocated in the previous phase to be built into the system) provide an unambiguous capability to detect or isolate a fault to the appropriate replacement unit at each level of maintenance. As a minimum, the design analysis shall be based on maintenance dependency models (or their equivalents) at the system level to quantify the degree of ambiguity at the lowest replaceable assembly. Also, the design analysis shall be based on an assessment of the inherent testability of the design, where there are large areas of ambiguity. In electronics assemblies consisting of digital logic, the dependency model analysis shall be augmented with 100T fault grading simulation for selected samples. In addition, the contractor shall prepare a worst case analysis of design or tolerance margins at each BIT sensor and test point for the system. As a result of these analyses, an assessment shall be provided of the capability of the built-in diagnostics to meet the fault coverage specification, identification of specific system areas where there is ambiguous fault detection or isolation, and assessment of the capability to limit the false alarms and false removals. The assessment shall identify the projected causes of false alarms and false removals and ambiguities in terms attributable to equipment design mechanization (including partitioning, test point placement, BIT limitations), transients or maintenance, and operational considerations. These analyses will be used to update the diagnostics functional allocation, where necessary, to resolve ambiguities or reduce overlap. These analyses shall be completed by the critical design review (CDR).

Assessment of external diagnostics. The contractor shall deliver at the CDR detailed requirements definition for: external for; test equipment (troubleshooting) approaches to be included in maintenance manuals, maintenance aids, and training requirements. These requirements shall each be supported by a diagnostics ambiguity analysis to be delivered at the same time. The analysis shall describe the degree to which diagnostics ambiguities are reduced and the areas where there is redundancy (overlap) of diagnostics capabilities. The analysis shall specifically highlight those areas where the combination of internal and external diagnostics cannot unambiguously detect/isolate a fault within the prescribed diagnostics time limits or false alarm and removal limits. The results, modified as necessary to resolve ambiguities, will be used to update requirements documents for external diagnostics.

Diagnostics maturation program. The contractor shall establish a diagnostics performance data collection system. Also, the contractor shall conduct diagnostics performance verification tests and demonstrations to evaluate the effectiveness of the diagnostics design. As a minimum, the diagnostics testing should include the insertion of a complete fault sample (approaching 100 percent) in customer-selected areas of the system, in order to evaluate the accuracy of the fault coverage prediction. Diagnostics capability requirement level should be specified at established milestones during FSD, prior to OT&E. In addition, diagnostics testing should include system operation throughout the specified environmental range, in order to evaluate false alarm and removal deficiencies. When external diagnostics capabilities are not available, substitutes shall be used to the maximum extent possible. In addition, the contractor shall monitor diagnostics performance whenever the system is operating, and perform an analysis to determine whether the diagnostics capabilities are operating in accordance with the design. Based on the maturation results, the contractor shall take corrective action in order to meet diagnostics capability requirements. The contractor shall provide a diagnostics maturation profile, periodic summary of diagnostics performance throughout the development cycle, and results of the diagnostics verification.

A22.5.3   Recommended list of deliverables

These items are to be identified in CDRL or applicable RFP package:

PRD/CDR embedded diagnostics design assessment results

External diagnostics capability detail specifications

Completion updated diagnostics analysis and simulation test results

Updated diagnostics capabilities field maturation plan

Updated notional diagnostics performance specification

A22.6   Production Phase

A22.6.1   Diagnostics maturation

The contractor shall mature the diagnostics capability in accordance with the established maturation plan to assure that the required improvements are made toward satisfying the updated Notional Diagnostics Performance Specification at each maintenance level. This includes:

1. Maintaining and utilizing the diagnostics data system to measure performance of the diagnostics capability and take required corrective action, in accordance with the incentive and warranty provisions

2. Planning the transitioning of the data system to the government

3. Demonstrating that the diagnostics capability satisfied the diagnostics requirements.

A22.6.2   Recommended list of deliverables

These items are to be identified in the CDRL of the applicable RFP proposal:

External diagnostic test results

Maturation results

RAM-LSA-Supply Support Dependencies

The bridge between system design and designing the system's support is a highly complex network of interdependent system engineering relationships. One role of the logistician is to act as integrator of the parts of this network and, through the use of tools and technology, orchestrate the integration process to achieve a supportable design.

An example of this integration process is the use of the LSAR to provide an estimate of the annual usage rates of the various assemblies, subassemblies, and piece parts that compose the end item and that are required to be provisioned. The usage rates for these parts are usually based on the annual number of expected failures requiring removal and replacement of the item. In the LSAR, the term for this usage rate is the *Maintenance Replacement Rate (MRR)*. The MRR is a major factor in determining initial sparing quantities. When researching data for inclusion in the LSAR, it is important to consider how the data will affect the logistics support goals, such as provisioning. The following example describes the relationship among the failure rate, failure mode ratio, task frequency, quantity per task, and the maintenance replacement rate (Fig. A23.1).

While each acquisition program is unique, factors such as system redundancy, acquiring agency, hardware/software configuration, and

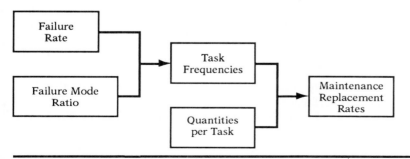

Figure A23.1 Data interdependencies.

type of system (mechanical or electronic) may affect the interrelationships of the preceding elements. On a specific program, therefore, the general interrelationships contained here would have to be appropriately modified; however, if the objective—to determine annual usage—is the final goal, the necessary modifications should become self-evident. The simplified hardware breakdown illustrated in Fig. A23.2 will be used throughout this discussion. A basic knowledge of the LSA process and the LSAR is assumed. All terms and definitions are consistent with the data element dictionary in MIL-STD-1388-2B, Appendix E.

A23.1   Failure Rates

Figure A23.2 shows a partial hardware breakdown tree of a typical end item consisting of more than one assembly. The first assembly, Assembly 1, is composed of three repair parts (piece parts) and one repairable subassembly. LSA Control Numbers (LCN) are assigned using the classical LCN assignment method.

For these items, the failure rates were determined during the performance of a Failure Mode, Effects, and Criticality Analysis (FMECA). For the purposes of the LSAR in determining support requirements (provisioning in this case), it is very important that the FMECA be performed against a hardware breakdown rather than a functional FMECA. This is necessary so that failures may be attributed to a particular component, since support requirements are identified against hardware rather than functions. A functional FMECA is insufficient because one hardware item may have more than one function or multiple hardware items may be required for one function. Although a functional FMECA may be needed to establish fault-isolation procedures, a program that has only performed a functional FMECA may at best have a difficult time tying failure rates to a particular piece of equipment and at worst the correct correlation may never be accurately established. It is important that the LSA manager be aware of this distinction between types of FMECAs. In one instance, on a relatively small acquisition program, a portable air-conditioner, only a functional FMECA was performed. The failure modes were allocated to the hardware items using "best engineering judgment," but all traceability to the reliability analysis was lost. If only a functional FMECA is called out by contract, the LSA manager will still have to have a hardware FMECA conducted and costed in order to properly perform the LSA.

The failure rates for LCNs A, A1, and A1D would be entered into the BD table ("Reliability, Availability and Maintainability Indicator

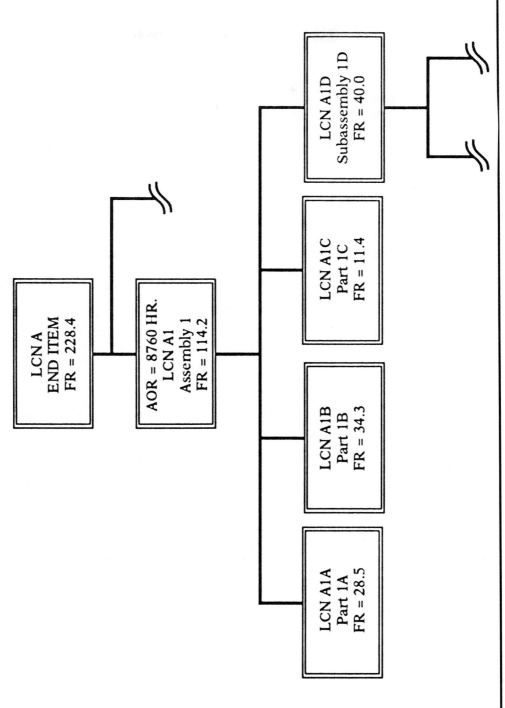

Figure A23.2 Hardware breakdown.

Characteristics") of the LSAR along with the other FMECA data. Failure rates are recorded in the LSAR as the mean number of failures per million hours. [*Note:* The mean time between failures (MTBF) is equal to the reciprocal of the failure rate times 10^6.] The three parts (LCN A1A, A1B, and A1C) shown in Fig. A23.2 are repair parts and would not be LSA candidates. LSA candidates are those items which require documentation of operational and logistics support parameters and are normally repairable assemblies. Only LSA candidates are documented in the LSAR's B tables. This means that failure rates of these three parts would not be included directly in the LSAR but rather indirectly as the failure mode ratios (FMR) of their next-higher assembly.

A23.2   Failure Mode Ratio

The next number to identify in the LSAR would be the FMR. The FMR is the fraction of a component's failure rate that may be attributed to the related failure mode. In the example, Assembly 1 would have four or more failure modes identified (at least one for each of its three repair parts and one subassembly). The sum of the FMRs related to failures of part 1A would be 0.25. This is 28.5 of Assembly 1's 114.2 failures in 1 million hours would result from a failure of part 1A. Thus the formula to calculate the FMR of a subassembly x is

$$\text{FMR}_x = \frac{\text{FR}_x}{\text{FR}_y} \qquad (A23.1)$$

FMR_x is the sum of all FMRs related to failures of x in y, and y is the next-higher assembly of x. Using Eq. (A23.1), the FMRs for the items in the example are

LCN A: $$\text{FMR}_{A1} = \frac{114.2}{228.4} = 0.50$$

LCN A1: $$\text{FMR}_{A1A} = \frac{28.5}{114.2} = 0.25$$

$$\text{FMR}_{A1B} = \frac{34.3}{114.2} = 0.30$$

$$\text{FMR}_{A1C} = \frac{11.4}{114.2} = 0.10$$

$$\text{FMR}_{A1D} = \frac{40.0}{114.2} = 0.35$$

Because the FMR is a percentage, the total of the FMRs for any one item will always equal 1.00. (*Note:* Items not shown in

Fig. A23.2 would account for the remaining 50 percent of LCN A's failure rate.)

A23.3 Task Codes

The task code in the LSAR is a seven-character code that uniquely identifies and provides information relative to the performance of the task itself. This information consists of task function, task interval, maintenance level, service designator, operability code, and a two-character task sequence code. The task code can convey a great deal of information concerning a particular task, and it is incumbent on the LSA manager to tailor the use of this code for his or her own particular program. The discussion in this paper will concern itself only with the first element of the task code, the task function code.

For the purposes of this example, the discussion will be limited to only two task function codes: H and J. The task function code is the first character of the LSAR task code. An H task identifies a remove/replace task. In an H task, a failure is corrected by removing the item under analysis (i.e., the item identified by the LCN associated with the task code) and replacing it with a like item. For example, line 1 of Table A23.1 would describe a task for the removal and replacement of Assembly 1.

A J task identifies a repair task. In a J task, the failure of the item under analysis is corrected by removal and replacement of a subordinate repair part or parts. Repair parts are not normally LSA candidates and in most cases are just piece parts. Line 2 of Table A23.1 would describe a task for the repair of Assembly 1 by removing and replacing one or more piece parts.

Although tasks described here will only be concerned with the actual "repair" of the failed item, it must be remembered that a repair task also must account for the time required to perform a fault-isolate/locate and a verification/test procedure to obtain the correct mean time to repair (MTTR). In most cases, the best way to accomplish this is by referencing these procedures at the beginning and end of the repair task narrative.

TABLE A23.1   Table CA (Task Requirements) R&M Example

	LCN LSACONXB	Task code TASKCDCA	Task AOR measurement base AORMSBCA	Task frequency TSKFRQCA	Task identification TASKIDCA
1	A1	HGOXBAA	H	1.000	R/R OF ASSEMBLY 1
2	A1	JGFXGAA	H	0.650	REPAIR OF ASSEMBLY 1
3	A1D	HGFXGAA	H	0.350	R/R OF SUBASSEMBLY 1D

A23.4   Task Frequency

The *task frequency* is the probable number of times per year that a task is performed. This number is a critical number in the LSAR. It is used to compute annual usage data for manpower and support equipment, as well as for spares and repair parts. The task frequency can include up to 3 decimal places. Therefore, a task with a frequency of 0.333 may be described correctly in any of these three ways: a task that is projected to be performed once every 3 years, a task that has a 33 percent probability of being performed during a year, or if there are 1000 end items in the field, 333 of them will require performance of this task during the year.

The task frequency for corrective maintenance actions is calculated using the formula in MIL-STD 1388-2B:

$$\text{TF} = \text{FMR} \cdot \left(\frac{\text{FR}}{10^6} + \frac{1}{\text{MTBM-IN}} + \frac{1}{\text{MTBM-ND}} \right) \cdot \text{AOR} \cdot \text{CF} \qquad (\text{A23.2})$$

In this example the mean time between maintenance—inherent (MTBM-IN) and the mean time between maintenance—no defect (MTBM-ND) will be assumed to be so great as to be negligible in Eq. (A23.2). Generally, these factors cannot be determined accurately until after the item is fielded for a new development item. The conversion factor (CF) is equal to 1. The factors above are therefore omitted from Eq. (A23.3) below. The annual operating requirement (AOR) is the estimated or required yearly rate of usage of an item. In this example the AOR is 8760 hours.

$$\text{TF} = \text{FMR} \cdot \frac{\text{FR}}{10^6} \cdot 8760 \qquad (\text{A23.3})$$

In this example, the sum of the FMRs of the end item accounted for by the failure of Assembly 1 is equal to 0.50. This is 114.2 of the end item's 228.4 failures per 1 million hours resulted from a failure of Assembly 1. Therefore, the task in line 1 of Table A23.1 (removal and replacement of Assembly 1) would have a task frequency of

$$\text{TF}_{\text{A1 HGOXBAA}} = .5 \cdot \frac{228.4}{10^6} \cdot 8760$$

$$= 1.000$$

Assembly 1 consists of three repair parts (part 1A, part 1B, and part 1C) and one subassembly (subassembly 1D). The failure modes of Assembly 1 caused by the failure of subassembly 1D would be corrected by the task shown in line 3 of Table A23.1. This task would have a task frequency of

$$TF_{A1D\ HGFXGAA} = FMR_{A1D} \cdot FR_{A1} \cdot AOR$$

$$= 0.35 \cdot \frac{114.2}{10^6} \cdot 8760$$

$$= 0.350$$

The remaining three failure mode ratios are a result of the failures of part 1A, part 1B, and part 1C, respectively. It may be possible that all three of these failure modes may be corrected by using the same task narrative description, i.e., "Remove cover from assembly, locate and remove faulty part, replace with like part from stores, and reseal assembly." A task of this type is identified in line 2 of Table A23.1 and has a task frequency of

$$TF_{A1\ JGFXGAA} = (FRM_{A1A} + FMR_{A1B} + FMR_{A1C}) \cdot FR_{A1} \cdot AOR$$

$$= (0.25 + 0.30 + 0.10) \cdot \frac{114.2}{10^6} \cdot 8760$$

$$= 0.650$$

It should be noted above that each time the assembly is removed at the organizational maintenance level (once per year in this example) there is some task to repair it at some higher level (TF_{A1D} $_{HGFXGAA}$ + $TF_{A1\ JGFXGAA}$ = $TF_{A1\ HGOXBAA}$ = 1.000). (*Note:* The repair level is indicated by the third character of the task code; O is organizational and F is intermediate.)

Lastly, from this example it can be seen clearly that the basis for the task frequency is the failure rates and FMRs obtained from the FMECA. For the task frequency to be calculated correctly, it is necessary for the LCN structure to be based on the end item's hardware breakdown and that the FMECA be conducted using this breakdown rather than a functional breakdown.

Even when this relationship is understood, if the analyst does not understand the use of the task frequency by other ILS elements (i.e., calculating annual manhour and repair part usage), he or she may misenter the data. In one instance, while an LSAR data-entry program was automatically calculating the task frequency, due to a programming error it was being entered at over five times the actual rate. When this error was pointed out, the analyst was unconcerned because this was just a "planning" number and the provisioners were determining sparing requirements using other data. However, a quick check with the provisioners established that they were in fact using the erroneous task frequency in their calculations.

A23.5   Quantity per Task

The CI tables (Task Provisioned Item) in the LSAR provide the logistician with space to enter a list of the items used in support of a particular task. Although the CI tables should contain all bulk items, spares, repair parts, etc., this example will be concerned only with the replacement parts. In addition to identifying the support items by reference number and name, other data [item category code (ICC), quantity per task (QTY/TASK) and unit of measure (UM)], must be entered to complete the record for each support item.

The ICC identifies a category that describes the support item, such as common tools, peculiar test equipment, technical manuals, etc. Using the description of the task codes given above, the ICC used to describe the replacement part in an H task (removal/replacement of an LSA candidate) will generally be an X (spare/assembly/repairable), while the ICC for the replacement part in a J task would be a Y (repair part). Further, the item being replaced in an H task will be the item identified by the LCN of that task.

The QTY/TASK is the quantity of the support item used in the performance of the task to the hundredth of a unit of measure. For items such as tools and test equipment, this will normally be 1.00 each. For bulk items, it may be something similar to "grease 0.25 pounds." For items that may be lost or damaged during performance of the task, the QTY/TASK may be "screw 0.10 each." This accounts for a screw that may be lost or stripped 1 in every 10 times the task is performed and would usually be an engineering estimate.

For an H task, the QTY/TASK of the replacement item will be 1.00 each (i.e., each time an item fails, it is replaced by one like item). The same is true in those cases where a J task describes the action requires to correct a failure that only involves the replacement of a single repair part.

However, in the J task described above, the failure may be corrected by replacing either part 1B, part 1C, *or* part 1D but not a combination of two or more of these parts. To include all these parts in the LSAR CI table and account for their usage properly, their QTY/TASK must be allocated in proportion to each part's failure contribution to the task frequency (i.e., weighted average). To accomplish this calculation, the following formula may be used:

$$\frac{QTY}{TASK} = \frac{\sum FMRs \text{ associated with the part's failure}}{\sum FMRs \text{ used to obtain the task frequency}} \qquad (A23.4)$$

Using this formula, the quantities for parts 1A, 1B, and 1C in the task shown in line 2 of Table A23.1 are

$$\frac{QTY}{TASK_{Part\ 1A}} = \frac{0.25}{0.65} = 0.38 \text{ each}$$

$$\frac{QTY}{TASK_{Part\ 1B}} = \frac{0.30}{0.65} = 0.65 \text{ each}$$

$$\frac{QTY}{TASK_{Part\ 1C}} = \frac{0.10}{0.65} = 0.15 \textit{ each}$$

$$\text{Total} = 1.00 \text{ each}$$

From the preceding example, the total usage of replacement parts is equal to 1 (allowing for rounding). Logically, this makes sense; that is, each time this task is performed, one part of some type will be replaced. However, if the logistics analyst does not understand the relationship between the task frequency and quantity/task, a common mistake is to list every repair part with a quantity of 1. The reasoning is that one cannot replace a partial quantity of a part, i.e., 0.5 of a resistor. Even though one cannot replace a fraction of a part, a fraction may indeed be the correct entry for this data field.

A23.6 Maintenance Replacement Rate (MRR)

The MRR establishes the number of replacements per maintenance cycle (1 year in this example) required for a particular usage of a part (defined by a unique part number–LCN combination) in an end item. This number is obtained by summing the products of the task frequencies and quantities per task for all tasks in which an item is required. The general formula for MRR is

$$MRR = \left(TF \cdot \frac{QTY}{TASK} \right) \tag{A23.5}$$

The MRRs for the parts in this example are

$$MRR_{Assembly\ 1} = TF_{A1\ HGOXBAA} \cdot \frac{QTY}{TASK_{Assembly\ 1}}$$

$$= 1.00 \cdot 1.00 = 1.000$$

$$MRR_{Part\ 1B} = TF_{A1\ JGFXGAA} \cdot \frac{QTY}{TASK_{Part1A}}$$

$$= 0.65 \cdot 0.38 = 0.2470$$

$$MRR_{Part\ 1C} = TF_{A1\ JGFXGAA} \cdot \frac{QTY}{TASK_{Part1B}}$$

$$= 0.65 \cdot 0.46 = 0.2990$$

$$\text{MRR}_{\text{Part 1D}} = \text{TF}_{\text{A1 JGFXGAA}} \cdot \frac{\text{QTY}}{\text{TASK}_{\text{Part1C}}}$$

$$= 0.65 \cdot 0.15 = 0.0975$$

$$\text{MRR}_{\text{Subassy 1A}} = \text{TF}_{\text{A1D HGFXGAA}} \cdot \frac{\text{QTY}}{\text{TASK}_{\text{Subassy 1A}}}$$

$$= 0.35 \cdot 1.00 = 0.3500$$

In this example, all the replacement parts were used in only one task. It must be remembered, however, that if a part is used in more than one task, the sum of those partial MRRs must be used to establish the total MRR for that part.

A23.7   Conclusion

In most cases, the tasks and replacement parts required to support those tasks involve only the items identified by the LCN of that task and those items in its next-lower LCN indenture level. In the foregoing example it has been shown that the failure rate, failure mode ratio, task frequency, quantity per task, and the maintenance replacement rate do not exist independently but are all interrelated. The logistician uses the LSAR as a data integrator among reliability, maintainability, and provisioning to establish the sparing requirements. However, the responsibility to accomplish this integration should not rest on the logistician alone if the LSAR is to be a useful data base. Provisioners should be aware of how the data for their requirements are developed and be able to identify eventual supply support problems early in program development. Likewise, the design and R&M engineers must not look at the FMECA data as an end product but also be aware of its impact throughout the logistics support analysis.

An Example of LSAR Task Documentation Conventions

While MIL-STD-1388-2B provides standardization to the LSAR data structure and element definitions, there is still latitude for variation within this standard. Examples of this latitude are which data element codes are acceptable to different customers and the depth of detail expected in the narrative data elements. Early agreement on these additional conventions is necessary to preclude needless rewrite and reentry of the data. These conventions also should be presented and discussed at the guidance conference, since the customer may require a different approach.

This appendix gives an example of the type of convention an ILS manager may want to provide to individuals entering task analysis/documentation into the LSAR. MIL-STD-1388-2B allows a large amount of flexibility both in selecting which data elements to document and in the level of detail that can be documented. The detail provided in these conventions ensures that everyone knows what is to be done and how. In a large organization these conventions also can provide consistency among analysts and possibly assist in encouraging compatibility among programs. In general, the more detailed the guidance provided early on, the more useful and clear the LSAR will be to both the internal users and customers. Some programs and contractors have gone as far as creating a guide with the unique data entry instructions for each specific data element field that was used in the particular program.

A24.1 Task Codes

When documenting tasks, one of the first considerations is assigning the task code. The task code is a data chain of six separate data elements that along with the LCN/ALC/UOC uniquely identify an operation/maintenance task. These codes, if used correctly, can provide a significant amount of information about the task and can be used by the LSAR ADP system to aggregate similar tasks for analysis. MIL-STD-1388-2B lists a number of codes, some of which will not be applicable to an individual program. A list of some of the more basic codes is provided below. This list should be reviewed in conjunction with the list in MIL-STD-1388-2B and the unique requirements of a particular program to identify the codes that will be used. The standard also provides definitions for these codes. This list may then be presented at the guidance conference.

Key point. *For each of the codes below, the standard provides many more choices. The purpose here is to eliminate codes that are not applicable to a particular program and to avoid the use of codes that may have overlap in definitions.*

A24.1.1 Task function codes

Task function codes identify specific maintenance, operator, or supporting functions necessary to the operation and maintenance of an item.

Access	W
Calibrate	F
Fault location	N
Inspect	A
Mission profile change	M
Operate	O
Remove and replace	H
Repair	J
Service	C
Test	B

Key point. *All tasks will be scheduled or unscheduled. The interval for the scheduled tasks will be apparent from the task frequency, so in most cases other interval codes are unnecessary.*

A24.1.2 Task interval codes

Task interval codes identify the timing of the task occurrence. In most situations it is convenient to use only the two codes below. The exact interval for scheduled tasks will be clarified by the task frequency.

Scheduled	B
Unscheduled	G

A24.1.3 Operation/maintenance level

The operation/maintenance-level code indicates the maintenance levels authorized to perform the maintenance function. This will identify the maintenance organization of the acquiring authority. In this example it is the Navy.

Operator/crew/unit-crew	C
Intermediate/direct support/afloat/ off-equipment	C
Organizational/on-equipment	O
Intermediate/direct support/afloat/ off-equipment	F
Intermediate/general support/ashore	H
Intermediate/ashore and afloat	G
Depot/shipyards	D

A24.1.4 Service designator

The service designator code identifies the major government agency (military branch) having jurisdiction over or executive management authority for the acquisition. *There will not normally be any flexibility in this particular data element.*

Navy	N

A24.1.5 Operability code

The operability code indicates the operational status and mission readiness of the item during the maintenance task.

System inoperable during equipment maintenance	A
System operable during equipment maintenance	B
Off-equipment maintenance	G
Turnaround	F

A24.1.6 Task sequence code

This is a two-position code assigned to each task. The code is used to differentiate between two or more tasks having the same LCN/ALC/UOC–task code combination in the first five characters of the task code.

A24.2 Task Narrative

The purpose of this section is to provide guidelines for the development of the LSAR task narratives. These narratives may be written by several authors. To achieve a uniformity in the LSAR, it is imperative that all writers become familiar with and follow the same conventions. Any questions and/or conflicts should be identified to and resolved by the LSA manager.

A24.2.1 General

Tasks are written against repairable items, not piece parts. Repairable items will be LSA candidates.

A24.2.2 Remove and replace tasks

A remove and replace task (H task) is a task that effects a repair by removing and replacing the item identified by the LCN of the task. This task is *not* documented as a repair of the next-higher assembly. Such tasks will be described in the task identification (TASKIDCA) field as "R/R of _____ ."

A24.2.3 Repair tasks

A repair task (J task) is a task that effects a repair by removing and replacing an item or items one indenture lower than the LCN of the task. These items will be piece parts (nonrepairables). These tasks will be described in the task identification (TASKIDCA) field as "Repair of _____ ." A task will only involve removal and replacement of the LCN of the task (H tasks) *or* nonrepairable items in its next lower indenture (J tasks); i.e., tasks in the LSAR will not normally involve items more than one level deeper than the LCN of the task.

A24.2.4 Task description sequence

Each task will begin with a fault-isolation procedure (subtasks) and end with a repair-verification (test) procedure (subtasks). That is, the task will consist of the following parts: (1) how the fault is detected/isolated, (2) what steps are taken to fix it, and (3) a test or verification procedure to ensure that it has been fixed. This corresponds to the steps needed to determine the mean time to repair (MTTR). In some cases the fault-isolation procedure may be the same as the repair-verification procedure. When the repair has been verified, include the procedures to return the equipment to current readiness condition, if necessary.

A24.2.5 Amount of detail

How much detail should be included in the task narrative? The maintenance engineer documenting the task should coordinate with the technical manual authors to ensure that the narratives contain enough detail to develop the technical manual. At a minimum, the step-by-step analysis is necessary of what needs to be done and what tools, test equipment, and parts are required. The tools, test equipment, parts, and bulk items are also documented in the CG and CI tables. There should be a one-to-one correspondence between the items used in the task narrative and those in the CG and CI tables.

A24.2.6   Warnings and cautions

Warnings/cautions should be included in the task narrative. Warnings are associated with procedures that could result in personnel injury, and cautions are associated with procedures that could result in damage to the equipment. Warnings and cautions should always precede the narrative describing the hazardous procedure. The SHSC (FMSHSCBI) and the HMPC (HAZMPCCA) should correspond to the warnings/cautions in the narrative. For example, if the SHSC or the HMPC is greater than D or 4, respectively, there should be either a warning or a caution in the task narrative. The format for warnings and cautions is to first state the hazard and then to provide avoidance information.

A24.2.7   Narrative rules

The following rules apply to the development of the actual narrative:

1. Use implied second person imperative mood.
2. Do not use articles unless required for clarity.
3. Use single action steps unless multiple simultaneous actions are required. Use a tiered breakdown (i.e., tasks, subtasks, elements). The beginning of each element will be identified by placing an E in the element indicator field (ELEMNTCC). Otherwise this field will be left blank. Elapsed times and manhours for the subtask will be recorded in the subtask requirement (CB) table and the subtask personnel requirement (CD) table, respectively.
4. Use notes for additional useful information.
5. If explanatory illustrations are required, they must be referenced in the narrative and included in the task support equipment (CG) table.
6. Each disassembly procedure must have a corresponding reassembly procedure.
7. Use full reference designation of components.
8. Identify hardware removed. If hardware is to be retained, state so. Otherwise, hardware removed must be included in the task provisioned item (CI) table.

9. Use generic names of tools except for specific requirements.

10. Use conventional names in procedures (i.e., cross-head screwdriver not screwdriver, cross-head).

11. Use panel markings as placarded. Since the LSAR is all capitalized, distinguish panel markings from text by use of quotes.

12. When a switch does not have a panel marking, give switch position [e.g., Set switch to on (up)].

13. Do not spell out numbers greater than twelve.

A24.2.8   Referencing tasks/subtasks

Procedures that will be performed multiple times can be referenced in subsequent task narratives. Most commonly these tasks involve fault isolation, testing, or access procedures. These tasks will be used exclusively as referenced tasks and will have a task frequency of zero. A procedure *must* be referenced by subtasks but may not include all the subtasks within a task. For example, the fault-isolation procedure may be referenced from the first subtask to the subtask at which the fault is determined for a particular task. Also, any tools or test equipment used in the referenced task must be included in the CG or CI tables of the referencing task as appropriate.

Subtasks may only be referenced one deep; i.e., a subtask referenced in a referencing subtask may not be merged in this system and will make tracking (auditing) more difficult. In order to reference "two deep," both tasks must be called out individually in the referencing task.

A24.2.9   Recording time

Times should be recorded in minutes for each subtask in a task both for elapsed time (SBMMETCB) and manhours (SUBMMMCD). These are the times required for only that subtask, *not* the accumulated times in the task. The times required for the entire task are recorded in hours in the CA table.

A24.3   Task Narrative Example

Tables A24.1 through A24.3 show the data entries for a remove and replace task. These figures illustrate the guidance provided above.

Note: The tables do not include all the data elements that are in that particular table but only those needed to illustrate the example. Also, the logistician will probably never see the data in this "table" format. Most programs will enter the data through a more user-

TABLE A24.1 Table CA (Task Requirement) Entries for Task Narrative Example

LCN LSACONXB	Task code TASKCDCA	Task AOR measurement base AORMSBCA	Task frequency TSKFRQCA	Task identification TASKIDCA	Predicted mean elapsed time PRDMETCA
A	NGONBAA	H	0	FI/FL END ITEM	18.2
A	BGONBAA	H	0	TEST END ITEM	2.0
AB	HGONAAA	H	1.250	R/R OF POWER SUPPLY	19.1

NOTE: The tasks are further defined in the following tables. Notice that the fault-isolation and test tasks have a zero task frequency.

521

TABLE A24.2 Table CB (Subtask Requirements) for Task Narrative Example

Index A24.	Subtask number SUBNUMCB	Subtask identification SUBTIDCB	Referenced subtask LCN RFDLCNCB	Referenced subtask task code REDTCDCB	Referenced subtask number RFDSUBCB	Subtask mean minutes of elapsed time SBMMETCB
.2.4*	001		A	NGONBAA	1	
.2.4*	002		A	NGONBAA	2	
.2.9	003	PREPARE TO OPEN POWER SUPPLY DRAWER				0.8
.2.7.7	004	EXTEND POWER SUPPLY DRAWER 4A3				0.5
	005	VERIFY POWER SUPPLY 4A3A4 IS FAULTY				1.5
.2.7.7	006	REMOVE POWER FROM UNIT 4				1.5
	007	REMOVE FAULTY POWER SUPPLY 4A3A4				3.0
.2.7.7	008	INSTALL NEW POWER SUPPLY				3.5
	009	VERIFY OPERATION OF P/S 4A3A4				1.5
.2.7.7	010	CLOSE POWER SUPPLY DRAWER 4A3				0.7
.2.4†	011		A	BGONBAA	1	
.2.4	012	RETURN TO READINESS CONDITION				0.1

*These two subtasks are the first two subtasks in the FI/FL task (to the point where the power supply is determined to be the fault). Their identification and elapsed time would be merged in LSA-019 by the LSAR software. The elapsed time for these two subtasks is 4 minutes.
†This subtask is the entire test task.

522

TABLE A24.3   Table CC (Sequential Subtask Description) for Task Narrative Example

Index A24.	Subtask number SUBNUMCB	Element indicator ELEMNTCC	Sequential subtask description text sequencing code TEXSEOCC	Sequential subtask description SUBNARCC
2.6	003		1	CAUTION
	003		2	THIS EQUIPMENT CONTAINS PARTS AND ASSEMBLIES SUBJECT TO DAMAGE BY ESD. OBSERVE ESD PRECAUTIONARY PROCEDURES.
2.7.3,.2.7.12,	003	E	3	SET "UNIT POWER" SWITCH 4A1S1 TO ON (UP).
2.7.7,.2.7.11,	003	E	4	SET "UNIT CONTROL" SWITCH 4A1S2 TO "LOCAL".
2.7.13	003	E	5	RELEASE FOUR LATCH FASTENERS THAT SECURE POWER SUPPLY 4A3 TO CABINET.
2.6	003		6	WARNING
	004		1	LETHAL VOLTAGES AND CURRENTS ARE PRESENT IN POWER SUPPLY DRAWER 4A3 WHEN POWER IS APPLIED TO CABINET. USE CARE NOT TO CONTACT LIVE CIRCUITS.
2.7.3	004		2	
	004	E	3	PULL OUT POWER SUPPLY DRAWER 4A3 UNTIL SLIDE ENGAGES
	004	E	4	RELEASE SLIDE LOCK.
	004	E	5	PULL OUT POWER SUPPLY DRAWER 4A3 UNTIL LOCKED IN FULLY EXTENDED POSITION.
	004	E	6	RESET POWER SUPPLY CIRCUIT BREAKER 4A3A4CB1.
.2.7.13	005	E	1	PRESS "RESET" PUSHBUTTON SWITCH 4A1S3.
	005	E	2	WAIT 30 SECONDS.
	005	E	3	OBSERVE "FAULT CODE" DISPLAY READOUT 4A1DS1.
	005	E	4	IF "FAULT CODE" DISPLAY READOUT 4A1DS1 INDICATES NO FAULT PROCEED TO SUBTASK 010, OTHERWISE PROCEED.
	005	E	5	SET "UNIT POWER" SWITCH 4A1S1 TO "OFF".
	006	E	1	AT POWER DISTRIBUTION UNIT 2, SET "UNIT 4 CB" (2CB3) TO OFF (DOWN).
	006	E	2	
	006	E	3	
	006	E	4	APPLY RED DANGER TAG TO "UNIT 4 CB" (2CB3) ACCORDING TO STANDARD TAGOUT PROCEDURES.
	006		5	
2.7.9,.2.7.10,	007	E	1	AT POWER SUPPLY 4A3A4, USING CROSS-HEAD SCREWDRIVER REMOVE AND RETAIN TWO SCREWS, LOCK WASHERS, AND FLAT WASHERS SECURING TERMINAL BOARD COVER AND REMOVE COVER.
	007		2	
	007		3	
.2.7.10	007	E	4	USING CROSS-HEAD SCREWDRIVER REMOVE AND RETAIN THREE SCREWS, LOCK WASHERS, AND FLAT WASHERS SECURING WIRES TO TERMINAL BOARD CONNECTORS.
	007		5	
	007		6	

TABLE A24.3   Table CC (Sequential Subtask Description) for Task Narrative Example (Continued)

Index A24.	Subtask number SUBNUMCB	Element indicator ELEMNTCC	Sequential subtask description text sequencing code TEXSEOCC	Sequential subtask description SUBNARCC
	007	E	7	USING CROSS-HEAD SCREWDRIVER REMOVE AND RETAIN SIX SCREWS, LOCK WASHERS, AND FLAT WASHERS SECURING POWER SUPPLY 4A3A4 TO BOTTOM OF DRAWER.
	007		8	
	007		9	
	007	E	10	REMOVE POWER SUPPLY 4A3A4 FROM DRAWER.
	007	E	11	USING CROSS-HEAD SCREWDRIVER REMOVE FOUR SCREWS, LOCK WASHERS, AND FLAT SECURING MOUNTING PLATE TO POWER SUPPLY.
	007		12	
.2.7.4	007		13	NOTE
	007		14	THE NEW POWER SUPPLY COMES WITH THE REQUIRED HARDWARE FOR ATTACHING TO THE MOUNTING PLATE.
	007		15	
	008	E	1	USING CROSS-HEAD SCREWDRIVER SECURE REPLACEMENT POWER SUPPLY TO MOUNTING PLATE WITH FOUR SCREWS, LOCK WASHERS, AND FLAT WASHERS.
	008		2	
	008		3	
	008	E	4	USING CROSS-HEAD SCREWDRIVER INSTALL POWER SUPPLY ASSEMBLY DRAWER USING RETAINED SIX SCREWS, LOCK WASHERS, AND FLAT WASHERS.
	008		5	
	008		6	
	008	E	7	USING CROSS-HEAD SCREWDRIVER CONNECT WIRES TO TERMINAL BOARD WITH RETAINED THREE SCREWS, LOCK WASHERS, AND FLAT WASHERS.
	008		8	
	008		9	
	008	E	10	USING CROSS-HEAD SCREWDRIVER INSTALL TERMINAL BOARD COVER USING RE-TAINED TWO SCREWS, LOCK WASHERS, AND FLAT WASHERS.
	008	E	11	ENSURE POWER SUPPLY CIRCUIT BREAKER 4A3A4CB1 IS SET TO ON.
	008	E	12	AT POWER DISTRIBUTION UNIT 2, REMOVE RED TAG.
	009	E	1	SET "UNIT 4 CB" (2CB3) TO ON (UP).
	009	E	2	SET "UNIT POWER" SWITCH 4A1S1 TO ON (UP).
	009	E	3	WAIT 30 SECONDS.
	009	E	4	OBSERVE "FAULT CODE" DISPLAY READOUT 4A1DS1.
	009	E	5	IF "FAULT CODE" DISPLAY READOUT 4A1DS1 INDICATES A FAULT
	009	E	6	REFER TO TROUBLESHOOTING INSTRUCTIONS. OTHERWISE PROCEED.
	009	E	7	
	010	E	1	RELEASE SLIDE LOCK AND PUSH POWER SUPPLY DRAWER 4A3 INTO CABINET.
	010		2	
	010	E	3	SECURE POWER SUPPLY DRAWER 4A3 TO CABINET WITH 4 LATCH FASTENERS.
	010		4	
	010	E	5	SET "UNIT CONTROL" SWITCH 4A1S2 TO "REMOTE".

524

friendly data entry screen. The "INDEX" field in Tables A24.2 and A24.3 is used to refer to subparagraph numbers in paragraph C above.

A24.4 Summary

The point of the conventions is to ensure that everyone is working from the same "playbook." Although the initial establishment of this guidance may appear to be tedious, the benefits to be derived will become apparent during the first review of the LSAR. In addition, after it is established on one program, it will become easier to develop for the next by modification. The example given is only one of many options and sets very narrow limits on the task writer. This does not imply that these particular conventions should be applied to all LSAR programs but rather to illustrate what was done on *one* program. Each requirer/LSA manager will probably have different and unique requirements.

Postproduction Support Checklist

A25.1 Supply Support

1. Continued producibility and availability of components and parts. [Every peculiar item within the system should be reviewed to the subcomponent level and NSN (see DODD 4005.16).]
   a. Are technical data available at a reasonable cost?
   b. Is stability of design a concern?
   c. Is competitive procurement appropriate?
   d. Is the production base adequate?
   e. What proprietary rights, if any, have been declared by the prime and subordinate contractors?
   f. Are rights in data procurable at a reasonable cost?
   g. What is life-of-type buy potential?
   h. Are repair facilities available?
   i. Is the component critical to system performance?
   j. What is the expected life of the system/subsystem?
   k. Is there foreign military sales support potential?
   l. Are work-around alternatives available?
   m. Are quality assurance requirements unique, difficult to duplicate?
   n. Is contract logistics support feasible?
   o. Will failure rates be high enough to sustain organized capability?
   p. Technological obsolescence. Is system replaceable with new technology?
   q. Will potential design changes eliminate the need for the part?
   r. Is engineering level-of-effort contract appropriate to ensure continued supportability?
2. What support equipment is required?
3. Will support of support equipment be available at a reasonable cost?
4. Is there an adequate organization to focus on and resolve postproduction problems?

A25.2 Engineering

1. Who has been designated to perform a quality acceptance inspection on technical data?
2. Will there be adequate field engineering support, configuration management, and Engineering Change Proposals (ECPs) support?

Will there be adequate support to update:

a. Technical manuals

b. Production drawings

c. Technical reports

d. Logistics support data

e. Operational and maintenance data

f. User's manuals

g. Data requirements

3. Will operational experience be considered in changes to the material system?

A25.3 Competitive Procurement

1. Is production rate tooling complex/cost significant; is it readily available or is there a long lead time for procurement?

2. Are all cost factors associated with a breakout/competitive procurement decision considered? Cost elements should encompass added tooling, special test equipment, qualification testing, quality-control considerations, rights in data procurement, etc. If performance specifications are applicable, the following additional costs pertain: cataloging, bin opening, item management, technical data, production and distribution variables, rate and ADP hardware/software augment costs, configuration control, and engineering requirement costs.

3. Are all potential customers included in the production requirements computations?

4. Do resources cover installs, spares, rejects, repairs, and FMS as applicable?

A25.4 ATE Support

1. Hardware

   a. Will hardware be supportable?

   b. Will mission/ECP changes be compatible?

   c. Will modifications be possible, supportable?

   d. Is system expandable?

2. Software

   a. Will diagnostic software changes be possible?

   b. Will the organizational structure allow for continuing software update?

   c. Will software changes caused by ECP/mission changes be incorporated?

A25.5   Storage and Handling

1. Will shelf life items be replaceable when they expire?
2. Will special shipping containers be replaceable/repairable?
3. Will peculiar manufacturing tools and dies be procured and stored?

A25.6   Technical Data

1. Will manufacturing shop standards and procedures be retained?
2. Will all changes that occur during the production phase be incorporated in the manufacturing shop drawings?

A25.7   Training

1. Will simulators and maintenance trainers be supportable in the out-years?
2. Will follow-on factory training be required?

A25.8   Maintenance

1. Will depot overhaul be required in the out-years? Organic or contract.
2. Will provisions be made in the front end to accommodate a service life extension program if required? (Most recent material systems have been extended past their original forecasted disposal date.)
3. Will components be available to support the depot overhaul program in the out-years?
4. Is it realistic to comingle manufacturing with repair on a single production line?

Front-End Cost Analysis (FECA) Model

This appendix presents a simple macro tradeoff analysis model designed to support front-end cost analysis (FECA). The user should have a working knowledge of LOTUS 1-2-3 or some other spreadsheet. The objective is to improve the LSA analyst's capability to estimate and compare cost projections for logistics support requirements.

The model herein is an adaptation of a MRSA initiative. It is designed for use with the LSA 200 and 300 series tasks prior to EMD. It uses a simple spreadsheet program which can be augmented/tailored, as necessary, for specific program applications. The model is convenient for updating and performing sensitivity analysis of "what if" questions. The resulting cost estimates can provide "bottom line" values for assessment of specific alternatives and/or excursions.

A26.1 Purpose

The prime purposes of the FECA model are to

Give visibility to the various costs associated with a logistics support concept.

Serve as a tradeoff analysis tool to measure the impact of supportability-related decisions.

Provide a framework for front-end cost analysis.

A26.2 Objectives

The objectives of the FECA model are to

Permit evaluation of a wide range of equipment, operations use, and support concepts and issues.

Be easy to tailor/customize.

Be readily usable, with minimal training, by an LSA analyst.

Be PC adaptable.

Minimize tedious, time-consuming calculations.

Provide an audit trail capability.

A26.3   Limitations

The following are limitations to the use of the FECA model:

1. The model provides approximations, with emphasis on being representative, broad in scope, generic, and consistent.
2. The model is designed for macro analyses early in the acquisition process rather than detailed analysis.
3. Outputs are only as good as the inputs.
4. Only current-year dollars are used.
5. The model makes no correction for unique cases or second effects.

A26.4   Model Description

A26.4.1   Model concept

The FECA model processes LSA task input data, including use (task 201, Use Study), system (task 202, Standardization), and baseline comparison system (BCS) data (task 203, Comparative Analysis), frequently developed to estimate logistics resources and A_o (see Fig. A26.1). Inputs must be at a consistent level of detail with each other and the issues to be addressed. BCS data can be used as the default set of data from which all tradeoffs are made. Figure A26.1 presents a conceptual view of this process. Tradeoffs are accomplished by making factor adjustment changes to represent the design concept or support concept being evaluated. The factors can be adjusted to represent different support concepts, standardization impact, new technologies, etc. Adjustment factors along with the default data become the basis for the adjusted data used in the resource calculations and resource outputs for a tradeoff analysis.

A26.4.2   Spreadsheet layout

The spreadsheet is constructed for convenience of data input; easy review of logistics resources, operational availability, and a mobility index; and ease of issue documentation. Figure A26.2 is a diagram of the spreadsheet showing the area where each type of data resides. Following is a description of these areas of the spreadsheet. The number following each data area name is its beginning cell address. Data area size depends on the number of functions being used for the system evaluated.

Data entry areas

Issue area (I1).   An input area for documentation of the issue being examined. A statement of the issue, opportunities for impact, factors ad-

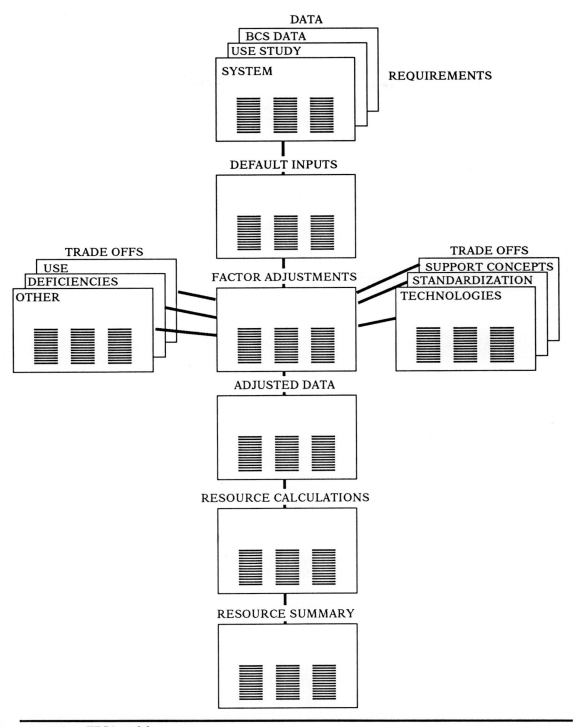

Figure A26.1 FECA model concept.

Introduction/Index				Standard Factors	Macros
Factor Adjustments			Issue Documentation		
Function 1	Function 2	Function 3			
System and Function Data Biased by Trade-off Factors for the issue					
Detailed Logistic Resources by Function For 1 Organization (Output)					
Intermediate Algorithms					
Summary of Logistic Resources by System for 1 Organization					
System Cost Over Life of System					
			Default Input Values For System and Functions Will Be Starting Point From Which to do Trade-offs For Various Support Concepts		

Figure A26.2   Spreadsheet.

justed, and observations can be entered. After a tradeoff is accomplished, summary results are automatically entered in this area. *Caution:*   Entering factors adjusted in this area does not change the value of the factor.

Adjustment factors area (A25).   The area where changes to the inputs are made when performing tradeoffs. The initial set of adjustment factors is set to 1. The user inputs a set of appropriate adjustment factors to do tradeoffs. These adjustment factors are then used to make changes across the entire set of spreadsheet functions.

System data (H58).   Common for all system functions. This is a set of input data for a system which is seldom changed during tradeoffs. The left side of Table A26.1 provides a listing of the system data.

Operations data (A74).   Data peculiar to the use or operation of a system such as the number of systems per organization and the number

TABLE A26.1   System Input Data

System-level data	Operations data
System name	Number of locations
System life	Quantity produced
Deployment requirements	Quantity required/organization
Required/organization	Operating hours/item/year, peace
Percent deployed in war	Crew operating size/item, operators
Percent deployed in peace/year	Remoteness of maintenance (h)
Surge rate (1–10)	Unit size (footprint)
Climate (operating) (1,2,3)	Mission operating capability
Threat level (1,2,3)	A_o objective
Crew hours/year operations	

of operating hours per year. These data are changed as system use changes but are usually static during the tradeoffs. The right side of Table A26.1 provides a listing of the operations data.

Default function data set (A325).   Consists of a data set for each function or subsystem. Examples of subsystems are propulsion, payload, common avionics, etc. Data input in this area are the baseline data for each function and become the static set of data on which all tradeoffs are based. Table A26.2 provides a list of the function input parameters.

TABLE A26.2   Input Data by Function: Function Characteristics

Percent operating h/yr/item (peace/normal)	Crew size (maint.)
Percent operating Crew appl. to function (0.1–1)	Training locations, maint.
Unit cost (M)	Training weeks, maint.
Unit weight, metric tons	Training locations, ops.
Number of subassemblies	Training weeks, ops.
MTBPM	SE complexity (1–3, 3 = complex)
MTBF (unscheduled) (h)	Lines of code
MPTM (scheduled) (h)	Software complexity (1–10, 10 = complex)
Unit	Software document quality (1–3, 1 = poor)
Intermediate	Software stability (1–5, 0 = no SW)
Depot	POL consumption (kg/h)
MTTR (unscheduled) h on equip.)	Consumable value/unit
MTTR (unscheduled) h off equip.)	kg/unit consumable
Diagnostic effectiveness (%)	Expenditure rate (units/event)
Equip. complexity (1–3, 3 = complex)	Events/operating hour % Sc5 h maint.
Performed at % NDI	
Organizational	Bat. dam. susceptibility index (1–10)
Intermediate	Bat. dam. repair diff. index (1–10)
Depot	% embedded training
% unscheduled repairs at	A_o objective
Units	
Intermediate	
Depot	

Output areas.   There are four sets of output from this model.

Issue area (I1).   Provides documentation germane to the issue. It is both an input and output area. It is described under Data Entry Area (I.a).

Detailed O&S cost output (A135).   Provides a detailed formated output of logistics resources by function for each logistics factor, allowing the analyst to see where major logistics resource contributions are located. This is where the majority of spreadsheet calculations are done. Table A26.3 provides a list of resources that are calculated and provided as output in the detailed O&S costs output report.

Production and O&S cost summary (A231).   Table A26.4 lists the resources provided in the formated logistics resources summary output. This output summarizes the logistics resource costs for the system. Most of the data for this output are calculated from the results of the detailed O&S costs output.

Logistics resources: Life-cycle cost estimate (A278).   This is a formated output that breaks out the logistics resource costs by development cost, production cost, and O&S cost for the life of the system. The detailed logistics resources output, logistics resources summary, and sys-

TABLE A26.3   Detailed O&S Cost Output (For One Organization in Millions of Dollars)

Equip. cost	Equipment supply cost/yr
Operations MH cost/yr	Item management cost/yr
Maint. MH cost/yr	Spares
Scheduled	Initial cost
Unit	Replenishment cost/yr
Intermediate	WARF cost
Depot	Support equipment cost
Total scheduled maint. MH	Unit
Unscheduled	Intermediate
Unit	Depot
Intermediate	SE support cost/yr
Depot	Unit
Total unscheduled maint. MH	Intermediate
Total maintenance MH cost/yr	Depot
Manpower maintenance cost/yr	Transportation cost/yr
Unit	Facilities cost
Intermediate	Facilities support cost/yr
Depot	Technical data cost
Total manpower maint. cost/yr	Technical data support cost/yr
Manpower operations cost/yr	Software cost
Manpower management cost/yr	Software support cost/yr
Training	POL cost/yr
Operations cost/yr	Consumables cost/yr
Maintenance cost/yr	A_o (peace)
Equipment qty. cost	A_o (war)
	Mobility index

TABLE A26.4   LCC Estimate (for One Organization in Millions of Dollars)

Production (cost)	O&S (cost/yr)
Equipment	Operations
Training equip. purchase	Sch. maint. (U,I,D)
Spares (initial and WARF)	Unsch. maint. (U.I.D.)
Support equipment (U,I,D)	Manpower maint. (U.I.D.)
Facilities	Manpower operation
Technical data	Manpower management
Software	Training (op., M., equip., sup.)
Total production cost	Item management
	Spares (replen.)
	SE support
	Transportation
	Facilities support
A_o (peace)	Technical data support
A_o (war)	Software support
	POL
Mobility index	Consumables
	Total O&S cost/yr

tem life are the main components for this output. Table A26.5 provides a list of the logistics resources that are calculated and output in the life-cycle cost estimates output

Spreadsheet documentation area (AI). A small area at the beginning of the spreadsheet which gives a brief description of the mode 1 and indices of the major input, output, and macros to simplify use of this model.

TABLE A26.5   Logistics Resources: LCC (for One Organization in Millions of Dollars)

Development cost	O&S cost
Production cost	Manpower operations
Equipment	Manpower maintenance
Training equipment	Manpower management
Spares (initial and WARF)	Training operations
Support equipment	Training maintenance
Facilities	Training equipment support
Technical data	Replenishment spares
Software	Support equipment support
	Transportation
Total production cost	Facilities support
	Technical data
	Software support
	POL
	Consumables
	Total O&S cost

Static data area. Spreadsheet data areas where system data are initially input. These data remain constant during the tradeoffs.

Default function data (A325).   Data sets for each function. Once the data set is entered, it becomes a static set to be used as the starting point for all tradeoffs.

General standard factors (AAl). Factors that seldom require changes such as depot repair cycle (DRC) time in days, cost of transportation per metric ton, and mean logistics delay time (MLDT) in hours.

Issues area (I.A.).   This is an input area for documentation of the issue being examined. A statement of the issue, opportunities for impact, factors adjusted, and observations can be entered. After a tradeoff is accomplished, the summary results are automatically entered in this area. *Caution:* Entering factors adjusted in this area does not change the value of the factor.

Interim calculations. Locations where a set of calculations is done between the input data step and the detailed logistics resources output.

Adjusted function data (A86).   Data calculated by taking the default function data (static data) multiplied by data factors adjusted to represent the concept or issue under consideration. The result is used in calculating the logistics resources for the concept or issue under analysis.

Interim algorithms (A215).   A subset of several more complex algorithms. This computation only needs to be done one time, and the results are input to several algorithms. (An example of an interim algorithm is the number of failures per year for a particular function. The number of failures is calculated and used in spares, maintenance, training, support equipment, and transportation algorithms.)

Macros (AGI). Contains macros to enable easy maneuvering around the spreadsheet and printing of sections of the output. Each macro is preceded by its name and followed by a description of its function.

A26.5   Algorithms

Table A26.6 contains the cell addresses of inputs, outputs, and algorithms for calculations that are used in this spreadsheet. All adjustment factors, system data, general standard factors, and macros are presented. The cell addresses and algorithms for one function are listed. Each function uses the same set of algorithms. Four functions have been used in this spreadsheet. The output summary of logistics resource estimates and cost over life of system algorithms reflect summary across the four functions of the system.

TABLE A26.6   Cell Addresses of Inputs, Outputs, and Algorithms for Calculations Used in This Spreadsheet

Symbols

Standard LOTUS symbols are used to designate math and system functions as follows:

Symbol	Meaning
/	Divided by
*	Multiplied by
'	Justification label (left)
"	Justification label (right)
~	Macro
@	Function at
+	Positive, add
^	Exponent
= < >	Equal, not equal
–	Negative, subtract
< >	Less than, greater than
#Not#	Logical NOT
#And#	Logical AND
#Or#	Logical OR
&	String combination
( )	Performed first
{?}	Pause in macro

Input Factors

Static Default System Data

Input factor 1: System name
  Definition: State name of system being analyzed.

Input factor 2: System life
  Definition: System life is entered as this factor.

Input factor 3: Deployment requirements
  Definition: Percent deployed in war, percent of systems deployed to combat location.

  or

  Definition: Percent redeployed in peace per year, percent of systems redeployed during peacetime per year.

Input factor 4: Required/organization
  Definition: Percent deployed in war and percent deployed in peacetime per year.

Input factor 5: Surge ratio (1–10)
  Definition: The ratio of peacetime use to wartime use.

Input factor 6: Climate operating (1, 2, 3)
  Definition: An index of climatic conditions.

Input factor 7: Threat level (1, 2, 3, 1 = low intensity)
  Definition: An index of combat action expected.

Input factor 8: Crew hr/yr operations
  Definition: Wartime hours per year for operator/crew members.

TABLE A26.6   Cell Addresses of Inputs, Outputs, and Algorithms for Calculations Used in This Spreadsheet (Continued)

Static default operations data

Input factor 1: Number of locations
  Definition: The number of separate operating locations requiring intermediate-level support.

Input factor 2: Quantity produced
  Definition: The total quantity that will be built.

Input factor 3: Quantity required/organization
  Definition: The quantity authorized for one organization.

Input factor 4: Op. hrs/item/yr peace
  Definition: Total operating hours in any mode for peacetime operations, including training hours.

Input factor 5: Crew size/item operators
  Definition: Number of personnel required to operate each system.

Input factor 6: Remoteness of maintenance
  Definition: Time in hours required to reach the equipment for maintenance beyond the operator capability when the unit is in the field.

Input factor 7: Unit size (footprint)
  Definition: The area size of the deployed unit in square meters.

Input factor 8: Mission capability
  Definition: The percent of total mission requirements that can be met by the system in normal operation.

Input factor 9: A_o objective
  Definition: The expected operational availability of the force to perform the mission in wartime. This would be taken from RAM rationale.

Static default function data: Data required for each system function

Input factor 1: % of op. hr/yr/item (peace/normal)
  Definition: Percent of system hours this function operates (may be less than or greater than 100 percent).

Input factor 2: % op. crew appl. to function (0.1–1)
  Definition: Percent of the operating crew supporting the function. Total for all functions must equal 1.

Input factor 3: Unit cost
  Definition: Cost of one unit of the function in millions of dollars.

Input factor 4: Unit weight, metric tons
  Definition: Weight of one unit of equipment in metric tons.

Input factor 5: Number of subassemblies
  Definition: Number of line-replaceable assemblies in one unit of the function (line-replaceable units × quantity per function).

Input factor 6: MTBPM
  Definition: Mean operating hours between preventive maintenance actions.

Input factor 7: MTBF (unscheduled)
  Definition: Mean operating hours between failures.

TABLE A26.6   Cell Addresses of Inputs, Outputs, and Algorithms for Calculations Used in This Spreadsheet (Continued)

Static default function data: Data required for each system function (Cont.)

Input factor 8: MTPM (scheduled)
   Definition: Mean time in hours to accomplish preventive maintenance.
     Unit
     Intermediate

Input factor 9: MTTR unsch. H, on equip.
   Definition: Mean time to repair actions accomplished on the equipment.

Input factor 10: MTTR unsch. H, off equip.
   Definition: Mean time to repair assemblies removed from the equipment.

Input factor 11: Diagnostic effectiveness
   Definition: Percent of corrective actions successful on the first attempt using all available means.

Input factor 12: Equip. complexity (1–3, 3 = complex)
   Definition: A complexity index of the function.

Input factor 13: % (scheduled) preventive maintenance performed at O, I, D
   Definition: Percent of preventive maintenance actions performed at each level by personnel assigned to that level.
     Unit
     Intermediate
     Depot

Input factor 14: % (unscheduled) corrective maint. repaired at:
   Definition: Percent of failures repaired at the level indicated. It is assumed that all failures result in the vehicle being returned to service at the unit level. Removed equipment is then repaired at DS/GS level. The model user is expected to assess the total distribution of repair actions required for each failure.
     Unit
     Intermediate
     Depot

Input factor 15: Crew size
   Definition: Number of persons required for system maintenance actions.

Input factor 16: Training locations, maint.
   Definition: Number of maintenance training locations required.

Input factor 17: Training weeks, maint.
   Definition: The number of weeks required to train each maintainer.

Input factor 18: Training location ops.
   Definition: The number of operator training locations per unit.

Input factor 19: Number ops. training weeks
   Definition: The number of training weeks required per operator.

Input factor 20: SE complexity (1–10)
   Definition: An index of complexity of support equipment.

Input factor 21: Lines of code
   Definition: The expected number of lines of code for end item software.

TABLE A26.6   Cell Addresses of Inputs, Outputs, and Algorithms for Calculations Used in This Spreadsheet (Continued)

Static default function data: Data required for each system function (Cont.)

Input factor 22: Software complexity (1–10, 1 = simple, 10 = complex)
   Definition: An index of software complexity.

Input factor 23: Software document quality (1–3, 3 = poor)
   Definition: An index of software quality.

Input factor 24: Software stability (0–3, 0 = no software)
   Definition: Expected rate of software change.

Input factor 25: POL consumption (kg/h)
   Definition: POL consumption rate for function.

Input factor 26: Consumable value/unit
   Definition: Cost per unit of consumables for ammunition, etc.

Input factor 27: Kg unit consumable
   Definition: Weight per unit of consumable.

Input factor 28: Expenditure rate (unit events)
   Definition: Consumption rate per event (i.e., miles/gallon).

Input factor 29: Events/opr. hour
   Definition: Events per operating hour (i.e., rounds/hour).

Input factor 30: % NDI
   Definition: Percent of the function that can be satisfied with existing equipment, not requiring development.

Input factor 31: Bat. dam. susceptibility index (1–10, 1 = not vulnerable)
   Definition: A vulnerability index of the function to battle damage.

Input factor 32: Bat. dam. repair diff. index (1–10, 1 = easy)
   Definition: An index of the difficulty of effecting battle damage repair.

Input factor 33: % Embedded training
   Definition: Percent of the functions that have embedded training.

Input factor 34: A_o objective
   Definition: The stated operational availability objective for the function.

Input factor 35: Condemnation rate, %
   Definition: Percent of failures resulting in disposal of the failed subassembly; 100 percent design for discard gives condemnation rate of 1.

Adjustments to factors (issues)

The adjustment factors are all set to one (1) unless modified by the analyst. To test a change or issue to the standard data, the following elements of data may be modified by the analyst using the FECA model.

TABLE A26.6   Cell Addresses of Inputs, Outputs, and Algorithms for Calculations Used in This Spreadsheet (Continued)

System factors
   System name
   System life
   Deployment requirements
   % deployed in war
   % redeployed in peace/yr
   Surge rate
   Climate
   Threat level
   Crew hr/yr ops.
   Manning policy

Operations factors
   Number of locations
   Quantity produced
   Quantity required/orgn.
   Op. hr/item/yr peace
   Crew size/items ops.
   Remoteness of maint. (hr)
   Unit size (footprint)
   Mission capability
   A_o objective

Function factors (for each function)
   % ops. h/year/item (peace/normal)
   % ops. crew appl. to function
   Unit cost (m)
   Unit weight (metric tons)
   Number of subassemblies
   MTBPM (h)
   MTBF (unsch.)(h)
   MPTM (sch.)(h) at:
      Unit
      Intermediate
      Depot
   MTTR (unsch.) h, on equip.
   MTTR (unsch.) h, off equip.
   Diagnostic effectiveness (%)
   Equip. complexity
   Condemnation rate (%)
   % sch. maint. performed at:
      Unit
      Intermediate
      Depot
   % unsch. maint. performed at:
      Unit
      Intermediate
      Depot
   Crew size (maint.)
   Training locations, maint.
   Training weeks, maint.
   Training locations, ops.

TABLE A26.6   Cell Addresses of Inputs, Outputs, and Algorithms for Calculations Used in This Spreadsheet (Continued)

Function factors (for each function) (Cont.)
   Training weeks, ops.
   SE complexity
   Lines of code
   Software complexity
   Quality of SE documentation
   SW stability
   POL consumption
   Consumable value/unit
   kg/unit consumable
   Expenditure rate (events/units)
   Events/operation hour
   % NDI
   Bat. dam. susceptibility index
   Bat. dam. rpr. diff. index
   % embedded training
   A_o objective

Output

Detailed output of logistics resources required for one organization in millions of dollars.

Output factor 1: Equipment cost
   Definition: Cost of all equipment purchased.
   Algorithm:
   Unit cost × adjustment factor × total quantity required

Output factor 2: Operations MH cost/yr
   Definition: Cost of operations manpower needed to operate the system for 1 year.
   Algorithm:
   Ops. h/yr/orgn. × crew size/item × yrs adjustment factor × % ops. crew applied to function × cost of MH operations × standard conversion factor to millions (0.000001)
   Default value: Cost of MH operations $12.32.
   Source: NAVCOMPT Notice 7041.

Output factor 3: Maintenance MH cost/yr
   Definition: Detailed costs associated with maintenance manhours at organization, intermediate, and depot levels for scheduled and unscheduled maintenance for 1 year.
   Algorithms:
   Scheduled maintenance MH cost
      Unit:
   Previous maintenance action/yr × (lst unit × additional units + adjustment factor × crew size × 2) × cost MH maint. (O) × conversion factor to millions (0.000001)
      Intermediate:
   Previous maintenance action/yr × (lst unit × additional units + adjustment factor × crew size × 2) × cost MH maint. (I) × conversion factor to millions (0.000001)
      Depot:

TABLE A26.6 Cell Addresses of Inputs, Outputs, and Algorithms for Calculations Used in This Spreadsheet (Continued)

Previous maintenance action/yr × (1st unit × additional units + adjustment factor × crew size × 2) × cost MH maint. (D) × conversion factor to millions (0.000001)
Total scheduled maintenance MH cost = unit + intermediate + depot
Unscheduled maintenance MH cost
   Unit:
Failures per year × unit unsch. × (MTTR unsch. × diagnostic effect factor + remoteness of ops. maint. hds. × crew maint. + size × 2) × cost of MH/unit × conversion to millions
   Intermediate:
Failures per year × intermediate unsch. × MTTR off equip. × diagnostic effect factor × cost of I MH × conversion to millions
   Depot:
Failures per year × depot unsch. × MTTR off equip. × diagnostic effect factor + cost of depot MH × conversion to millions
Total unscheduled maint. MH = O + I + D above
Total maint. MH = total sch. + total unsch.
Default factors: See default factors for output factor 4.

Output factor 4: Manpower maint. cost/yr
  Definition: Provides a method to compute manpower requirements based on manhours requirements developed in output factors 2 and 3 above.
  Algorithms:
   Unit: 2000/[1000 × (2 + % deployed in war)] × (sch. unit MH + unsch. MH) × total war force factor + bat. damage hd × crew size maint. × cost per unit MH factor × conversion to millions
   Intermediate: 2000/[1000 × (2 + % deployed in war)] × (sch. I MH + unsch. I MH) × total war force factor + bat. damage hd × crew size maint. × cost per I MH factor × conversion to millions
   Depot: (2000/3000) × (depot sch. + depot unsch. MH) × total war force factor + bat. damage hd × crew size maint. × cost per D MH factor × conversion to millions
Total unscheduled maint. manpower cost
Total = unit + intermediate + depot
Total maint. manpower costs
Total maintenance power costs = total sch. + total unsch.
  Default values: Cost of MMH O = 12.32 × 2
                Cost of MMH I = 15.02 × 3
                Cost of MMH D = 15.02 × 3
  Source: NAVCOMPT Notice 7041

Output factor 5: Manpower operations
  Definition: Computes the operational manpower requirements.
  Algorithm: (Operating hours/unit/yr ÷ % deployed in war) × total war force factor × 2000 × cost MH operations × conversion to millions

Output Factor 6: Manpower management
  Definition: Computes the management manpower requirements.
  Algorithm: (Number of subassemblies ÷ items per manager) × (1 − % of NDI) × 2000 × cost/h mgt. × conversion to millions × 0.1

TABLE A26.6 Cell Addresses of Inputs, Outputs, and Algorithms for Calculations Used in This Spreadsheet (Continued)

Output factor 7: Operations training cost
  Definition: Computes the operations training costs for the program.
  Algorithm: [Operations manpower ÷ (2000 × cost/MH maint.)] × attrition rate %/ yr × training weeks per operator × cost/training week

Output factor 8: Maintenance training cost
  Definition: Computes the maintenance training costs for the program
  Algorithm: (Maintenance manpower O, I, and D) ÷ (2000 × cost maint. MH) × attrition rate % × trng. wks./maint. position × cost/training week

Output factor 9: Item management
  Definition: Computes the management of items at the wholesale level (item manager).
  Algorithm: (Number of subassemblies ÷ items per manager) × (1 − % of NDI) × 2000 × cost/h to manage item × conversion to millions

Output factor 10: Spares initial cost
  Definition: Provides a cost for initial spares support of the program.
  Algorithm: $(1 - A_o \text{ objective}^2)$ × failures per year × [unsch. I% maint. × I RCT days + unsch. depot % maint. (depot RC days + O&S days) × unit cost ÷ (365 × number of subassemblies)] × diagnostic effect factor + failures per year × 2 × avg. cost repair part × conversion to millions
  Default values:
    I repair cycle days = general factor
    D repair cycle days = general factor
    Avg. cost repair part = general factor
  Source: General factors

Output factor 11: Replenishment spares
  Definition: Computes the cost of replenishment spares to support the program for 1 year.
  Algorithm:
Failures per year × (avg. cost per repair × conversion to millions) + condemnation rate × unit cost ÷ number of subassemblies
  Default value: Condemnation rate = function factor
  Source: Functions input data

Output factor 12: War cost factor
  Definition: Computes the cost of the surge to wartime rates.
  Algorithm:
(Failures per year ÷ 365) × war surge/threat factor × (unit cost ÷ number of assemblies) × MLDT hours × diagnostic effect factor + threat level (1,2,3) × battle damage susceptibility index × MLDT hours × qty. reqd. per orgn. × % deployed in war ÷ 1000 × unit cost ÷ number of assemblies
  Default value: See above
  Source: General factors/function inputs

TABLE A26.6   Cell Addresses of Inputs, Outputs, and Algorithms for Calculations Used in This Spreadsheet (Continued)

Input factor 13: Support equipment (SE)
  Definition: Computes the support equipment required to support the system at O, I, and D.
  Algorithms:
    Unit: If O, go to I [(Failures per year × MTTR (unsch.) h, on equip. × surge rate × threat level × diagnostic effect factor + battle damage index) ÷ 4000 + number of locations] × SE complexity factor × unit cost ÷ 8
    Intermediate: [(Failures per year × MTTR (unsch.) h, off equip. × I level % unsch. maint. performed × surge rate × threat level × diagnostic effect factor + battle damage index) ÷ 4000 + number of locations] × SE complexity × unit cost ÷ 8
    Depot: [(Failures per year × MTTR (unsch.) h, off equip. × depot level % unsch. maint. performed × surge rate × threat level × diagnostic effect factor + battle damage index) ÷ 4000 + number of locations] × SE complexity × unit cost ÷ 8
  Default value: SE complexity factor
  Source: LSA Task 202

Output factor 14: Support of SE
  Definition: Computes the cost to support SE for 1 year at O, I, and D.
  Algorithms:
    Unit: Unit SE cost × SE support E. cost ratio
    Intermediate: I level SE cost × SE support E. cost ratio
    Depot: D level SE cost × SE support E. cost ratio
  Default value: SE support cost ratio
  Source: General factor

Output factor 15: Transportation cost
  Definition: Cost of transportation for 1 year to and from depot.
  Algorithm: [Operation hours per year per organization × (% unsch. depot maint. ÷ MTFB (unsch.) h) × diagnostic effect factor × unit weight (metric tons) × 2 ÷ number of subassemblies + (events per operating hour × expenditure rate (units/events) × kg/unit consumable ÷ 1000) + POL consumption (kg/h) ÷ 1000 + 2 × % sch. maint. at depot × unit weight ÷ (MTPM × number of assemblies)] × $ per metric ton × conversion to millions
  Default values: Unit weight (metric tons), number of subassemblies, events per operating hour, expenditure rate, kg/unit consumable, POL consumption, and unit weight
  Sources: Unit weight (metric tons) = system level
    Number of subassemblies = system level
    Events per operating hour = system level
    Expenditure rate = system level
    Kg/unit consumable = system level
    Unit weight = general factors

TABLE A26.6   Cell Addresses of Inputs, Outputs, and Algorithms for Calculations Used in This Spreadsheet (Continued)

Output factor 16: Facilities
   Definition: Approximate facilities cost for program at all locations.
   Algorithm:
Unit size (footprint) × qty. required/organization ÷ 10 × cost of facilities/M^2
   average × equip. complexity × climate × threat level × conversion to millions
   Default values: Unit footprint, qty. required, cost of facilities, equipment complexity, climate, threat level, and conversion
   Sources: Unit footprint = general factor
      Quantity required = system factor
      Equip. complexity = system factor
      Climate = system factor
      Threat level = system factor
      Conversion factor = general factor

Output factor 17: Facilities support
   Algorithm:
Product from factor 16 × 0.01

Output factor 18: Technical data
   Definition: Cost of technical data to support the program.
   Algorithm: [Unit cost (M) × unit weight (metric tons) × equip. complexity ÷ (unit size (footprint) × 150)] × 0.1 × cost/page tech. data × conversion to millions
   Default value: Cost per page of tech. data
   Source: General factors

Output factor 19: Technical data support
   Definition: Cost to maintain technical data for 1 year.
   Algorithm:
Result from output factor 18 × 0.1 × 0.5

Output factor 20: Software (SW)
   Definition: Computes the cost of software for the system for 1 year.
   Algorithm: (Software complexity × lines of code × quality of SW documentation ÷ 3) × cost per line of SW code × conversion to millions
   Default value: Software complexity, lines of code, quality of SW documentation, and cost per line of SW code
   Source: Lines of code = system factor
      Software complexity = system factor
      Cost per line of SW code = general factor
      Quality of SW documentation = system factor

Output factor 21: Software support
   Definition: Cost to maintain SW for 1 year.
   Algorithm: [Quality of SW documentation = 0, 0, (software complexity × lines of code × SW stability) ÷ (quality of SW documentation × 0.5)] × conversion to millions
   Default value: Quality of SW documentation = system factor
      Software stability = system factor
   Source: System factors

TABLE A26.6   Cell Addresses of Inputs, Outputs, and Algorithms for Calculations Used in This Spreadsheet (Continued)

Output factor 22: POL
   Definition: Cost of POL used per year.
   Algorithm:
Opr h/yr/orgn. × POL consumption (kg/h) × cost/kg of POL × conversion to millions
   Default value: POL consumption (kg/h)
      Cost of POL (per kg)
   Source: POL consumption = system factor
      Cost of POL = general factor

Output factor 23: Consumables
   Definition: Cost of consumables to support 1 year of operation.
   Algorithm:
Opr. h/yr × events/opr. h × expenditure rate × consumable value/unit × conversion to millions
   Default values: Events per opr. h
      Expenditure rate
   Source: System factors

   Output factor 24: Peace Time A_o
   Definition: Peacetime A_o to support the program for 1 year.
   Algorithm:
(Peace uptime − peace downtime) ÷ total peace time
   Default value: N/A

Output factor 25: War Time A_o
   Definition: Wartime A_o for program.
   Algorithm:
(War uptime − war downtime) ÷ total war uptime

Output factor 26: Cost over life of system
   Systems cost: B283 (sys. name)

Output factor 27: Development costs (B285)
   Algorithm:
Equip. cost × (1 − % NDI) × equip. complexity × 0.1

Logistics resources summary

This output summarizes the logistics resource costs for the system. Most of the data for this output are calculated from the results of the detailed logistics resources output. This is a formulated output that breaks out the logistics resource costs by development cost, production cost, and O&S for the life of the system.

General standard factors

Attrition rate: %/yr: Annual rates of personnel turnover in given unit or force to determine the number of new accessions to be trained.
h/shift: The typical length of a day during a wartime mission.
Days/week: The typical number of days per week during a wartime mission.
$/metric ton transportation: The cost in current year dollars (CYD) to ship a metric ton of equipment from the United States to Europe.
Avg. cost/repair part: The average cost in CYD of a typical repair part.
Cost/h mgt.: Average personnel cost per hour for managing the system at the commodity command level.
Cost/MH maint.: Average cost per manhour for maintenance personnel at all levels (U, I, and D).

TABLE A26.6   Cell Addresses of Inputs, Outputs, and Algorithms for Calculations Used in This Spreadsheet (Continued)

Cost/tng. week: The average weekly cost to train a person; includes trainee pay, instructor cost, facility cost, and cost of training materials less ammunition.

Cost/page of tech. data: The cost in CYD to write and produce one page of technical data.

Cost/line of software code: The cost in CYD to write, program, and debug a line of software code.

Cost of facilities/m² avg.: The cost in CYD to build and maintain a square meter of maintenance facility.

Cost/kg of POL: The average cost in CYD for 1 kg of fuel.

SE support cost ratio (%): Ratio of cost to maintain support equipment for 1 year to the acquisition cost of the support equipment.

Training equipment ratio (%): Ratio of the cost to maintain training equipment for 1 year to the acquisition cost of training equipment.

Items/mgr.: The average number of items assigned to an item manager at the commodity command.

IRCT days: Intermediate repair cycle time. The average time required to cycle a part to an intermediate maintenance activity, repair it, and return it to supply.

DRC days: Depot repair cycle. Average time to repair an item at the depot and return it to supply. Does not include transportation time.

O&S days: Order and ship days. The average time required to requisition and ship a spare or repair part.

MLDT hours: Mean logistics delay time. The average total downtime, not including active maintenance time K (includes O&S days and repair time).

WARF days: Wartime replacement factors days. The wartime operating duration in days for which spares are maintained with a unit.

Conversion to millions factor: Standard factor used to convert to a value expressed in millions.

Intermediate Algorithms for Each Function

This area contains algorithms that are a subset of several more complex algorithms. Thus the computation only needs to be done once and the results input to several other algorithms in paragraph 3 (output).

Output factor 1: Operating hours per year per organization.
   Definition: Self-explanatory.
   Algorithm:
Operating hours per item per year × percent operating hours per year per item (peace/normal) × quantity required per organization
   Default value: N/A (*Note:* Default value will be omitted on future factors if N/A.)

Output factor 2: Failures per year
   Definition: Total of all failures for 1 year.
   Algorithm: (Product from factor 1 above) ÷ MTBF (unsch.)

Output factor 3: Previous maint. actions per year
   Definition: Comparison of previous maintenance actions per year.
   Algorithm: (Product from factor 1 above) × (MTBPM × MTBPM h)
Output factor 4: Diagnostic effect factor
   Definition: Computes a diagnostic effect factor from the system default factor constant and the adjustment factor (as used).
   Algorithm: [1 + (1 − diagnostic effectiveness %)]
   Default value: Diagnostic effectiveness
   Source: Function factors

TABLE A26.6   Cell Addresses of Inputs, Outputs, and Algorithms for Calculations Used in This Spreadsheet (Continued)

Output factor 5: Uptime
   Definition: Total time all systems are considered available for use.
   Algorithm: 24 h per day × number of systems

Output factor 6: Downtime, peace
   Definition: To provide a computation of total downtime in peacetime.
   Algorithm: (Operating hours per year per orgn. ÷ 365 days) × [(MTTR (unsch.) h, on equip. + O&S days. + remoteness of maint.) ÷ MTBF (unsch.) h + MPTM (sch.) h, unit ÷ MTBPM h]
   Default value: O&S days
   Source: General factors

Output factor 7: Downtime, war
   Definition: Calculates the total system downtime during wartime.
   Algorithm: (Operating hours per year per orgn. ÷ 365) × % NDI × expenditure rate × [(MPTM h, unit ÷ MPTM h) + (2 × MTTR (unsch.) h + MLDT h + 2 × remoteness of maint. h) ÷ (2 × MTBF (unsch.) h)] + qty. required per orgn. × operating climate, exponential 2 × battle dam. exp. × (2 × MTTR (unsch.) on equip. × battle dam. rep. diff. index + 2 × remoteness of maint. + MLDT h) ÷ 2000

Output factor 8: A_o objective squared
   Definition: Computation of the square of the A_o.
   Algorithm: (A_o objective × A_o objective)

Output factor 9: War surge threat factor
   Definition: Computation of composite war surge threat factor.
   Algorithm:
Consumable value/unit × expenditure rate (units/events) × % NDI

Output factor 10: Total war force factor
   Definition: Compute the total war force factor.
   Algorithm:
War surge threat factor + (1 − consumable value/unit)

Logistics Resource Cost Algorithms

Production cost equipment B240
   (add B144 through E144)
Training equipment purchase B241
   (add B177 through E177)
Spaces (init. + warf.) B242
   (add B183 through E183 and B185 through E185)
Support equipment (O, I, D) B243
   (add B188 through E190)
Facilities B244
   (add B199 through E199)
Technical data B245
   (add B202 through E202)
Software B246
   (add 205 through E205)
Total production costs B248
   (add B240 through B246)

TABLE A26.6   Cell Addresses of Inputs, Outputs, and Algorithms for Calculations Used in This Spreadsheet (Continued)

O&S cost/yr B250
Operations MH B251
  (add B145 through E145)
Sched. maint. (O, I, D) B252
  (add B152 through E152)
Unsched. maint. (O, I, D) B253
  (add B159 through E159)
Manpower maint. (O, I, D) B254
  (add B168 through E168)
Manpower ops. B255
  (add B170 through E170)
Manpower management B256
  (add B172 through E172)
Training (op., M, equip. sp.) B257
  (add B175 through E175 and B178 through E178)
Item management B258
  (add B180 through E180)
Spares (replen.) B259
  (add B184 through E184)
SE support B260
  (add B193 through E193)
Transportation B261
  (add B197 through E197)
Facilities support B262
  (add B200 through E200)
Technical data support 263
  (add B203 through E203)
Software support 264
  (add B206 through E206)
POL B265
  (add B208 through E208)
Consumables B266
  (add B209 through E209)
Total O&S cost B268
  (add B254 + B255 + B257 through B267)
A_o peace B270
  (sum B221 through E221 − sum B222 through E222) ÷ sum B221 through E221
A_o war B271
  (sum B221 through E221 − sum B223 through E233) ÷ sum B221 through E221
Mobility index B272
  (sum B213 through E213)

TABLE A26.6   Cell Addresses of Inputs, Outputs, and Algorithms for Calculations Used in This Spreadsheet (Continued)

Macros

This section contains macros to enable easy maneuvering around the spreadsheet and printing of sections of the output. Each macro is preceded by a description of its function:

Description	Macro
Data entry	\E
Setup print area	\P
Adjust funct. data	\A
Sys. life cycle	\C
Detail log res.	\D
Gen. std. factors	\G
Issues	\I
Adjust K factors	\K
Sum log res.	\L
Mission input data	\M
Default input data	\O
System input data	\S
Print issue data	RI1.M20 ~ GP
Print issue factors	RA25.E72 ~ GP
Print system data	RH58.I69 ~ GP
Print cost of LCC	RA278.B321 ~ PGQ

A26.6   Recommendations for Tailoring the FECA Model

Recommendations for tailoring the FECA model include

1. The number of functions [or subsystem(s)] used in the model may be altered to meet the requirements of the system being addressed.

2. The algorithms may be altered to represent more closely the specific functions of concern on a specific system. It is not optimized for a specific function or system.

3. The model makes no correction for unique cases or second-order effects.

PPS Solution Options

The tables in this appendix provide solution options to the various problem causes cited in the Chap. 5 discussion of the LSA post-production task.

TABLE A27.1 PPS Source Problem

Problem cause	Solution options
Out of production: no advance notice	Obtain tooling/data rights; develop alternate source(s) Develop alternate part Reinstitute production for periodic buy
Projected demand exceeds production capability	Develop alternate source Increase production capability Develop alternate part
Low rate of production	Increase production rate Develop alternate source Reduce demand Develop alternate part

TABLE A27.2 PPS Procurement Problem

Problem cause	Solution options
Projected demand exceeds order stock	Reduce demand Identify substitute Accelerate procurement Increase orders/budget Improve demand forecasting
Increasing production and production lead times	Larger periodic buys Reduce demand Breakout to develop alternate source(s) Improve demand forecasting Identify a substitute
Decreasing sources	Rotate buys with product sources Breakout to new source(s) Reduce demand Lifetime buy Increase the orders Negotiate product warranties

TABLE A27.3   PPS Forecasted Change

Problem cause	Solution options
Production terminating (advanced notice)	Develop a new source; maintain production capability Substitute a part Lifetime buy
Equipment change (ECP, other)	Develop an interchangeable replacement part Develop a retrofit Recover/retain change-out assets
Predicted failure trend increase	Apply a new technology Update demand forecast Substitute a part Increase frequency of buys

TABLE A27.4   PPS Problem: Insufficient Spares and Repair Parts for System Life

Problem cause	Solution options
Demands by other programs	Locally procure immediate requirements Upgrade priority status Increase insurance quantity
Unpredicted failure rate increase	Substitute a part Apply new technology Increase reorder quantity

TABLE A27.5   PPS Support Problems

Problem cause	Solution options
Problem in supporting element (S&TE)	Apply new technology Substitute/equipment Revise/update procedures

Projected demand exceeds stock on hand	Critical part:
	  Expedite procurement locally
	  Substitute a part
	  Procure next-higher assembly
	  Cannabalize existing assets
	  Improve forecasting
	Noncritical part:
	  Maintain production capability
	  Update demand forecast
	  Increase frequency of buys
	  Improve forecasting

Acronyms

ACIM	Availability-Centered Inventory Model
ADP	Automated Data Processing
AF	Air Force
AFLC	Air Force Logistics Command
AFSC	Air Force Systems Command
AI	Artificial Intelligence
ALC	Air Logistics Center
AMC	Army Materiel Command/Acquisition Method Code
A_o	Operational Availability
APL	Allowance Parts List
BCS	Baseline Comparison System
CAE	Computer-Aided Engineering
CALS	Computer-Aided Acquisition and Logistics
CASA	Cost Analysis Strategy Assessment
CASREP	Casualty Report
CDR	Critical Design Review
CDRL	Contract Data Requirements List
CE	Concept Exploration
CEEM	Cost-Effectiveness Evaluation Model
CELSA	Cost-Estimating Methodology for LSA
CFE	Contractor-Furnished Equipment
CITIS	Contractor-Integrated Technical Information System
CMRS	Calibration Measurement Requirements Summary
COMOCK	Computer Mock-up
CRLCMP	Computer Resource Life-Cycle Management Plan
DBMS	Data Base Management System
DDN	Defense Data Network
DED	Data Element Definition
DEMVAL	Demonstration Validation
DI	Data Item
DID	Data Item Description
DLSC	Defense Logistics Support Center

DOD	Department of Defense
DODD	DOD Directive
DOP	Degree of Protection/Depot Overhaul Point
DS	Direct Support
DSMC	Defense Systems Management College
DT	Development Testing/Down Time
DTC	Design to Cost
DTLCC	Design to Life-Cycle Cost
DVAL	Demonstration Validation
EBCDIC	Extended Binary Code Decimal Interchange Code
ECP	Engineering Change Proposal
EMD	Engineering Manufacturing Development
ENG	Engineering
EW	Electronic Warfare
FEA	Front-End Analysis
FECA	Front-End Cost Analysis
FGC	Functional Group Code
FH	Flight Hours
FI	Fault Isolation
FIPS	Federal Information Processing Standard
FMEA	Failure Modes Effects Analysis
FMECA	Failure Modes Effects and Criticality Analysis
FMS	Foreign Military Sales
FSC	Federal Supply Classification
FSD	Full-Scale Development
GFE	Government-Furnished Equipment
GFI	Government-Furnished Information
GIDEP	Government-Industry Data Exchange Program
GS	General Support
GSE	Ground Support Equipment
HARDMAN	Hardware/Military Manpower Integration
HDBK	Handbook
HF	Human Factors
HMPC	Hazardous Maintenance Procedure Code
HSI	Human Systems Integration
HW	Hardware
IAW	In Accordance With
ICC	Item Category Code

ICP	Inventory Control Point
IGES	Initial Graphics Exchange Standard
ILS	Integrated Logistics Support
ILSMT	ILS Management Team
ILSP	ILS Plan
IMIP	Industrial Modernization Incentives Program
IRLA	Item Repair-Level Analysis
I/R	Interchangeability/Repairability
ISG	Industry Support Group
ITO	Instruction to Offerers
JIT	Just in Time
JMSNS	Justification for Major Systems New Start
JSTAR	Joint Surveillance and Target Attack Radar System
JVX	Joint Services Advanced Vertical Lift Aircraft
LAPL	Lead Allowance Parts List
LCC	Life-Cycle Cost
LCN	Logistics Control Number
LCOM	Logistics Composite Model
LEM	Logistics Element Manager
LLTIL	Long-Lead-Time Item List
LM	Logistics Manager
LOR	Level of Repair
LORA	Level-of-Repair Analysis
LORAP	LORA Plan
LRA	Line-Replaceable Assembly
LRU	Line-Replaceable Unit
LSA	Logistics Support Analysis
LSAP	LSA Plan
LSAR	LSA Record
MAC	Maintenance Allocation Chart
MAMDT	Mean Active Maintenance Downtime
MANPRINT	Manpower and Personnel Integration
MCLOR	Marine Corps LORA Model
MCM	Material Change Management
MDTD	Mean Downtime Documentation
MDTOA	Mean Downtime for Outside Assistance
MDTOR	Mean Downtime for Other Reasons
MDTT	Mean Downtime for Training

MEC	Mission Essentiality Code
MEL	Maintenance Expenditure Limit
MIP	Maintenance Index Page
MLDT	Mean Logistics Delay Time
MMH	Maintenance Manhours
MPA	Maintenance Planning Analysis
MPT	Manpower, Personnel, and Training
MRC	Maintenance Requirements Card
MR	Maintenance Ratio
MRRI	Maintenance Replacement
MRSA	Material Readiness Support Agency
MSRT	Mean Supply Response Time
MTBF	Mean Time between Failures
MTBFS	MTBF Software
MTBM	Mean Time between Maintenance
MTBMA	MTBM Actions
MTBMCF	MTBM Critical Failure
MTBR	Mean Time between Removals
MTD	Maintenance Task Distribution
MTTR	Mean Time to Repair
MTTRS	Mean Time to Reboot
NAVAIR	Naval Air Systems Command
NAVSEA	Naval Sea Systems Command
NDI	Nondevelopmental Item
NDT	Nondestructive Testing
NHA	Next-Higher Assembly
NRLA	Network Repair-Level Analysis
NSIA	National Security Industrial Association
NSN	National Stock Number
OJT	On-the-Job Training
OLSP	Operational Logistics Support Plan
OSD	Office of the Secretary of Defense
OSI	Open System Interconnection
OT	Operating Time
OTSA	Off-the-Shelf Analysis
PCO	Procurement Contracting Officer
PCS	Permanent Change of Station
PDR	Preliminary Design Review

PHS&T	Packaging, Handling, Storage, and Transportation
PLISN	Provisioning List Item Sequence Number
PMS	Planned Maintenance System
POC	Point of Contact
POL	Petroleum, Oil, and Lubricants
POM	Program Objectives Memorandum
POSTPROD	Postproduction
PPL	Provisioning Parts List
PPLI	PPL Index
PPS	Postproduction Support
P^3I	Preplanned Product Improvement
QA	Quality Assurance
RAM	Reliability Availability Maintainability
RCM	Reliability-Centered Maintenance
RFP	Request for Proposals
RIL	Repair Items List
RIW	Reliability Incentive Warrantee
RLA	Repair-Level Analysis
RT	Repair Time/Relocation Time
RTD	Repair Task Distribution
SEMP	System Engineering Management Plan
SERD	Support Equipment Recommendation Data
SGML	Standard Generalized Mark-up Language
SMR	Source, Maintenance, and Recoverability Code
SOLELETTER	Society of Logistics Engineers Newsletter
SQL	Structured Query Language
SRO	Systems Readiness Objective
SRU	Shop-Replaceable Unit
SSA	Source Selection Authority
SSAC	SSA Council/Committee
SSC	Source Selection Chairman/Skill Specialty Code
ST	Standby Time
SW	Software
SSEB	Source Selection Evaluation Board
SSP	Source Selection Plan
SYS	System
SYSCOM	Systems Command
TALDT	Total Administrative and Logistics Downtime

TCM	Total Corrective Maintenance Time
TM	Technical Manual
TMDE	Test, Measurement, and Diagnostic Equipment
TP	Test Program
TPM	Total Preventive Maintenance Time
TPS	Test Program Set
TQM	Total Quality Management
TT	Total Time
TTEL	Tools and Test Equipment List
UAV	Unmanned Aerial Vehicle
UMMIPS	Uniform Material Movement Issue Priority System
UOC	Usable on Code
UUT	Unit under Test
VAMOSC	Visibility and Management of O&S Costs
VE	Value Engineering
WBS	Work Breakdown Structure
WRA	Weapon Replaceable Assembly
WUC	Work Unit Code

References

Part A identifies documents specifically referenced in this book. Part B consists of additional references of possible interest to the reader. Part C lists various related DOD specifications and standards.

A. References cited in text

1. Linda Green, *Logistics Engineering,* Wiley, New York, 1991.
2. MIL-STD-1388-1A, "Logistic Support Analysis."
3. James V. Jones, *LSA Handbook,* McGraw-Hill, New York, 1989.
4. Benjamin Blanchard, *Logistics Engineering and Management,* Prentice-Hall, New York, 1989.
5. MIL-STD-1388-2A, "DOD Requirements for a Logistic Support Analysis Record."
6. AMC-P700-4, LSA Techniques Guide.
7. MRSAP 700-11, Cost Estimating Methodology for LSA, December 1988.
8. Defense Systems Management College, Systems Engineering Guide, December 1986.
9. Benjamin Blanchard, *System Engineering Management,* Wiley, New York, 1989.
10. Defense Systems Management College, Cost Analysis Strategy Assessment, September 1986.
11. Navy LSA Applications Guide, May 1986.
12. DOD DIR 5000.1, "Defense Acquisition," February 1991.
13. DOD INST 5000.2, "Defense Acquisition Management Policies and Procedures," February 1991.
14. Ferguson and Hertz, "Requirements planning," *Airpower Journal,* 1991.
15. Gansler, "Changes in the defense engineering/business environment for the 1990s," *The Military Engineer,* Vol. 81, July 1989.
16. HARDMAN Publication 84-01, HARDMAN Methodology: Equipment/Systems/Subsystem, May 1985.
17. Draft SPAWAR MPA-LSA Desk Guide, February 1991.
18. DOD TQM Guide, February 1991.
19. DOD Handbook, "DOD CALS Program Implementation Guide," September 1990.
20. Report: "Application of CALS to the UAV Family," April 1991.
21. IDA Report R-338, "The Role of Concurrent Engineering in Weapon System Acquisition," December 1988.
22. SOLE Report, National Workshop on Software Logistics, August 1989.
23. Betts, Keene, and Keene, Software Logistics, Presentation at 1990 SOLE Symposium.
24. Ferens, Software Logistics Tutorial, Presentation at 1991 SOLE Symposium.
25. EWCM LSA Plan, May 1987; and Biedenbender and Klingensmith, EWCM—Integrating Design, R&M, and LSA, 1988 R&M Symposium.
26. Draft UAV LSA Applications Guide, Block One, July 1991.
27. Draft DOD Standard, "RCM Requirements for DOD Weapon Systems and Equipment," May 1991.
28. NSIA Report, "Guidelines for Preparation of Diagnostic Requirements," July 1986.
29. NAVSEA TO300-ACQ-PRO-010, "Systems and Equipment LSA Procedures Manual," 1991.
30. Draft for SPAWAR, "Write Improved SOWs Easily (WISE) Manual," October 1986.
31. Draft, UAV Family FECA Model, August 1991.

32. MRSA draft, "LSA During the CE Phase," derived from draft Coogan Report, Guide to LSA During CE, 1991.
33. Army Logistics Management Center LSA Course.
34. Navy Operational Availability Handbook, June 1986.
35. AMC-P700-27, "LORA Procedures Guide," February 1991.
36. NAVSEA TL081-AB-PRO-010, "NAVSEA LORA Procedures Manual," December 1990.
37. AFLC/AFSC Pamphlet 800-28, "Repair Level Analysis Program," 1991
38. Naval Air Systems Command, "LORA Default Data Guide," August 1991.
39. Draft DOD LORA Standard, March 1991.
40. MRSA OTSA 84-01, "Graphical Repair/Discard Analysis Procedure Handbook," 1991.
41. TRADOC/AMC Pamphlet 70-11, "RAM Rationale Report Handbook," July 1987.
42. Emanuel Parzen, *Modern Probability Theory and Its Application,* Wiley, New York, 1991.
43. William Feller, *An Introduction to Probability Theory and Its Applications,* Wiley, New York, 1991.
43. Donald Isaacson, LCC tutorial, 1991 SOLE Symposium.

B. Other references

LSA guidance

44. Draft, "NAVSEA LSA Implementation Procedures," November 1984.
45. NAVSEA TO300-ACQ-PRO-020, "Ships LSA Procedures Manual," March 1988.
46. DARCOM-C 700-4, "LSA/LSAR Tailoring Procedures Guide," February 1980.
47. JVX LSA Plan, February 1985.
48. Draft, "NAVELEX LSA/LSAR Tailoring Procedures Guide," 1983.
49. ERC Pamphlet, "Government Performance of Selected LSA Requirements,"
50. Draft, "NAVAIR Common Support Equipment LSA Tailoring Guide," April 1985.
51. "Army LOGPARS Planning and Requirements Simplification Users Handbook," October 1988.
52. DARCOM P 750-16, "DARCOM Guide TO LSA," June 1979.
53. "AF LSA Users Guide," March 1989.
54. "U.S. Marine Corps Guide for Tailoring LSA/LSAR," March 1986.
55. "LSA Guidance Conference Handbook," March 1989.
56. Biedenbender, A reexamination of the LSA/LSAR and R&M interface, 1985 SOLE Symposium.

NDI

57. "OSD Handbook SD-2, Buying NDI," October 1990.
58. Dept. Navy, "Guidelines for Implementation of Policies for NDI," February 1987.
59. MRSA Draft, "LSA Guide for NDI," 1991.

Integrated diagnostics

60. MIL-STD-1814, "Integrated Diagnostics," April 1991.
61. Air Force Guide Specification, "Integrated Diagnostics," April 1991.

Miscellaneous

62. IDA Paper P-2306, "The Relationship Between CALS and CE," March 1990.
63. MIL-HDBK-248, "Acquisition Streamlining," February 1989.
64. "ILS Planning Guide for DOD Systems and Equipments," October 1968.
65. "ILS Implementation Guide for DOD Systems and Equipments," March 1972.
66. Mary Eddings Earles, *Factors, Formulas, and Structure,*

C. Military specifications and standards

67. MIL-STD-2173, "Reliability-Centered Maintenance to Naval Aircraft Weapon Systems and Support Equipment."

68. MIL-I-8500D, "Interchangeability and Replaceability."
69. MIL-Q-9858A, "Quality Program Requirements."
70. MIL-H-46855B, "Human Engineering Requirements for Military Systems, Equipment and Facilities."
71. MIL-S-52779A, "Software Quality Assurance Program Requirements."
72. MIL-T-3100, "Drawings, Engineering and Associated Lists."
73. MIL-T-81821, "Trainers, Maintenance, Equipment and Services, General Specification for."
74. MIL-M-85337A (NAVY), "Manuals, Technical, Quality Assurance Plan."
75. DOD-STD-100C, "Engineering Drawings Practices."
76. MIL-STD-470A, "Maintainability Requirements (for Systems and Equipment)."
77. DOD-STD-480A, "Configuration Control Engineering Changes, Deviations, and Waivers."
78. MIL-STD-780F, "Work Unit Codes for Aeronautical Equipment, Uniform Numbering System."
79. MIL-STD-785B, "Reliability Program for Systems and Equipment Development and Production."
80. MIL-STD-499, "Engineering Management."
81. MIL-STD-881A, "Work Breakdown Structure for Defense Material Items."
82. MIL-STD-965A, "Parts Control Program."
83. MIL-STD-1365A, "General Design Criteria and Handling Equipment Associated with Weapons and Weapon Systems."
84. MIL-STD-1390B, "Level of Repair."
85. MIL-STD-1561B, "Provisioning Procedures, Uniform DOD."
86. MIL-STD-1629A, "Procedures for Performing a Failure Mode, Effects and Criticality Analysis."
87. MIL-STD-1840A, "Computer-Aided Acquisition Logistics System."
88. MIL-STD-2073-1A, "DOD Materiel Procedures for Development and Application of Packaging Requirements."
89. MIL-STD-2076 (AS), "Unit Test Compatibility with Automatic Test Equipment, General Requirements for."
90. MIL-STD 2077A (AS), "Test Program Sets; General Requirements for."
91. DOD-STD-2165, "Defense System Software."
92. DOD-STD-2167A, "Defense System Software Development."

Index

ABOUT THE AUTHORS

Dick Biedenbender is an independent consultant on ILS, LSA, and CALS. He previously was a senior logistics engineer with Logistic Systems Architects, providing support to the Unmanned Aerial Vehicles (UAV) Joint Project Office. Prior to that, he was technical director of Evaluation Research Corporation's Logistics Division. Before his retirement from the Office of the Secretary of Defense, he was the program manager for the development of the current LSA standard. He is now active in teaching and consulting with both contractors and government agencies.

Florence Vryn has worked in support of programs such as Modular Automatic Test Equipment for the U.S. Air Force and Navigation Sub-system for Strategic Systems and for defense contractors Fairchild/Republic, Airborne Instruments Laboratories, and Sperry Systems Management. She also served as chairperson for Support Systems Group, Logistics Management Committee of the National Security Industrial Association (NSIA). Ms. Vryn recently retired from her position as manager, Integrated Logistics Support, Tactical Systems, Unisys.

John Eisaman has been working in the acquisition logistics field since 1984, primarily in the LSA arena. He has provided support and taught LSA/LSAR courses to the Navy, Marine Corps, National Security Agency, General Electric, Ford Aerospace, Rockwell, Magnavox, Unisys, and other government and private organizations. He is currently working for the U.S. Postal Service supporting its national automation program.

LSA Task/Subtask Index

Task/Subtask	Title
101	Development of an Early Logistic Support Analysis Strategy
101.2.1	LSA Strategy
101.2.2	Update
102	Logistic Support Analysis Plan
102.2.1	LSA Plan
102.2.2	Update
103	Program and Design Reviews
103.2.1	Establish Review Procedures
103.2.2	Design Reviews
103.2.3	Program Reviews
103.2.4	LSA Review
201	Use Study
201.2.1	Supportability Factors
201.2.2	Quantitative Factors
201.2.3	Field Visits
201.2.4	Use Study Reports and Updates
202	Mission Hardware, Software, and Support System Standardization
202.2.1	Supportability Constraints
202.2.2	Supportability Characteristics
202.2.3	Recommended Approaches
202.2.4	Risks
203	Comparative Analysis
203.2.1	Identify Comparative Systems
203.2.2	Baseline Comparison System

Task/Subtask	Title
203.2.3	Comparative System Characteristics
203.2.4	Qualitative Supportability Problems
203.2.5	Supportability, Cost, and Readiness Drivers
203.2.6	Unique System Drivers
203.2.7	Updates
203.2.8	Risks and Assumptions
204	Technological Opportunities
204.2.1	Recommended Design Objectives
204.2.2	Update
204.2.3	Risks
205	Supportability and Supportability Related Design Factors
205.2.1	Supportability Characteristics
205.2.2	Supportability Objectives and Associated Risks
205.2.3	Specified Requirements
205.2.4	NATO Constraints
205.2.5	Supportability Goals and Thresholds
301	Functional Requirements Identification
301.2.1	Functional Requirements
301.2.2	Unique Functional Requirements
301.2.3	Risks
301.2.4	Operations and Maintenance Tasks
301.2.5	Design Alternatives